ALUMINUM STRUCTURES

A Guide to Their Specifications and Design

Second Edition

J. Randolph Kissell
Robert L. Ferry
The TGB Partnership

JOHN WILEY & SONS, INC.

Library of Congress Cataloging-in-Publication Data:

ISBN: 0-471-01965-8

Printed in the United States of America

10 9 8 7 6 5 4 3 2 1

CONTENTS

PREFACE TO THE FIRST EDITION

The purpose of this book is to enlighten humanity and contribute to the general betterment of this orb that we call home. Failing that, we will settle for giving engineers enough guidance in the use of aluminum that they will feel confident designing with it. The Aluminum Association, an industry association of aluminum producers, publishes the *Specifications for Aluminum Structures* (hereafter called the Aluminum *Specifications*), which are the generally accepted criteria for the design of aluminum structures. Our book is keyed to the sixth edition of the Aluminum *Specifications*, and readers should have access to it.

Structural engineering may be regarded as the practice of analyzing and designing structures. The analysis process resolves the loads applied to the structure into the resulting forces and moments in the components of the structure. Structural design is, then, the sizing of the structure's components to safely sustain these forces and moments. Academic curricula typically train students in structural analysis, as well as in the design methods appropriate to common materials of construction (i.e., steel, concrete, and perhaps timber), and many excellent texts on these subjects are available. We assume that the reader is already well versed in structural analysis and acquainted with steel design. Our objective is to expand readers' design capability beyond steel, and to present aluminum as another material of construction.

While this text is keyed to the Aluminum *Specifications*, it is also organized to parallel steel design practice. We compare the requirements of the Aluminum *Specifications* to the provisions for the design of steel structures found in the American Institute of Steel Construction (AISC) *Manual of Steel Construction*. Those design requirements and considerations that are particular to aluminum, then, are presented in the context of the steel design background that we assume on the part of the reader.

In addition to bridging the gap between the familiar old state of steel and the exciting new realm of aluminum, we also seek to bridge the gap between the theoretical and the real worlds. We recognize that one of the greatest difficulties in the transition from student to practitioner is knowing how to apply the design methods in "the book" to real-life problems. Whether that book is a text or an industry specification, it often seems that the problem at hand does not neatly fit into any of the categories given. We include a step-by-step design process for real-world applications. If our steps do not spare readers from a 12-step program, then their problems are beyond the scope of this text.

J. Randolph Kissell
Robert L. Ferry

The TGB Partnership
Hillsborough, North Carolina

PREFACE TO THE SECOND EDITION

We were frankly surprised by the reaction to the first edition of this book. While it never threatened to reach the *New York Times* best seller list, the favorable comments were more numerous and heartfelt than we had expected. When a reader wrote that "you will be pleased to know that your book is rapidly becoming dog-eared as it is one of the most popular books in our library," we knew we had achieved our goal. What may have been the most surprising was the international notice the book received, including a Japanese translation and very favorable European reviews. All this almost made up for the work it took to write it.

Once we'd milked the acclaim for all we could, it was time to think about a second edition. The Aluminum Association forced our hand when it revised the *Specification for Aluminum Structures* in the 2000 edition of the *Aluminum Design Manual*. Since this book is a guide to the *Specification*, an update was due. The changes to the *Specification* are more than cosmetic, such as changing the title to the singular "*Specification*." They include changes to tension limit states, design compressive strengths for yielding, design bearing stresses, slip-critical connections, screw pull-out strengths, and others, as well as metrication of mechanical properties. We've revised our text accordingly and metricated it, too, although we haven't been pedantic about metrication in order to preserve readability. We've also added the benefit of what is, we hope, additional wisdom gained from experience since the first edition. Since the *Specification* continues to be a living document, we're dealing with a moving target, but that keeps life interesting.

We welcome readers' comments—this time with slightly less trepidation than before. It's also easier now since this time we have an e-mail address: tgb@mindspring.com. Thanks for your interest in aluminum and our book.

J. RANDOLPH KISSELL
ROBERT L. FERRY

The TGB Partnership
Hillsborough, North Carolina

PART I
Introduction

Double-layer aluminum space frame under construction. This is one of the two structures pictured in Figure 2.3. (Courtesy of Conservatek Industries, Inc.)

1 What's in This Book?

Our book is about the use of aluminum as a material of construction for structural components.

Our major themes are:

- The suitability of aluminum as a structural material,
- How to design aluminum structural components in accordance with the Aluminum Association's *Specification for Aluminum Structures,*
- How to apply the design methods to actual structures.

We begin by introducing you to aluminum, and we hope that by the end of Part I you are sufficiently well acquainted to be ready to get serious about the relationship. In Part II we explain the design requirements of the 2000 edition of the *Specification for Aluminum Structures* (hereafter called the Aluminum *Specification*), published by the Aluminum Association in its *Aluminum Design Manual* (4). Those of you who can't wait to plug and chug may want to jump right ahead to Part III, and refer back to Part II only when you want to know "Where did that come from?"

We assume that you have already had ample exposure to methods of load determination and structural analysis, so we do not replow that ground. We do, however, include in Part II a discussion on local buckling since this is a limit state (i.e., failure mode to you old-timers) that you may have been sheltered from if your design experience has been primarily with hot-rolled steel.

As we discussed in the Preface, we have keyed the discussion of design requirements to the Aluminum *Specification.* In Part II we compare these design provisions to the more familiar requirements for steel buildings published by the American Institute of Steel Construction (AISC) in the *Specification for Structural Steel Buildings* (hereafter called the Steel *Specification*) (38, 39). The Aluminum *Specification* is primarily intended for building structures; thus, we focus on these applications.

Throughout the book we give attention to those features of aluminum that differentiate it from other structural materials, particularly steel. Perhaps the most significant feature that distinguishes aluminum from steel is its *extrudability.* Extruding is the process of forming a product by pushing it through an opening called a *die.* The cross section of the resulting product is determined by the shape of the die. You may simply prepare a drawing of the

3

cross section that you desire for a certain application, then have the mill make a die for producing that shape. This is not the case for steel.

We know from personal experience that while custom extrusions enable designers to exercise a great deal of creativity, the process of sizing a unique shape can be very tedious. When designing with steel, engineers often restrict their choices to those shapes listed in tables of *compact sections,* where the section properties and dimensions are all provided, and the *slenderness* of the cross-sectional *elements* have already been checked to confirm that they are not governed by *local buckling.* While this approach may be safe, it is not very creative. When we create our own shape, however, we assume responsibility for determining its section properties and checking the slenderness of the cross-sectional elements. Furthermore, we may find that our new section is not compact, and we must then determine the local buckling stress limits. As mentioned previously, Part II includes a comprehensive explanation of the behavior of these slender (light gauge) shapes, which is also pertinent to the design of cold-formed steel structures. Although your task does become more complicated when you venture beyond using off-the-shelf shapes, we will guide you through it.

Your first reaction may be that the chore of performing these additional calculations poses too large a cost to pay for obtaining your creative license. We have made it easier, however, by presenting in Part III a straightforward method of performing the design checks required by the Aluminum *Specification.* We also provide some simple tables to make the process easier. Thus, if you pay attention, you can achieve maximum design freedom with minimal computational burden.

We presented the design checks required for individual structural components in Part III, and in Part IV we illustrate the application of these design requirements to actual structures. These include an example of cold-formed construction to demonstrate design with slender shapes, and we demonstrate the checks for beams, columns, and combined stresses in the design of a triangulated dome frame.

We present the design requirements and examples in the Allowable Stress Design (ASD) format because it is still the method in widest use. In Part V, however, we remove the shroud of mystery from Load and Resistance Factor Design (LRFD), so that when you do encounter it, you need not fear it.

Finally, we have compiled useful data in the Appendices, including a cross-reference in Appendix H of the provisions of the Aluminum *Specification* indexed to where they are discussed in this book. There is also a glossary of technical terms.

2 What Is Aluminum?

This chapter does not deal with the origins of aluminum or how it is refined from bauxite, although the ruins at Les Baux de Provence in southern France are certainly worth a visit. There is an ingot of aluminum in the museum at La Citadelle des Baux as a tribute to the metal that is produced from the nearby red rock, which the geologist Berthier dubbed "bauxite" in honor of this ancient fortress in 1821 (135). The ruins of the medieval stronghold, though, are the real attraction. We'll defer to Fodor's and Frommer's on the travel tips, and to Sharp on a discussion of the history, mining, and production of aluminum (133). Our purpose in this chapter is to discuss aluminum's place in the families of structural metals.

2.1 METAL IN CONSTRUCTION

We include aluminum with steel and reinforced concrete as a metal-based material of construction. While our basis for this grouping may not be immediately obvious, it becomes more apparent when considered in an historical context (103).

Prior to the development of commercially viable methods of producing iron, almost all construction consisted of gravity structures. From the pyramids of the pharoahs to the neoclassical architecture of Napoleonic Europe, builders stacked stones in such a way that the dead load of the stone pile maintained a compressive state of force on each component of the structure (see Figure 2.1). The development of methods to mass-produce iron, in addition to spawning the Industrial Revolution in the nineteenth century, resulted in iron becoming commercially available as a material of construction. Architecture was then freed from the limitations of the stone pile by structural components that could be utilized in tension as well as compression. American architect Frank Lloyd Wright observed that with the availability of iron as a construction material, "the architect is no longer hampered by the stone beam of the Greeks or the stone arch of the Romans." Early applications of this new design freedom were the great iron and glass railway stations of the Victorian era. Builders have been pursuing improvements to the iron beam ever since.

An inherent drawback to building with iron as compared to the old stone pile is the propensity of iron to deteriorate by oxidation. Much of the effort to improve the iron beam has focused on this problem. One response has

Figure 2.1   Pont du Gard in southern France. An aqueduct that the ancient Romans built by skillfully stacking stones.

been to cover iron structures with a protective coating. The term *coating* may be taken as a reference to paint, but it is really much broader than that. What is reinforced concrete, for example, but steel with a very thick and brittle coating? Because concrete is brittle, it tends to crack and expose the steel reinforcing bars to corrosion. One of the functions served by prestressing or posttensioning is to apply a compressive force to the concrete in order to keep these cracks from opening.

While one approach has been to apply coatings to prevent metal from rusting, another has been to develop metals that inherently don't rust. Rust may be roughly defined as that dull reddish-brown stuff that shiny steel becomes as it oxidizes. Thus, the designation of "stainless" to those iron-based metals that have sufficient chromium content to prohibit rusting of the base metal in atmospheric service. The "stain" that is presented is the rust stain. *Stainless steel* must have been a term that originated in someone's marketing department. The term confers a quality of having all the positive attributes of steel but none of the drawbacks.

If we were to apply a similar marketing strategy to aluminum, we might call it "light stainless steel." After all, it prevents the rust stain as surely as stainless steel does, and it weighs only about one-third as much. Engineers who regard aluminum as an alien material may be more favorably disposed toward "light stainless steel."

For the past century and a half, then, structural engineers have relied on metals to impart tension-carrying capability to structural components. Technical development during that time has included improvement in the properties of the metals available for construction. One of the tasks of designers is to determine which metal best suits a given application.

2.2 MANY METALS FROM WHICH TO CHOOSE

Structural metals are often referred to in the singular sense, such as "steel," "stainless steel," or "aluminum," but, in fact, each of these labels applies to a family of metals. The label indicates the primary alloying element, and individual alloys are then defined by the amounts of other elements contained, such as carbon, nickel, chromium, and manganese. The properties of an alloy are determined by the proportions of these alloying elements, just as the characteristics of a dessert are dependent on the relative amounts of each ingredient in the recipe. For example, when you mix pumpkin, spices, sugar, salt, eggs, and milk in the proper quantities, you make a pumpkin pie filling. By adding flour and adjusting the proportions, you can make pumpkin bread. Substituting shortening for the pumpkin and molasses for the milk yields ginger cookies. Each adjustment of the recipe results in a different dessert. Whereas the addition of flour can turn pie filling into bread, adding enough chromium to steel makes it stainless steel.

While this is a somewhat facetious illustration, our point is that just as the term *dessert* refers to a group of individual mixtures, so does the term *steel*. Steel designates a family of iron-based alloys. When the chromium content of an iron-based alloy is above 10.5%, it is dubbed *stainless steel* (136). Even within the stainless steel family, dozens of recognized alloys exist, each with different combinations of alloying ingredients. Type 405 stainless steel, for example, contains 11.5% to 14.5% chromium and 1.0% or less of several other elements, including carbon, manganese, silicon, and aluminum. Should the alchemist modify the mixture, such as by switching the relative amounts of iron and aluminum, substituting copper for carbon and magnesium for manganese, and then leaving out the chromium, the alloy might match the composition of aluminum alloy 2618. As this four-digit label implies, it is but one of many aluminum alloys. Just as with desserts, there is no one best metal mixture, but rather different mixtures are appropriate for different occasions. The intent of this text is to add aluminum-based recipes to the repertoire of structural engineers who already know how to cook with steel.

2.3   WHEN TO CHOOSE ALUMINUM

2.3.1   Introduction

Today aluminum suffers from a malady similar to that which afflicted tomatoes in the eighteenth century: many people fail to consider it out of superstition and ignorance. Whereas Europeans shunned tomatoes for fear that they were poisonous, engineers seem to avoid aluminum for equally unfounded reasons today.

One myth is that aluminum is not sufficiently strong to serve as a structural metal. The fact is that the most common aluminum structural alloy, 6061-T6, has a minimum yield strength of 35 ksi [240 MPa], which is almost equal to that of A36 steel. This strength, coupled with its light weight (about one-third that of steel), makes aluminum particularly advantageous for structural applications where dead load is a concern. Its high strength-to-weight ratio has favored the use of aluminum in such diverse applications as bridge rehabilitation (Figure 2.2), large clear-span dome roofs (Figure 2.3), and fire truck booms. In each case, the reduced dead load, as compared to conventional materials, allows a higher live or service load.

Aluminum is inherently corrosion-resistant. Carbon steel, on the other hand, has a tendency to self-destruct over time by virtue of the continual conversion of the base metal to iron oxide, commonly known as rust. Although iron has given oxidation a bad name, not all metal oxides lead to progressive deterioration. Stainless steel, as noted previously, acquires its feature of being rust-resistant by the addition of chromium to the alloy mixture. The chromium oxidizes on the surface of the metal, forming a thin transparent film. This chromium oxide film is passive and stable, and it seals the base

Figure 2.2 Installation of an aluminum deck on aluminum beams for the Smithfield Street Bridge in Pittsburgh, Pennsylvania. (Courtesy of Alcoa)

metal from exposure to the atmosphere, thereby precluding further oxidation. Should this film be scraped away or otherwise damaged, it is self-healing in that the chromium exposed by the damage will oxidize to form a new film (136).

Aluminum alloys are also rendered corrosion-resistant by the formation of a protective oxide film, but in the case of aluminum it is the oxide of the base metal itself that has this characteristic. A transparent layer of aluminum oxide forms on the surface of aluminum almost immediately upon exposure to the atmosphere. The discussion on coatings in Section 3.2 describes how color can be introduced to this oxide film by the anodizing process, which can also be used to develop a thicker protective layer than one that would occur naturally.

Corrosion-prone materials are particularly problematic when used in applications where it is difficult or impossible to maintain their protective coating. The contacting faces of a bolted connection or the bars embedded in reinforced concrete are examples of steel that, once placed in a structure, are not accessible for future inspection or maintenance. Inaccessibility, in addition to preventing repair of the coating, may also prevent detection of coating

Figure 2.3   Aerial view of a pair of aluminum space frames covered with mill finish (uncoated) aluminum sheeting. (Courtesy of Conservatek Industries, Inc.)

failure. Such locations as the seam of a bolted connection or a crack in concrete tend to be places where moisture or other agents of corrosion collect.

Furthermore, aluminum is often used without any finish coating or painting. The cost of the initial painting alone may result in steel being more expensive than aluminum, depending on the quality of coating that is specified. Coatings also have to be maintained and periodically replaced. In addition to the direct cost of painting, increasing environmental and worker-safety concerns are associated with painting and paint preparation practices. The costs of maintaining steel, then, give aluminum a further advantage in life-cycle cost.

2.3.2 Factors to Consider

Clearly, structural performance is a major factor in the selection of structural materials. Properties that affect the performance of certain types of structural members are summarized in Table 2.1.

For example, the strength of a stocky compression member is a function of the yield strength of the metal, while the strength of a slender compression member depends on the modulus of elasticity. Since the yield strength of aluminum alloys is frequently comparable to those of common carbon and stainless steels, aluminum is very competitive with these materials when the application is for a stocky column. Conversely, since aluminum's modulus of elasticity is about one-third that of steel's, aluminum is less likely to be competitive for slender columns.

Strength is not the only factor, however. An example is corrosion resistance, as we noted above. Additional factors, such as ease of fabrication (extrudability and weldability), stiffness (modulus of elasticity), ductility (elongation), weight (density), fatigue strength, and cost are compared for three common alloys of aluminum, carbon steel, and stainless steel in Table 2.2.

While cost is critical, comparisons based on cost per unit weight or unit volume are misleading because of the different strengths, densities, and other properties of the materials. Averaged over all types of structures, aluminum components usually weigh about one-half that of carbon steel or stainless steel members. Given this and assigning carbon steel a relative cost index of 1 results in an aluminum cost index of 2.0 and stainless cost index of 4.7. If initial cost were the only consideration and carbon steel could be used without coatings, only carbon steel would be used. But, of course, other factors come

TABLE 2.1 Properties That Affect Structural Performance of Metals

Structural Performance of	Property
tensile members	yield strength, ultimate strength, notch sensitivity
columns (compression members)	yield strength, modulus of elasticity
beams (bending members)	yield strength, ultimate strength, modulus of elasticity
fasteners	ultimate strength
welded connections	ultimate strength of filler alloy; ultimate strength of heat-affected base metal

TABLE 2.2 Comparing Common Structural Shapes and Grades of Three Metals

Property	Aluminum 6061-T6	Carbon steel A36	Stainless steel 304, cold-finished
extrudability (see Section 3.1)	very good	not practical	very limited
weldability	fair, but reduces strength	good, no strength reduction	good
corrosion resistance	good	fair	very good
tensile yield strength	35 ksi	36 to 50 ksi	45 ksi
modulus of elasticity	10,000 ksi	29,000 ksi	27,000 ksi
elongation	8% to 10%	20%	30%
density	0.098 lb/in.^{3}	0.283 lb/in.^{3}	0.284 lb/in.^{3}
fatigue strength (plain metal, 5 million cycles)	10.2 ksi	24 ksi	
relative yield strength-to-weight ratio	2.8	1.0 to 1.4	1.2
cost by weight	$1.20/lb	$0.30/lb	$1.40/lb
cost by volume	$0.12/in.^{3}	$0.084/in.^{3}	$0.42/in.^{3}
cost index (see text)	2.0	1.0	4.7

into play, such as operation and maintenance costs over the life of the structure. Also, in specific applications, the rule of thumb that an aluminum component weighs one-half that of a steel member doesn't always hold true. For example, an aluminum component might weigh considerably less when a corrosion allowance must be added to the steel. In other cases, the low material cost of steel is offset by higher fabrication costs, such as applications requiring complex cross sections (for example, curtainwall mullions). In such cases, the cost of steel is much more than just the material cost since the part must be machined, cold-formed, or welded to create the final shape, while the costs of aluminum fabrication are almost nonexistent (the material cost includes the cost to extrude the part to its final shape).

Because of stainless steel's high cost, it is used only when weight is not a consideration and finish and weldability are. In fact, when stainless steel is used in lieu of aluminum, the reason is often only concern about welding aluminum.

The families of structural metals, and the individual alloys within each, then, offer a wide range of choices for designers. Each recipe or alloy designation results in certain characteristics that serve specific purposes. When corrosion resistance, a high strength-to-weight ratio, and ease of fabrication are significant design parameters, aluminum alloys merit serious consideration.

2.4 ALUMINUM ALLOYS AND TEMPERS

2.4.1 Introduction

While sometimes it is appropriate to bake flour mixed with nothing but water, such as when one is hurrying out of Egypt with a pharoah in hot pursuit, baked goods are generally improved by the judicious addition of other ingredients. Whether the base is bran flour or corn flour, transforming the flour into muffins requires throwing in a pinch of this or that. So it is for alloys. Whether the base metal is iron or aluminum, it is rarely used in its pure form. Small amounts (often less than 1%) of other elements, which are sometimes called *hardeners,* are required to attain more useful properties.

One of the properties of critical interest for structural metals is their strength. Unalloyed aluminum has an ultimate tensile strength of about 13 kips/in.^{2} (ksi) [90 MPa]. This value can be increased by more than 30 ksi [200 MPa], however, by adding a dash of zinc, then throwing in a pinch or two of copper and magnesium and just a smidgen of chromium. Putting this recipe in the oven and heating it at the prescribed temperature and duration can bring the strength up to more than 80 ksi [550 MPa]. Variations on the ingredients and heating instructions can yield alloys to meet almost any engineering appetite.

Aluminum alloys are divided into two categories: *wrought alloys*, those that are worked to shape, and *cast alloys*, those that are poured in a molten state into a mold that determines their shape. The Aluminum Association maintains an internationally recognized designation system for each category, described in ANSI H35.1, *Alloy and Temper Designation Systems for Aluminum* (42). The wrought alloy designation system is discussed in the next section and the cast alloy system in Section 3.1.4. While strength and other properties of both wrought and cast products are dependent on their ingredients, or the selective addition of alloying elements, further variations on these properties can be achieved by *tempering*. Tempering refers to the alteration of the mechanical properties of a metal by means of either a mechanical or thermal treatment. Temper can be produced in wrought products by the strain-hardening that results from cold working. Thermal treatments may be used to obtain temper in cast products, as well as in those wrought alloys identified as *heat-treatable*. Conversely, the wrought alloys that can only be strengthened by cold work are designated *non-heat-treatable*.

2.4.2 Wrought Alloys

The Aluminum Association's designation system for aluminum alloys was introduced in 1954. Under this system, a four-digit number is assigned to each alloy registered with the Association. The first number of the alloy designates the primary alloying element, which produces a group of alloys with similar properties. The Association sequentially assigns the last two digits.

The second digit denotes a modification of an alloy. For example, 6463 is a modification of 6063 with slightly more restrictive limits on certain alloying elements, such as iron, manganese, and chromium, to obtain better finishing characteristics. The primary alloying elements and the properties of the resulting alloys are listed below and summarized in Table 2.3:

1xxx: This series is for *commercially pure aluminum,* defined in the industry as being at least 99% aluminum. Alloy numbers are assigned within the 1xxx series for variations in purity and which elements compose the impurities; the main ones are iron and silicon. The primary uses for alloys of this series are electrical conductors and chemical storage or processing because the best properties of the alloys of this series are electrical conductivity and corrosion resistance. The last two digits of the alloy number denote the two digits to the right of the decimal point of the percentage of the material that is aluminum. For example, 1060 denotes an alloy that is 99.60% aluminum.

2xxx: The primary alloying element for this group is *copper,* which produces high strength but reduced corrosion resistance. These alloys were among the first aluminum alloys developed and were originally called *duralumin.* Alloy 2024 is, perhaps, the best known and most widely used alloy in aircraft. Most aluminum-copper alloys fell out of favor, though, because they demonstrated inadequate corrosion resistance when exposed to the weather without protective coatings and are difficult to weld.

3xxx: *Manganese* is the main alloying element for the 3xxx series, increasing the strength of unalloyed aluminum by about 20%. The corrosion resistance and workability of alloys in this group, which primarily consists of alloys 3003, 3004, and 3105, are good. The 3xxx series alloys are well suited to architectural products, such as rain-carrying goods and roofing and siding.

TABLE 2.3   Wrought Alloy Designation System and Characteristics

Series Number	Primary Alloying Element	Relative Corrosion Resistance	Relative Strength	Heat Treatment
1xxx	none	excellent	fair	non-heat-treatable
2xxx	copper	fair	excellent	heat-treatable
3xxx	manganese	good	fair	non-heat treatable
4xxx	silicon	—	—	varies by alloy
5xxx	magnesium	good	good	non-heat-treatable
6xxx	magnesium and silicon	good	good	heat-treatable
7xxx	zinc	fair	excellent	heat-treatable

4xxx: *Silicon* is added to alloys of the 4xxx series to reduce the melting point for welding and brazing applications. Silicon also provides good flow characteristics, which in the case of forgings provide more complete filling of complex die shapes. Alloy 4043 is commonly used for weld filler wire.

5xxx: The 5xxx series is produced by adding *magnesium*, resulting in strong, corrosion resistant, high welded strength alloys. Alloys of this group are used in ship hulls and other marine applications, weld wire, and welded storage vessels. The strength of alloys in this series is directly proportional to the magnesium content, which ranges up to about 6% (Figure 2.4).

6xxx: Alloys in this group contain *magnesium* and *silicon* in proportions that form magnesium silicide (Mg_2Si). These alloys have a good balance of corrosion resistance and strength. 6061 is one of the most popular of all aluminum alloys and has a yield strength comparable to mild carbon steel. The 6xxx series alloys are also very readily extruded, so they constitute the majority of extrusions produced and are used extensively in building, construction, and other structural applications.

7xxx: The primary alloying element of this series is *zinc*. The 7xxx series includes two types of alloys: the aluminum-zinc-magnesium alloys, such as 7005, and the aluminum-zinc-magnesium-copper alloys, such as 7075 and

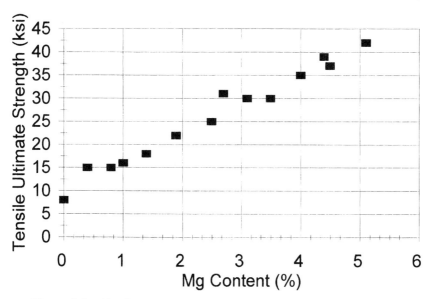

Figure 2.4 Aluminum alloy tensile strength versus magnesium content.

7178. The alloys of this group include the strongest aluminum alloy, 7178, which has a minimum tensile ultimate strength of 84 ksi [580 MPa] in the T6 temper, and are used in aircraft frames and structural components. The corrosion resistance of those 7xxx series alloys alloyed with copper is less, however, than the 1xxx, 3xxx, 5xxx, and 6xxx series, while the corrosion resistance of the 7xxx alloys alloyed without copper is fairly good. Some 7xxx alloys without copper, such as 7008 and 7072, are used as cladding to cathodically protect less corrosion resistant alloys.

8xxx: The 8xxx series is reserved for alloying elements other than those used for series 2xxx through 7xxx. *Iron* and *nickel* are used to increase strength without significant loss in electrical conductivity, and so are useful in such conductor alloys as 8017. Aluminum-lithium alloy 8090, which has exceptionally high strength and stiffness, was developed for aerospace applications (see Section 9.3.3).

9xxx: This series is not currently used.

Experimental alloys are designated in accordance with the above system, but with the prefix "X" until they are no longer experimental. Producers may also offer proprietary alloys to which they assign their own designation numbers.

This wrought alloy designation system had 357 registered alloys by 2001 (19), but only 81 appear in the Aluminum Association's *Aluminum Standards and Data 2000* (11) and 22 in the Association's *Specification for Aluminum Structures* (4), an indication that only a small percentage are commonly used. The signatories to the international accord on the designation system include organizations in the U.S., Russia, the United Kingdom, South Africa, Germany, Brazil, Belgium, Italy, Australia, Spain, China, Austria, France, Argentina, Mexico, Poland, Japan, Peru, Romania, Norway, Netherlands, and Switzerland, and the European Aluminum Association, so the system enjoys nearly global recognition. While the international accord has done much to standardize designations, given the number of signatories, it is perhaps not surprising the registration of so-called "national variations" has compromised uniformity. (There were 55 of these at last count in 2001.) Such variations are assigned a capital letter following the numerical designation (for example, 6005A, is used in Europe and is a variation on 6005). The chemical composition limits for national variations are similar to the Aluminum Association limits but vary slightly. Also, old habits die hard, and often the previous designations used in various European countries are used informally there.

Don't be alarmed if you see yet other designations for aluminum alloys that use the system described above but with a prefix. The Unified Numbering System (UNS), which the Society of Automotive Engineers and ASTM in conjunction with other technical societies, U.S. government agencies, and trade associations developed to identify metals and alloys, includes aluminum alloys. The UNS number for wrought aluminum alloys uses the same number as the Aluminum Association designation but precedes it with "A9" (for example, UNS A95052 for 5052) in order to differentiate aluminum alloys from other metal alloys covered by the UNS. The UNS number for cast aluminum alloys also uses the same number as the Aluminum Association designation (discussed below) but precedes it with "A" and a number "0" or higher (for example, UNS A14440 for A444.0).

You can also still come across the pre-1954 designations, usually when someone who doesn't know much about aluminum quotes an old reference. To help in these instances, as well as when you're confronted with modifying an historic aluminum structure, a cross reference between the old system designations and the new is given in Appendix A. What's probably even more amazing than the fact that these designations still occasionally appear is that many of the old alloys, such as 24S (now 2024), 43S (4043), and 61S (6061), are still popular. You'll also still occasionally hear reference to duralumin, the commercial name given to the original aluminum-copper alloys. Unfortunately these alloys proved to be the least durable of aluminum, and use of the name has faded.

Informal appellations are also given to aluminum alloy series. The 2xxx and 7xxx series are sometimes referred to as "aircraft alloys," but they are also used in other applications, including bolts and screws used in buildings. The 1xxx, 3xxx, and 6xxx series alloys are sometimes referred to as "soft," while the 2xxx, 5xxx, and 7xxx series alloys are called "hard." This description refers to the ease of extruding the alloys: hard alloys are more difficult to extrude, requiring higher-capacity presses, and are, thus, more expensive.

2.4.3 Tempers

Aluminum alloys are tempered by *heat treating* or *strain hardening* to further increase strength beyond the strengthening effect of adding alloying elements. Alloys are divided into two groups based on whether or not their strengths can be increased by heat treating. Both *heat treatable* and *non-heat treatable* alloys can be strengthened by strain hardening, which is also called *cold-working*. The alloys that are not heat-treatable may only be strengthened by cold-working. Whether or not an alloy is heat treatable depends on its alloying elements. Alloys in which the amount of alloying element in solid solution in aluminum increases with temperature are heat treatable. In general, the 1xxx, 3xxx, 4xxx, and 5xxx series wrought alloys are not heat treatable, while the 2xxx, 6xxx, and 7xxx wrought series are, but minor exceptions to this rule exist.

Non-heat treatable alloys may also undergo a heat treatment, but this heat treatment is used only to stabilize properties so that strengths do not decrease over time—behavior called *age softening*—and is required only for alloys with an appreciable amount of magnesium (the 5xxx series). Heating to 225°F to 350°F [110°C to 180°C] causes all the softening to occur at once and, thus, is used as the stabilization heat treatment.

Before tempering, alloys begin in the *annealed* condition, the weakest but most ductile condition. Tempering, while increasing the strength, decreases ductility and, therefore, decreases workability.

Strain hardening is achieved by mechanical deformation of the material at ambient temperature. In the case of sheet and plate, this is done by reducing its thickness by rolling. As the material is worked, it becomes resistant to further deformation and its strength increases.

Two heat treatments can be applied to annealed condition, heat treatable alloys. First, the material can be *solution heat treated*. This allows soluble alloying elements to enter into solid solution; they are retained in a super-saturated state upon *quenching*, a controlled rapid cooling usually performed using air or water. Next, the material may undergo a *precipitation heat treat-ment*, which is also called *artificial aging*. Here, constituents are precipitated from solid solution to increase the strength. An example of this process is the production of 6061-T6 sheet. From its initial condition, 6061-O annealed material is heat treated to 990°F [530°C] as rapidly as possible (solution heat treated), then cooled as rapidly as possible (quenched), which renders the temper T4. The material is then heated to 320°F [160°C] and held for 18 hours (precipitation heat treated); upon cooling to room temperature, the tem-per is T6.

Solution heat treated aluminum may also undergo *natural aging*. Natural aging, like artificial aging, is a precipitation of alloying elements from solid solution, but because it occurs at room temperature, it occurs much more slowly (over a period of days and months rather than hours) than artificial aging. Both aging processes result in an increase in strength and a corre-sponding decrease in ductility. Material that will be subjected to severe form-ing operations, such as cold heading wire to make rivets or bolts, is often purchased in a T4 temper, formed, and then artificially aged or allowed to naturally age. Care must be taken to perform the forming operation before too long a period of time elapses, or natural aging of the material will cause it to harden and decrease its workability. Sometimes T4 material is refriger-ated to prevent natural aging if cold forming required for fabrication into a product, such as a fastener or a tapered pole, will not be performed shortly after solution heat treatment.

The temper designation system is the same for both wrought and cast alloys, although cast alloys are only heat treated and not strain hardened, with the exception of some 85x.0 casting alloys. The temper designation follows the alloy designation, the two being separated by a hyphen, for example, 5052-H32. Basic temper designations are letters. Subdivisions of the basic tempers are given by one or more numbers following the letter.

The basic temper designations are:

F As fabricated.   Applies to the products of shaping processes in which no special control over thermal conditions or strain hardening is employed. For wrought products, no mechanical property limits exist.

O Annealed.   Applies to wrought products that are annealed to obtain the lowest strength temper, and to cast products that are annealed to improve ductility and dimensional stability. The "O" may be followed by a number other than zero.

H Strain-hardened   (wrought products only). Applies to products that have their strength increased by strain-hardening, with or without supplementary thermal treatments, to produce some reduction in strength. The "H" is always followed by two or more numbers.

W Solution heat-treated.   An unstable temper applicable only to alloys that spontaneously age at room temperature after solution heat-treatment. This designation is specific only when the period of natural aging is indicated, for example, W $\frac{1}{2}$ hour.

T Thermally treated to produce stable tempers other than F, O, or H.
Applies to products that are thermally treated, with or without supplementary strain-hardening, to produce stable tempers. The "T" is always followed by one or more numbers.

Strain-Hardened Tempers   For strain-hardened tempers, the first digit of the number following the "H" denotes:

H1 Strain-hardened only.   Applies to products that are strain-hardened to obtain the desired strength without supplementary thermal treatment. The number following this designation indicates the degree of strain-hardening, for example, 1100-H14.

H2 Strain-hardened and partially annealed.   Applies to products that are strain-hardened more than the desired final amount and then reduced in strength to the desired level by partial annealing. For alloys that age-soften at room temperature, the H2 tempers have the same minimum ultimate tensile strength as the corresponding H3 tempers. For other alloys, the H2 tempers have the same minimum ultimate tensile strength as the corresponding H1 tempers and slightly higher elongation. The number following this designation indicates the strain-hardening remaining after the product has been partially annealed, for example, 3005-H25.

H3 Strain-hardened and stabilized.   Applies to products that are strain-hardened and whose mechanical properties are stabilized either by a low-temperature thermal treatment or as a result of heat introduced during fabrication. Stabilization usually improves ductility. This designation is applicable only to those alloys that, unless stabilized, gradually age-soften at room temperature. The number following this designation indicates the degree of strain-hardening remaining after the stabilization has occurred, for example, 5005-H34.

H4 Strain-hardened and lacquered or painted.   Applies to products that are strain-hardened and subjected to some thermal operation during subsequent painting or lacquering. The number following this designation indicates the degree of strain-hardening remaining after the product has been thermally treated as part of the painting or lacquering curing. The corresponding H2X or H3X mechanical property limits apply.

The digit following the designation H1, H2, H3, or H4 indicates the degree of strain-hardening. Number 8 is for the tempers with the highest ultimate tensile strength normally produced and is sometimes called *full-hard.* Number 4 is for tempers whose ultimate strength is approximately midway between that of the O temper and the HX8 temper, and so is sometimes called *half-hard.* Number 2 is for tempers whose ultimate strength is approximately midway between that of the O temper and the HX4 temper, which is called *quarter hard.* Number 6 is for tempers whose ultimate strength is approximately midway between that of the HX4 temper and the HX8 temper called *three-quarter hard.* Numbers 1, 3, 5, and 7 similarly designate intermediate tempers between those defined above. Number 9 designates tempers whose minimum ultimate tensile strength exceeds that of the HX8 tempers by 2 ksi [15 MPa] or more. An example of the effect of the second digit is shown in Table 2.4.

The third digit, when used, indicates a variation in the degree of temper or the mechanical properties of a two-digit temper. An example is pattern or embossed sheet made from the H12, H22, or H32 tempers; these are assigned H124, H224, or H324 tempers, respectively, since the additional strain hardening from embossing causes a slight change in the mechanical properties.

TABLE 2.4   HX1 through HX9 Temper Example

| Temper | Ultimate Tensile Strength | | Description |
	ksi	MPa	
5052-O	25	170	annealed
5052-H32	31	215	$\frac{1}{4}$-hard
5052-H34	34	235	$\frac{1}{2}$-hard
5052-H36	37	255	$\frac{3}{4}$-hard
5052-H38	39	270	full-hard
5052-H39	41	285	

Heat-Treated Tempers   For heat-treated tempers, the numbers 1 through 10 following the "T" denote:

T1 Cooled from an elevated temperature shaping process and naturally aged to a substantially stable condition.   Applies to products that are not cold-worked after cooling from an elevated temperature shaping process, or in which the effect of cold work in flattening or straightening may not be recognized in mechanical property limits, for example, 6005-T1 extrusions.

T2 Cooled from an elevated temperature shaping process, cold-worked, and naturally aged to a substantially stable condition.   Applies to products that are cold-worked to improve strength after cooling from an elevated temperature shaping process, or in which the effect of cold work in flattening or straightening is recognized in mechanical property limits.

T3 Solution heat-treated, cold-worked, and naturally aged to a substantially stable condition.   Applies to products that are cold-worked to improve strength after solution heat treatment, or in which the effect of cold work in flattening or straightening is recognized in mechanical property limits, for example, 2024-T3 sheet.

T4 Solution heat-treated and naturally aged to a substantially stable condition.   Applies to products that are not cold-worked after solution heat treatment, or in which the effect of cold work in flattening or straightening may not be recognized in mechanical property limits, for example, 2014-T4 sheet.

T5 Cooled from an elevated temperature shaping process and then artificially aged.   Applies to products that are not cold-worked after cooling from an elevated temperature shaping process, or in which the effect of cold work in flattening or straightening may not be recognized in mechanical property limits, for example, 6063-T5 extrusions.

T6 Solution heat-treated and then artificially aged.   Applies to products that are not cold-worked after solution heat treatment, or in which the effect of cold work in flattening or straightening may not be recognized in mechanical property limits, for example, 6063-T6 extrusions.

T7 Solution heat-treated and then overaged/stabilized.   Applies to wrought products that are artificially aged after solution heat treatment to carry them beyond a point of maximum strength to provide control of some significant characteristic, for example, 7050-T7 rivet and cold heading wire and rod. Applies to cast products that are artificially aged after solution heat treatment to provide dimensional and strength stability.

T8 Solution heat-treated, cold-worked, and then artificially aged. Applies to products that are cold-worked to improve strength, or in which the effect of cold work in flattening or straightening is recognized in mechanical property limits, for example, 2024-T81 sheet.

T9 Solution heat-treated, artificially aged, and then cold-worked. Applies to products that are cold-worked to improve strength after artificial aging, for example, 6262-T9 nuts.

T10 Cooled from an elevated temperature shaping process, cold-worked, and then artificially aged. Applies to products that are cold-worked to improve strength, or in which the effect of cold work in flattening or straightening is recognized in mechanical property limits.

Additional digits may be added to designations T1 through T10 for variations in treatment. Stress-relieved tempers follow various conventions, which are described below.

Stress relieved by stretching:

T—51 Applies to plate and rolled or cold-finished rod or bar, die or ring forgings, and rolled rings when stretched after solution heat treatment or after cooling from an elevated temperature shaping process. The products receive no further straightening after stretching, for example, 6061-T651.

T—510 Applies to extruded rod, bar, profiles, and tubes, and to drawn tube when stretched after solution heat treatment or after cooling from an elevated temperature shaping process.

T—511 Applies to extruded rod, bar, profiles, and tubes, and to drawn tube when stretched after solution heat treatment or after cooling from an elevated temperature shaping process. These products may receive minor straightening after stretching to comply with standard tolerances.

These stress-relieved temper products usually have larger tolerances on dimensions than other products of other tempers.

Stress relieved by compressing:

T—52 Applies to products that are stress relieved by compressing after solution heat treatment or cooling from an elevated temperature shaping process to produce a permanent set of 1% to 5%.

Stress relieved by combined stretching and compressing:

T—54 Applies to die forgings that are stress relieved by restriking cold in the finish die.

For wrought products heat-treated from annealed or F temper (or other temper when such heat treatments result in the mechanical properties assigned to these tempers):

T42   Solution heat-treated from annealed or F temper and naturally aged to a substantially stable condition by the user (as opposed to the producer), for example, 2024-T42.

T62   Solution heat-treated from annealed or F temper and artificially aged by the user (as opposed to the producer), for example, 6066-T62.

While the temper designations for wrought alloys are grouped according to whether the strength level was obtained by strain-hardening or by thermal treatment, artificial aging is the more important distinction for structural design purposes because it affects the shape of the stress-strain curve. The Aluminum *Specification* gives one set of buckling constant formulas for the artificially aged tempers (T5 through T10), and another set of formulas that applies to both the non-heat-treated alloys (H), as well as to the heat-treated alloys that have not been artificially aged (T1 through T4). These buckling constants are discussed in detail in Section 5.2.1.

You'll occasionally encounter tempers that don't match any of the ones described above, such as 6063-T53. Individual producers register these with the Aluminum Association, which publishes their properties in *Tempers for Aluminum and Aluminum Alloy Products* (25).

2.5   STRUCTURAL APPLICATIONS OF ALUMINUM

2.5.1   Background

Aluminum's markets have developed gradually over the 100-year history of commercial production of the metal. Its first use was cooking utensils in the 1890s, followed by electrical cable shortly after 1900, military uses in the 1910s, and aircraft in the 1930s. Aluminum's use in construction began around 1930, when landmark structures, such as the national Botanic Garden Conservatory in Washington, DC, and the Chrysler and Empire State Buildings in New York City were erected with aluminum structural components. Aluminum didn't really crack the construction market until after World War II, when aluminum was first used to clad buildings. This was done with the advent of extrusions and curtainwall technology, discussed further below. The 1960s saw the rapid expansion of aluminum's largest market, packaging. In the 1990s, aluminum use in transportation has grown markedly, especially in automobiles and light trucks.

Aluminum's most recent markets are its largest—transportation, packaging, and construction, in that order—and together they account for two-thirds of U.S. aluminum consumption (Table 2.5).

Applications of aluminum can be divided into two classes: structural and nonstructural. Structural applications are those for which the size of the part is driven primarily by the load which it must support; nonstructural applications are the rest. About half the transportation and building and construction

TABLE 2.5   Aluminum's Big Markets

Market	1999 Consumption (billion lb)	Portion of Total Aluminum Shipments
Transportation	7.8	32.2%
Containers and Packaging	5.0	20.7%
Building and Construction	3.2	13.1%
Total	16.0	66.0%

applications of aluminum are structural, amounting to roughly 6 billion lb per year, or about a quarter of all aluminum produced in the U.S. annually.

Structural applications for aluminum can be further divided into aerospace and nonaerospace applications. Aerospace applications tend to use specialized alloys, such as aluminum-lithium alloys used for the space shuttle's external fuel tanks, and most (but not all) 2xxx and 7xxx series alloys, often called "aerospace alloys." Because the aerospace market is so specialized, we'll concentrate our attention on the other, broader structural applications for aluminum, especially the construction market.

Of the semi-fabricated aluminum products used in construction, rolled products (sheet and plate) and extruded shapes, make up roughly two-thirds and one-third, respectively, of the semi-fabricated products consumption in this market, as shown in Table 2.6.

The transportation industry also uses aluminum in significant structural applications, including cars, trucks, buses, tractor trailers, rail cars, and ships.

2.5.2   Building and Construction Applications

At a rate of about 3 billion lb per year in the U.S. and an assumed installed price of $5 per lb, the building market represents about $15 billion annually. What's being built with all this money?

One of the largest segments of this market is the aluminum curtain wall industry. Beginning in earnest in the 1950s with the United Nations Building

TABLE 2.6   Construction Use of Semi-Fabricated Aluminum

Product	1999 Consumption (billion lb)	Portion of Construction Consumption
Sheet and Plate	1.18	62%
Extruded Shapes	0.62	33%
Total	1.80	95%

in New York City and the Alcoa Building in Pittsburgh (Figure 2.5), aluminum curtain walls began to play a significant role in modern building construction. These walls act like large curtains hung from the building frame, serving to maintain a weather-tight envelope while resisting wind loads and transmitting them to the frame. Vertical and horizontal extruded aluminum mullions serve as the structural members (Figure 2.6). Aluminum extrusions also enjoy wide use as frames for doors and windows and in storefronts.

Recently, standing seam aluminum roof sheeting has become a popular architectural product. Some of these standing seam products are used as structural members to span between roof purlins. Aluminum sheet is employed for roofing and siding for corrosive applications or for architectural appeal, as well as routine use for flashing, gutters, siding, soffit, fascia, and down-

Figure 2.5   Aluminum curtain wall on the Alcoa Building. (Courtesy of Alcoa)

Figure 2.6 Custom-extruded aluminum framing supporting a glass wall. (Courtesy of Kawneer)

spouts on buildings. Patio and pool enclosures and canopies and awnings are also frequently constructed of aluminum for its ease of fabrication and corrosion resistance.

Welded aluminum tanks are used for industrial storage and process vessels and pipe for corrosive liquids and are well suited to cryogenic applications due to aluminum's good low-temperature properties (Figure 2.7). Large, clear-span aluminum roofs with bolted frames, clad with aluminum sheet, cover tanks and basins for water storage, wastewater treatment, and petrochemical and bulk storage (Figure 2.8). Floating roofs with aluminum pontoons and aluminum sheet decks are used to minimize evaporation of volatile liquids in tanks. Aluminum handrails and pedestrian bridges are also used in these industries for their corrosion resistance (Figure 2.9). Aluminum's non-sparking properties are preferred over steel in potentially flammable atmospheres.

Aluminum signs and sign structures, light poles, and guard rails are used for highways and railroads (Figure 2.10), and aluminum has been used for a number of bridge decks and bridge structures in North America and Europe. Culverts made of large diameter corrugated aluminum pipe are used for bridges, liners, and retaining walls (Figure 2.11). Aluminum landing mats for

Figure 2.7 Double-wall cryogenic tanks storing liquified natural gas at −260°F [−160°C]. The inner tanks are aluminum. (Courtesy of Chicago Bridge & Iron)

Figure 2.8 Clear-span aluminum dome covering a petroleum storage tank. (Courtesy of Conservatek Industries, Inc.)

Figure 2.9 Aluminum walkway and handrail at a municipal wastewater treatment plant. (Courtesy of Conservatek Industries, Inc.)

Figure 2.10 Aluminum structure supporting a highway sign at a railroad crossing.

Figure 2.11 Aluminum culvert. (Courtesy of Contech Construction Products Inc.)

aircraft are tough and portable. Aluminum is also used for portable bridges for military vehicles.

Aluminum's corrosion resistance lends itself to marine applications, including gangways and floating docks. To minimize the weight above the waterline to enhance stability, aluminum is often used in structures and decks for offshore oil platforms.

A striking example of aluminum's use as construction equipment is the 37 miles of aluminum pipe used as scaffolding for the renovation of the Washington Monument in 1999. (This seemed only fitting, since a 9 inch tall aluminum pyramid had been used to cap the structure, then the tallest in the world, upon its completion in 1884.) Aluminum is also used for ladders, trench shoring, and concrete forms because of its strength, durability, and light weight.

3 Working with Aluminum

Now that you've heard about aluminum, you may want to know what it looks like. We'll describe it, beginning with the forms in which it is produced, how these forms are shaped and altered to become structural components, and how these components can be dressed up with coatings. We will include some comments on how this process differs from the preparation of steel for duty and will conclude with a few suggestions on how to put the structural components into place. This chapter, then, presents the product forms in which aluminum is most commonly used for structural components, and how these product forms are fabricated and erected.

3.1 PRODUCT FORMS

The forms of aluminum used in structural components include extrusions, flat-rolled products, castings, and forgings. The most widely used of these forms are extrusions and the flat-rolled products, sheet and plate. Castings typically have less reliable properties than the wrought product forms, and forgings are often more expensive to produce than other wrought forms. Castings and forgings do, however, lend themselves to more complex shapes than extrusions and flat-rolled products.

3.1.1 Extrusions

Introduction What do aluminum and Play-Doh have in common? They can both be extruded, of course (Figure 3.1). *Extrusions* are produced by pushing solid material through an opening called a *die* to form parts with complex cross sections. Aluminum is not the only metal fabricated this way, but it is the most readily and commonly extruded. (*Stainless steel* can also be extruded, but it requires such great pressures that only small and simple stainless *shapes* can be made). The extrusion process makes aluminum an extremely versatile material for structural design. Rather than being limited to the standard rolled shapes, designers can concoct their own cross sections, putting material where it is needed. Solid and hollow cross sections, even sections with multiple hollows, can be readily extruded (Figure 3.2).

While extrusions dominate applications for parts with a constant cross section, bar and rod are also produced by rolling, and tubes and wire by drawing,

Figure 3.1   Engineer of the future extruding a Play-Doh I-beam. (Play-Doh is a registered trademark of Kenner.)

a process by which material is pulled (as opposed to pushed) through a die to change the cross section or harden the material. Cold-finishing may be used to improve surface finish and dimensional tolerances. Sometimes a combination of methods is used; for example, tube may be extruded and then drawn; bar may be rolled and then cold-finished. Products that have been cold-finished or drawn are held to tighter tolerances on cross-sectional dimensions than extruded products.

Standard Extruded Shapes   Before World War II, most aluminum shapes were produced by rolling, like steel, and so had cross sections similar to those of steel. Many of these shapes had sloped flanges that facilitated rolling but complicated connection details. Wartime and postwar demand for aluminum products prompted better production techniques, especially extrusions, which eventually displaced much of the rolled production. Since extrusions are not subject to the limitations of the rolling process, the need for sloped flanges

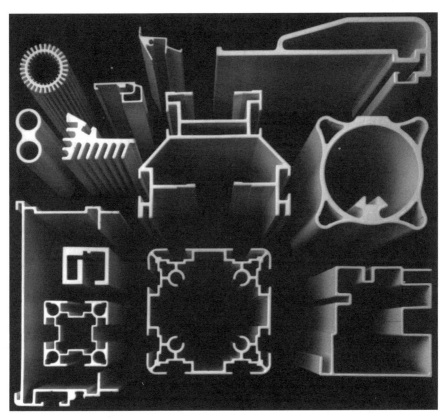

Figure 3.2 Examples of complex extruded shapes. (Courtesy of Cardinal Aluminum Co.)

was gone. Extrusions continued to be produced in shapes that looked like rolled products, however, because these shapes were standard. Around 1970 the Aluminum Association introduced standard channel and I-beam shapes designed to be extruded, with constant thickness flanges and optimum dimensions for strength (Figure 3.3). Today, almost no aluminum shapes are produced by rolling. Many of the old cross sections suited to production by rolling are still shown in catalogs, however, even though today they are extruded, not rolled.

A number of common extruded shapes are shown in extruders' catalogs. Some of these shapes, as well as the Aluminum Association standard shapes, are included among those listed in the *Aluminum Design Manual*, Part VI, Section Properties. (See Section 6.2 for a warning on availability; 15 in. [380 mm] deep channels are about the deepest shapes extruded for general use.) Extruders usually maintain an inventory of dies. Some are proprietary and,

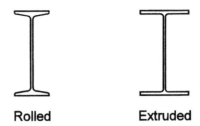

Rolled   **Extruded**

Figure 3.3 Rolled shape with sloped flanges, and extruded shape with constant thickness flanges.

thus, are for the exclusive use of a particular customer; others are called *open dies*, and these are the shapes shown in catalogs and available to any paying customer.

The Aluminum Association has established standard extruded I-beam and channel shapes, as mentioned above, in depths from 2 to 12 in. (They are not available in hard metric sizes.) These sections are efficient for structural applications and are produced by a number of extruders. Appendix B contains dimensions and section properties for these shapes, called Aluminum Association standard channels and I beams. Design data for these shapes is given in Appendices D and E.

Aluminum *pipe*, with the same dimensions as steel pipe of the same nominal size and schedule, is also extruded, in diameters up to 12 in. Dimensions and section properties are listed in Table 3.1. Aluminum pipe is usually 6061-T6 alloy, but it is also produced in 3003-H18 (but only under 1 in. nominal pipe size), 3003-H112, and 6063-T6.

Custom Extruded Shapes Sometimes the Aluminum Association or other standard shapes are impractical or inefficient for a specific application. In these cases, users may design their own shapes. Keep in mind that some limitations exist:

1) *Minimum Thickness:* The minimum thickness is a function of a number of factors, including the circle size (larger shapes require larger wall thickness) and whether a shape is hollow or solid. Table 3.2 provides approximate guidelines, but actual limits depend on the shape and the extruder. Elongation testing is not performed for shapes less than 0.062 in. [1.6 mm] thick.

2) *Maximum Length:* Extrusions can be produced up to 100 ft [30 m] long, but 40 ft [12 m] is generally the practical limit for extrusions shipped by truck. Structural shapes available from metal supply warehouses usually are stocked in 20 ft or 25 ft [6 m or 7.5 m] maximum lengths.

3) *Maximum Circle Size:* Extrusions are produced from *billets*, which are usually cylindrically shaped. Because of this, extrusion cross-section size is

TABLE 3.1 Diameters, Wall Thicknesses, Weights—Pipe

Nominal Pipe Size ① (in.)	Schedule Number	Outside Diameter (in.)			Inside Diameter (in.)	Wall Thickness (in.)			Weight per Foot (lb)	
		Nom. ①	Min ② ④	Max. ② ④	Nom.	Nom. ①	Min. ②	Max. ②	Nom. ③	Max. ② ③
$\frac{1}{8}$	40	0.405	0.374	0.420	0.269	0.068	0.060	..	0.85	0.091
	80	0.405	0.374	0.420	0.215	0.095	0.083	..	0.109	0.118
$\frac{1}{4}$	40	0.540	0.509	0.555	0.364	0.088	0.077	..	0.147	0.159
	80	0.540	0.509	0.555	0.302	0.119	0.104	..	0.185	0.200
$\frac{3}{8}$	40	0.675	0.644	0.690	0.493	0.091	0.080	..	0.196	0.212
	80	0.675	0.644	0.690	0.423	0.126	0.110	..	0.256	0.276
$\frac{1}{2}$	5	0.840	0.809	0.855	0.710	0.065	0.053	0.077	0.186	..
	10	0.840	0.809	0.855	0.674	0.083	0.071	0.095	0.232	..
	40	0.840	0.809	0.855	0.622	0.109	0.095	..	2.294	0.318
	80	0.840	0.809	0.855	0.546	0.147	0.129	..	0.376	0.406
	160	0.840	0.809	0.855	0.464	0.188	0.164	..	0.453	0.489
$\frac{3}{4}$	5	1.050	1.019	1.065	0.920	0.065	0.053	0.077	0.237	..
	10	1.050	1.019	1.065	0.884	0.083	0.071	0.095	0.297	..
	40	1.050	1.019	1.065	0.824	0.113	0.099	..	0.391	0.422
	80	1.050	1.019	1.065	0.742	0.154	0.135	..	0.510	0.551
	160	1.050	1.019	1.065	0.612	0.219	0.192	..	0.672	0.726
1	5	1.315	1.284	1.330	1.185	0.065	0.053	0.077	0.300	..
	10	1.315	1.284	1.330	1.097	0.109	0.095	0.123	0.486	..
	40	1.315	1.284	1.330	1.049	0.133	0.116	..	0.581	0.627
	80	1.315	1.284	1.330	0.957	0.179	0.157	..	0.751	0.811
	160	1.315	1.284	1.330	0.815	0.250	0.219	..	0.984	1.062

TABLE 3.1 Diameters, Wall Thicknesses, Weights—Pipe (*continued*)

Nominal Pipe Size ① (in.)	Schedule Number	Outside Diameter (in.)			Inside Diameter (in.)	Wall Thickness (in.)			Weight per Foot (lb)	
		Nom. ①	Min ② ④	Max. ② ④	Nom.	Nom. ①	Min. ②	Max. ②	Nom. ③	Max. ② ③
$1\frac{1}{4}$	5	1.660	1.629	1.675	1.530	0.065	0.053	0.077	0.383	..
	10	1.660	1.629	1.675	1.442	0.109	0.095	0.123	0.625	..
	40	1.660	1.629	1.675	1.380	0.140	0.122	..	0.786	0.849
	80	1.660	1.629	1.675	1.278	0.191	0.167	..	1.037	1.120
	160	1.660	1.629	1.675	1.160	0.250	0.219	..	1.302	1.407
$1\frac{1}{2}$	5	1.900	1.869	1.915	1.770	0.065	0.053	0.077	0.441	..
	10	1.900	1.869	1.915	1.682	0.109	0.095	0.123	0.721	..
	40	1.900	1.869	1.915	1.610	0.145	0.127	..	0.940	1.015
	80	1.900	1.869	1.915	1.500	0.200	0.175	..	1.256	1.357
	160	1.900	1.869	1.915	1.338	0.281	0.246	..	1.681	1.815
2	5	2.375	2.344	2.406	2.245	0.065	0.053	0.077	0.555	..
	10	2.375	2.344	2.406	2.157	0.109	0.095	0.123	0.913	..
	40	2.375	2.351	2.399	2.067	0.154	0.135	..	1.264	1.365
	80	2.375	2.351	2.399	1.939	0.218	0.191	..	1.737	1.876
	160	2.375	2.351	2.399	1.687	0.344	0.301	..	2.581	2.788
$2\frac{1}{2}$	5	2.875	2.844	2.906	2.709	0.083	0.071	0.095	0.856	..
	10	2.875	2.844	2.906	2.635	0.120	0.105	0.135	1.221	..
	40	2.875	2.846	2.904	2.469	0.203	0.178	..	2.004	2.164
	80	2.875	2.846	2.904	2.323	0.276	0.242	..	2.650	2.862
	160	2.875	2.846	2.904	2.125	0.375	0.328	..	3.464	3.741

3	5	3.500	3.469	3.531	3.334	0.083	0.071	0.095	1.048	..
	10	3.500	3.469	3.531	3.260	0.120	0.105	0.135	1.498	..
	40	3.500	3.465	3.535	3.068	0.216	0.189	..	2.621	2.830
	80	3.500	3.465	3.535	2.900	0.300	0.262	..	3.547	3.830
	160	3.500	3.465	3.535	2.624	0.438	0.383	..	4.955	5.351
$3\frac{1}{2}$	5	4.000	3.969	4.031	3.834	0.083	0.071	0.095	1.201	..
	10	4.000	3.969	4.031	3.760	0.120	0.105	0.135	1.720	..
	40	4.000	3.960	4.040	3.548	0.226	0.198	..	3.151	3.403
	80	4.000	3.960	4.040	3.364	0.318	0.278	..	4.326	4.672
4	5	4.500	4.469	4.531	4.334	0.083	0.071	0.095	1.354	..
	10	4.500	4.469	4.531	4.260	0.120	0.105	0.135	1.942	..
	40	4.500	4.455	4.545	4.026	0.237	0.207	..	3.733	4.031
	80	4.500	4.455	4.545	3.826	0.337	0.295	..	5.183	5.598
	120	4.500	4.455	4.545	3.624	0.438	0.383	..	6.573	7.099
	160	4.500	4.455	4.545	3.438	0.531	0.465	..	7.786	8.409
5	5	5.563	5.532	5.625	5.345	0.109	0.095	0.123	2.196	..
	10	5.563	5.532	5.625	5.295	0.134	0.117	0.151	2.688	..
	40	5.563	5.507	5.619	5.047	0.258	0.226	..	5.057	5.461
	80	5.563	5.507	5.619	4.813	0.375	0.328	..	7.188	7.763
	120	5.563	5.507	5.619	4.563	0.500	0.438	..	9.353	10.10
	160	5.563	5.507	5.619	4.313	0.625	0.547	..	11.40	12.31
6	5	6.625	6.594	6.687	6.407	0.109	0.095	0.123	2.624	..
	10	6.625	6.594	6.687	6.357	0.134	0.117	0.151	3.213	..
	40	6.625	6.559	6.691	6.065	0.280	0.245	..	6.564	7.089
	80	6.625	6.559	6.691	5.761	0.432	0.378	..	9.884	10.67
	120	6.625	6.559	6.691	5.501	0.562	0.492	..	12.59	13.60
	160	6.625	6.559	6.691	5.187	0.719	0.629	..	15.69	16.94

TABLE 3.1 Diameters, Wall Thicknesses, Weights—Pipe (continued)

Nominal Pipe Size ① (in.)	Schedule Number	Outside Diameter (in.)			Inside Diameter (in.)	Wall Thickness (in.)			Weight per Foot (lb)	
		Nom. ①	Min ② ④	Max. ② ④	Nom.	Nom. ①	Min. ②	Max. ②	Nom. ③	Max. ② ③
8	5	8.625	8.594	8.718	8.407	0.109	0.095	0.123	3.429	..
	10	8.625	8.594	8.718	8.329	0.148	0.130	0.166	4.635	..
	20	8.625	8.539	8.711	8.125	0.250	0.219	..	7.735	8.354
	30	8.625	8.539	8.711	8.071	0.277	0.242	..	8.543	9.227
	40	8.625	8.539	8.711	7.981	0.322	0.282	..	9.878	10.67
	60	8.625	8.539	8.711	7.813	0.406	0.355	..	12.33	13.31
	80	8.625	8.539	8.711	7.625	0.500	0.438	..	15.01	16.21
	100	8.625	8.539	8.711	7.437	0.594	0.520	..	17.62	19.03
	120	8.625	8.539	8.711	7.187	0.719	0.629	..	21.00	22.68
	140	8.625	8.539	8.711	7.001	0.812	0.710	..	23.44	25.31
	160	8.625	8.539	8.711	6.813	0.906	0.793	..	25.84	27.90
10	5	10.750	10.719	10.843	10.482	0.134	0.117	0.151	5.256	..
	10	10.750	10.719	10.843	10.420	0.165	0.144	0.186	6.453	..
	20	10.750	10.642	10.858	10.250	0.250	0.219	..	9.698	10.47
	30	10.750	10.642	10.858	10.136	0.307	0.269	..	11.84	12.79
	40	10.750	10.642	10.858	10.020	0.365	0.319	..	14.00	15.12
	60	10.750	10.642	10.858	9.750	0.500	0.438	..	18.93	20.45
	80	10.750	10.642	10.858	9.562	0.594	0.520	..	22.29	24.07
	100	10.750	10.642	10.858	9.312	0.719	0.629	..	26.65	28.78

12	5	12.750	12.719	12.843	12.438	0.156	0.136	0.176	7.258	..
	10	12.750	12.719	12.843	12.390	0.180	0.158	0.202	8.359	..
	20	12.750	12.622	12.878	12.250	0.250	0.219	..	11.55	12.47
	30	12.750	12.622	12.878	12.090	0.330	0.289	..	15.14	16.35
	40	12.750	12.622	12.878	11.938	0.406	0.355	..	18.52	20.00
	60	12.750	12.622	12.878	11.626	0.562	0.492	..	25.31	27.33
	80	12.750	12.622	12.878	11.374	0.688	0.602	..	30.66	33.11

① In accordance with ANSI/ASME Standards B36.10M and B36.19M.

② Based on standard tolerances for pipe .

③ Based on nominal dimensions, plain ends, and a density of 0.098 lb per cu in., the density of 6061 alloy. For alloy 6063, multiply by 0.99, and for alloy 3003 multiply by 1.011.

④ For schedules 5 and 10, these values apply to mean outside diameters.

TABLE 3.2a   Approximate Minimum Thicknesses for 6061 and 6063 Solid Extrusions

Circle Size (in.)	6063 Minimum Thickness (in.)	6061 Minimum Thickness (in.)	Circle Size (mm)	6063 Minimum Thickness (mm)	6061 Minimum Thickness (mm)
2 to 3	0.039	0.045	50 to 75	1.0	1.1
3 to 4	0.045	0.050	75 to 100	1.1	1.3
4 to 5	0.056	0.062	100 to 125	1.4	1.6
5 to 6	0.062	0.062	125 to 150	1.6	1.6
6 to 7	0.078	0.078	150 to 175	2.0	2.0
7 to 8	0.094	0.094	175 to 200	2.4	2.4
8 to 10	0.109	0.109	200 to 250	2.8	2.8
10 to 11	0.125	0.125	250 to 275	3.2	3.2
11 to 12	0.156	0.156	275 to 300	4.0	4.0
12 to 20	0.188	0.188	300 to 500	4.8	4.8

usually limited to that which fits within a circle. Larger extrusion presses use larger circle size dies. There are numerous 10 in. [250 mm] and smaller presses, about a dozen mills with presses larger than 12 in. [300 mm], while the largest in North America is 31 in. [790 mm]. Because of production limitations, an extrusion typically cannot fill the full area of the circle, but rather only part of the circle area (Figure 3.4).

4) *Maximum Area:* The largest cross-sectional area that can be extruded is about 125 in^2 [80,600 mm^2].

5) *Maximum Weight:* The maximum total weight of an extrusion is limited by the weight of a billet to about 4,300 lb [2,000 kg].

TABLE 3.2b   Approximate Minimum Thicknesses for 6061 and 6063 Hollow Extrusions

Circle Size (in.)	6063 Minimum Thickness (in.)	6061 Minimum Thickness (in.)	Circle Size (mm)	6063 Minimum Thickness (mm)	6061 Minimum Thickness (mm)
2 to 3	0.062	0.078	50 to 75	1.6	2.0
3 to 4	0.078	0.094	75 to 100	2.0	2.4
4 to 5	0.094	0.109	100 to 125	2.4	2.8
5 to 6	0.109	0.125	125 to 150	2.8	3.2
6 to 7	0.125	0.156	150 to 175	3.2	4.0
7 to 8	0.156	0.188	175 to 200	4.0	4,8
8 to 10	0.188	0.250	200 to 250	4.8	6.3

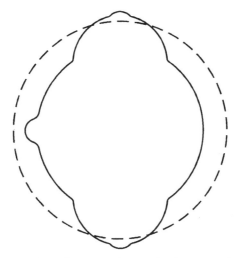

Figure 3.4   Useable portion of an extrusion circle size. Extrusion profiles must generally fit within the area indicated by the dotted line.

Before these limitations are reached, however, cost and availability may be overriding considerations. A number of techniques can be employed to minimize extrusion cost:

1) Minimize the size of the smallest circle that encloses the section. Sometimes this can be done by extruding a folded version of the section desired and then unfolding it after extruding by rolling or bending.

2) Avoid hollows where possible. They require more complex dies and are more difficult to extrude.

3) Avoid large differences in wall thickness in different parts of the cross section. For 6063, the largest ratio of maximum to minimum wall thickness should be 3:1; for 6061-T6, use 2:1.

4) Keep the perimeter-to-cross-sectional area ratio as low as possible. Some extruders estimate costs based on this ratio.

5) Avoid sharp corners, using generous fillets or rounding where possible.

6) Ask extruders which changes to a proposed section would reduce cost.

Extruded structural shapes can perform additional functions by incorporating some of these features:

1) *Interlocking Sections:* Extrusions can be designed to interlock with other extrusions to facilitate connections. Several kinds of extrusion interlocks exist. Some are designed to act like hinges (Figure 3.5a), with one piece being slid into the other from one end. Other extrusions are designed with nested

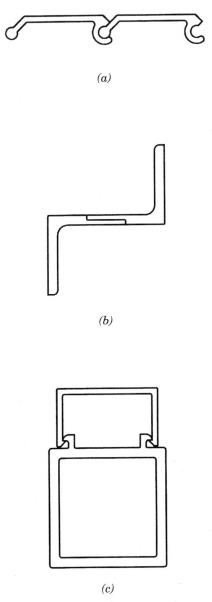

(a)

(b)

(c)

Figure 3.5   Examples of extruded interlocks.

interlocks that align parts, which are then fastened together (Figure 3.5b). Still others are designed as friction fit or snap-lock interlocks, which require the parts to deform elastically to fit together (Figure 3.5c). Unlike the first two joints, the friction fit joint cannot be disassembled without destroying the parts. None of these interlocks are considered to transmit enough longitudinal shear to enable the parts to act as a unit structurally without other means of fastening (133). To prevent longitudinal slippage at significant loads requires a fit too tight to make with parts extruded to standard tolerances.

2) *Skid-Resistant Surfaces:* Although the extrusion process is only capable of producing longitudinal features in a cross section, skid-resistant surfaces are routinely made with extrusions. This is achieved by extruding small triangular ribs on the surface (Figure 3.6), and then notching the ribs in the transverse direction after extruding. Extruders may perform the notching economically by running a ribbed roller over the ridges.

3) *Indexing Marks:* Shallow grooves extruded in parts can indicate where a line of holes is to be punched or drilled. These marks (raised or grooved) can also be used to identify the extruder, distinguish an inside from an outside surface, or distinguish parts otherwise similar in appearance. Dimensions for a typical indexing mark are shown in Figure 3.7.

4) *Screw Chases:* A *ribbed slot* (also called a *screw chase* or *boss*) can be extruded to receive a screw. This may be used, for example, to attach a batten bar to a frame member (Figure 3.8). A method for calculating the amount of lateral frictional resistance to sliding of a screw in a chase is given in AAMA TIR-A9-1991 (34). This calculation must be performed if the design considers

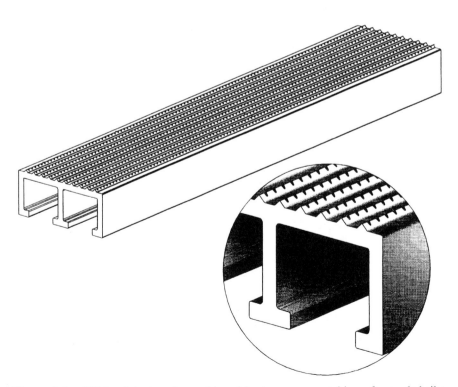

Figure 3.6 Skid-resistant surface achieved by transverse notching of extruded ribs.

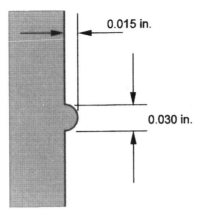

Figure 3.7 Typical dimensions for an extruded indexing or identification mark.

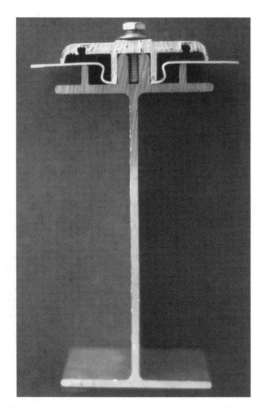

Figure 3.8 Use of a screw chase to connect extruded parts. (Courtesy of Conservatek Industries, Inc.)

both of the connected parts to act as a continuous section. The dimensions of the screw chase are a function of screw dimensions and strength. If the screw chase is too narrow, the head of the screw may break off before the screw is fully driven. If the chase is too wide, the ribs of the chase may strip during fastener installation. The maximum screw torque before stripping is fairly sensitive to the chase width. The narrow dimension of the chase should be less than the root diameter of the screw. The dimensions used successfully for a screw chase for a $\frac{1}{4}$-20 UNC screw are shown in Figure 3.9.

5) *Gasket Retaining Grooves or Guides:* In curtain wall, fenestration, and other applications, elastomeric gaskets are often required between parts. By extruding a groove in the metal that matches a protrusion on a gasket, you can eliminate the need for field assembly or adhesives. Usually the gasket is press-fit into the extrusion in the shop. Dimensions for such a detail are shown in Figure 3.10. Care must be taken to avoid stretching the gasket during installation to prevent contraction of the gasket in the field to a length shorter than required.

6) *Extrusions as a Substitute for Plate:* Plate costs about $1\frac{1}{2}$ times the cost of extrusions, so it's desirable to utilize extrusions rather than plate wherever possible. Extruding to final width dimensions also eliminates the need to cut plate to the desired width, thereby saving fabrication costs. Extruded bars are available from a number of extruders through about 18 in. [457 mm] widths or more, depending on thickness.

7) *Non-prismatic Extrusions:* Extrusions may have different cross sections along their length when stepped extrusion methods are used. The smallest section is extruded first, the die is changed, and a larger section that contains the full area of the smaller section is extruded next. This method is used on aircraft wings to minimize the amount of machining needed to produce tapered members. Other tapered members, such as light poles, may be produced by *spinning*. Set-up costs for these methods are high, so they tend to be used only on parts that will be produced in quantity.

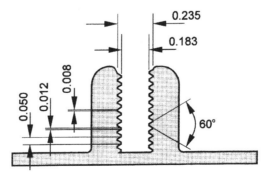

Figure 3.9   Dimensions for a screw chase for a $\frac{1}{4}$ in. diameter screw.

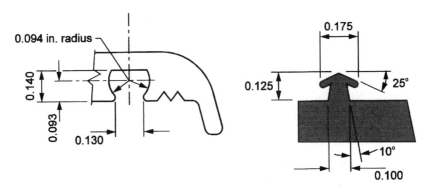

Figure 3.10 Dimensions to allow a gasket to be press fit into an extruded slot.

8) *Grooves for Fasteners:* Grooves can be extruded to permit screw heads to be flush with the surface of an extrusion to avoid the need for countersinking the fastener hole. Groove widths sized to a bolt head flat width can also be used to prevent rotation of the bolt during tightening of the nut. Another use for grooves is to reduce the loss of cross-sectional area that occurs at holes (Figure 3.11).

9) *Integral Backing for Welds:* As shown in Figure 3.12, built-in backing for longitudinal welds along an extrusion edge can be provided, eliminating the need for separate backing and the need to hold it in place during welding.

Hollow Extruded Shapes If you're like most people, you may need a moment of head scratching to imagine how hollow extruded shapes are possible. Extruders use three methods:

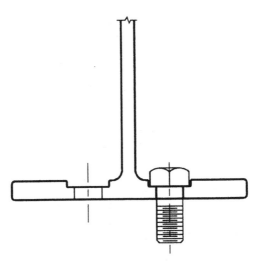

Figure 3.11 Use of an extruded groove in a line of bolts to secure the bolt heads from spinning, and to reduce the loss of effective net area in the cross section.

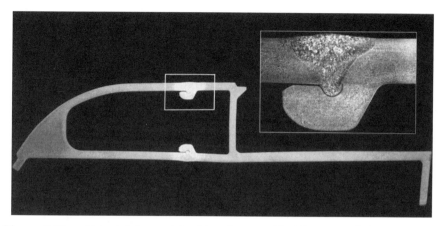

Figure 3.12 Extruded integral backing for a weld. (Courtesy of the Aluminum Association)

1) *Solid Billet and a Porthole Die:* In this method, the metal is divided into two or more streams by the supports for the mandrel that forms the hollow portion of the shape. The metal must reunite (sometimes referred to as "welding") behind the supports before it flows through the die, outlining the perimeter of the shape.

2) *Solid Billet and a Piercer Operating Through a Hollow Ram:* This method avoids the seams inherent in the porthole die approach and, thus, produces a seamless extrusion.

3) *Hollow Billet with a Die and Mandrel:* This method also produces a seamless extrusion, but it is rarely used.

The material specification identifies if hollow extrusions must be seamless (for example, ASTM B241 *Aluminum and Aluminum-Alloy Seamless Pipe and Seamless Extruded Tube*) (53). Do seamless hollow shapes have better structural properties than shapes with seams? The answer is no, not with respect to material properties; the minimum strengths for porthole die produced hollow extrusions are the same (all other things being equal) as those for seamless extrusions. For properly extruded shapes, seams do not appear to have any adverse effect on structural performance, such as burst pressure or fatigue.

Extruded Structural Alloys For purposes of extruding, the 1xxx, 3xxx, and 6xxx series alloys are considered "soft," which means they are more easily and economically extruded than other alloys. The 6xxx series alloys are the most useful for structural applications because of their combination of strength, corrosion resistance, and weldability. Specific extrusion alloys in this series that merit consideration are:

- 6061, the granddaddy of aluminum structural extrusions, usually provides the best combination of strength, economy, and corrosion resistance of aluminum alloys. It is widely used and available.
- 6063 has long been used for architectural applications as a lower strength and slightly lower-cost alloy compared to 6061. This alloy is well suited to such finishes as anodizing. 6463 has properties very similar to 6063. When chemically brightened, it gives a reflective high luster finish.
- 6005 and 6105, in the T5 temper, have the same strength in the unwelded condition as 6061-T6, but they are easier to extrude and less quench sensitive and so are more suitable for complex shapes. When welded, 6005-T5 minimum strengths are 85% those of welded 6061-T6.
- 6066 has higher strength than 6061 but is less corrosion resistant and is harder to extrude.
- 6070 has considerably higher strength than 6061 and is generally deemed corrosion resistant. Its minimum extruded thickness is slightly more limited than 6061, and it requires larger radii to bend. When ordered in sufficient quantity, it costs about 10% to 15% more than 6061 but is 25% stronger.
- 6262-T6 has the same strength and corrosion resistance as 6061-T6. It contains 0.4% to 0.7% lead, which provides it with better machinability than the other 6xxx alloys.
- 6351-T5 has the same strength as 6061-T6, while 6351-T6 is slightly stronger. 6351 enjoys slightly better corrosion resistance than 6061, and its fracture toughness is more reliably achieved than for 6061.
- 6082, a relatively new alloy, is discussed in Section 9.3.2.
- 7005 is as corrosion resistant as 6061 but with much higher strength; its disadvantage is that it is less easily extruded than 6061.

All of these alloys have nearly the same density, modulus of elasticity, and coefficient of thermal expansion. Where strength is an important factor, 6061-T6, 6351-T5 and -T6 and 6005-T5 and 6105-T5 should be considered. If higher strength is needed, 6066, 6070, and 7005 are candidates. Where strength requirements are less demanding, as is the case when deflection or fatigue considerations govern the design, 6063 is well suited. 6463 is used for applications where a high luster finished appearance is critical.

Table 3.3 summarizes information on these alloys. The corrosion ratings are from the Aluminum Association's A through E scale with A being best (see the *Aluminum Design Manual*, Part IV, Materials), and strengths are minimums. Costs are approximate and vary by supplier. In quantities sufficient to be extruded on order, solid shapes of 6063-T6 cost about $1.20/lb in 2001; hollow shapes cost about $0.10/lb more.

Extrusion alloys not listed in Table 3.3, such as the 2xxx series (e.g., 2014, 2024, and 2219), 5xxx series (e.g., 5083, 5086, 5154, and 5454), and 7xxx

TABLE 3.3 Properties of Some Commonly Extruded Aluminum Alloys

Alloy Temper	F_{tu} (ksi)	F_{tu} (MPa)	F_{ty} (ksi)	F_{ty} (MPa)	General Corrosion Resistance	Relative Cost
6063-T5*	22	150	16	110	A	100%
6063-T6	30	205	25	170	A	100%
6463-T6	30	205	25	170	A	104% to 105%
6005-T5	38	260	35	240		103%
6105-T5	38	260	35	240		103%
6061-T6	38	260	35	240	B	103%
6162-T6	38	260	35	240		
6262-T6	38	260	35	240	B	109%
6351-T5	38	260	35	240	A	
6351-T6	42	290	37	255	A	
6066-T6	50	345	45	310	C	
6070-T6	48	330	45	310	B	110% to 115%
7005-T53	50	345	44	305		117%

*up through 0.500 in. [12.50 mm] thick.

series with zinc and copper (e.g., 7050, 7075, 7178), are more difficult to extrude. Correspondingly, their costs are higher than those for the 6xxx series. The 5xxx series alloy extrusions are generally only used in marine applications where their corrosion resistance justifies their cost. While the 2xxx and 7xxx alloys are generally very strong, they have serious drawbacks, such as reduced corrosion resistance (still usually better than steel, though) and poor weldability. Alloys 2014 and 2024 are not fusion welded because of reduced strength and corrosion resistance at the welds and often require coating to resist corrosion.

Extruders and Their Capabilities   The Aluminum Extrusion Council (AEC) (www.aec.org) makes available a list of aluminum extruders and their capabilities. Many of these companies provide additional services for extruded products, such as finishing, including anodizing and painting; and fabricating, such as drilling, punching, welding, and bending. Minimum mill orders vary by extruder, circle size, and shape (solid or hollow). For smaller solid shapes (6 in. [150 mm]), minimum orders range around 500 lb [200 kg]; for larger or hollow shapes, minimum orders are about 1,000 to 2,000 lb [500 to 1,000 kg]. When smaller quantities are needed, they can be obtained from distributors or warehouses. Die charges range around several hundred dollars with typically at least six week lead times.

Extrusion Tolerances   Standard extrusion cross-sectional dimensional *tolerances* have been established and are published by the Aluminum Association. They are given in Table 3.4. For other tolerances (such as length, straightness, twist, flatness of flat surfaces, surface roughness, contour of curved surfaces, squareness of cut ends, corner and fillet radii, and angularity) consult *Aluminum Standards and Data* (11). Determining extrusion tolerances can be more complicated than you'd think, so you may want to consult your extruder. Closer tolerances can be held at increased cost; most extrusions can be ordered to tolerances that are one-half the *Aluminum Standards and Data* values. The 5xxx series alloys have larger tolerances than other extruded alloys.

Fabrication   For many applications, extrusions require little additional fabrication once they've been shipped from the mill. Typical processes involve cutting to length and fabricating holes. Extrusions may also be rolled to take on curvature for camber or other purposes. However, if strains required by the rolling process are great enough, they may produce cracking. This situation can sometimes be avoided by rolling before the material is *artificially aged*. For example, 6061 extrusions can be rolled in the -T4 temper and then artificially aged at the mill to the -T6 temper, which is usually needed for strength in the finished member. The -T4 temper has a minimum elongation of 16% versus 8% for the -T6 temper, so cracking is more readily avoided this way.

Extrusion Quality Assurance   Extrusion mills will generally supply certification meeting Aluminum Association requirements at no additional charge if this is requested when the material is ordered. The number of samples for mechanical tests is determined as follows: for extrusions weighing less than 1 lb/ft [1.7 kg/m], one sample is taken for each 1,000 lb. [500 kg], or fraction thereof, in a *lot*; for extrusions weighing 1 lb/ft [1.7 kg/m] or more, one sample is taken for each 1,000 ft [300 m], or part thereof, in a lot. Tension tests are made for the longitudinal direction only, unless other prior agreement is made with the supplier. The *tensile yield strength*, *tensile ultimate strength*, and *elongation* are reported and must meet or exceed the minimum values for these properties established in *Aluminum Standards and Data*.

Purchasers of extrusions ordered in mill quantities for building structures often do not require identification marking. (Indexing, or identification, marks [see above] extruded into the shape are, therefore, particularly useful to keep track of producer and alloy.) Extruded structural shapes purchased from a warehouse usually have identification marking consisting of the producer and the alloy and temper. Suppliers may provide either continuous (at intervals no greater than 40 in. [1,000 mm] along the length) or spot (once per piece) marking.

TABLE 3.4 Cross-Sectional Dimension Tolerances—Profiles①

Except for T3510, T4510, T6510, T73510, T76510 and T8510 Tempers⑦

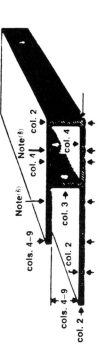

	Metal Dimensions				Tolerance ② ③—in. Plus and Minus								
					Space Dimensions								
					Allowable Deviation from Specified Dimension Where More Than 25 Percent of the Dimension is Space ⑥ ⑧								
Specified Dimension	Allowable Deviation from Specified Dimension Where 75 Percent or More of The Dimension is Metal ⑨ ⑩		Wall Thickness ④ Completely ⑤ Enclosing Space 0.11 sq. in. and Over (Eccentricity)		At Dimensioned Points 0.250–0.624 Inches from Base of Leg		At Dimensioned Points 0.625–1.249 Inches from Base of Leg		At Dimensioned Points 1.250–2.499 Inches from Base of Leg		At Dimensioned Points 2.500–3.999 Inches from Base of Leg		At Dimensioned Points 4.000–5.999 Inches from Base of Leg		At Dimensioned Points 6.000–8.000 Inches from Base of Leg
	All Except Those Covered by Column 3													
in.	Col. 2		Col. 3		Col. 4		Col. 5		Col. 6		Col. 7		Col. 8		Col. 9
Col. 1	Alloys 5083 5086 5454	⑪ Other Alloys	Alloys 5083 5086 5454	⑪ Other Alloys	Alloys 5083 5086 5454	⑪ Other Alloys	Alloys 5083 5086 5454	⑪ Other Alloys	Alloys 5083 5086 5454	⑪ Other Alloys	Alloys 5083 5086 5454	⑪ Other Alloys	Alloys 5083 5086 5454	⑪ Other Alloys	Alloys 5083 5086 5454	⑪ Other Alloys

(Diagram annotations: cols. 4–9; Note⑥; cols. 4–9; col. 2; Note⑻; col. 2; col. 3; col. 2; col. 1; col. 2; col. 4)

TABLE 3.4 Cross-Sectional Dimension Tolerances—Profiles① (continued)

Circumscribing Circle Sizes Less Than 10 In. in Diameter

Column-group tolerances: ±15% of specified dimension; ±.090 max, ±.015 min. • ±10% of specified dimension; ±.060 max, ±.010 min.

Cross-Sectional Dimension														
Up thru 0.124	.009	.006	.013	.010	.015	.014	⋮	⋮	⋮	⋮	⋮	⋮	⋮	⋮
0.125–0.249	.011	.007	.016	.012	.018	.014	.020	.016	⋮	⋮	⋮	⋮	⋮	⋮
0.250–0.499	.012	.008	.018	.014	.020	.016	.022	.018	.024	.020	⋮	⋮	⋮	⋮
0.500–0.749	.014	.009	.021	.016	.023	.018	.025	.020	.027	.022	⋮	⋮	⋮	⋮
0.750–0.999	.015	.010	.023	.018	.025	.020	.027	.022	.030	.025	.035	.030	⋮	⋮
1.000–1.499	.018	.012	.027	.021	.029	.023	.032	.026	.036	.030	.041	.035	⋮	⋮
1.500–1.999	.021	.014	.031	.024	.033	.026	.038	.031	.043	.036	.049	.042	.057	.050
2.000–3.999	.036	.024	.046	.034	.050	.038	.060	.048	.069	.057	.080	.068	.092	.080
4.000–5.999	.051	.034	.061	.044	.067	.050	.081	.064	.095	.078	.111	.094	.127	.110
6.000–7.999	.066	.044	.076	.054	.084	.062	.104	.082	.121	.099	.142	.120	.162	.140
8.000–9.999	.081	.054	.091	.064	.101	.074	.127	.100	.147	.120	.182	.145	.197	.170

Circumscribing Circle Sizes 10 In. in Diameter and Over

Column-group tolerances: ±15% of specified dimension; ±.090 max, ±.025 min. • ±15% of specified dimension; ±.090 max, ±.015 min.

Cross-Sectional Dimension														
Up thru 0.124	.021	.014	.025	.018	.027	.020	⋮	⋮	⋮	⋮	⋮	⋮	⋮	⋮
0.125–0.249	.022	.015	.026	.019	.029	.022	.035	.028	⋮	⋮	⋮	⋮	⋮	⋮
0.250–0.499	.024	.016	.028	.020	.032	.024	.038	.030	.058	.050	⋮	⋮	⋮	⋮
0.500–0.749	.025	.017	.030	.022	.035	.027	.049	.040	.068	.060	⋮	⋮	⋮	⋮
0.750–0.999	.027	.018	.031	.023	.039	.030	.057	.050	.079	.070	.099	.090	⋮	⋮
1.000–1.499	.028	.019	.033	.024	.043	.034	.069	.060	.089	.080	.109	.100	⋮	⋮
1.500–1.999	.036	.024	.046	.034	.056	.044	.082	.070	.102	.090	.122	.110	.182	.170
2.000–3.999	.051	.034	.061	.044	.071	.054	.097	.080	.117	.100	.137	.120	.197	.180
4.000–5.999	.066	.044	.076	.054	.086	.064	.112	.090	.132	.110	.152	.130	.212	.190
6.000–7.999	.081	.054	.091	.064	.101	.074	.127	.100	.147	.120	.167	.140	.227	.200
8.000–9.999	.096	.064	.106	.074	.116	.084	.142	.110	.162	.130	.182	.150	.242	.210
10.000–11.999	.111	.074	.121	.084	.131	.094	.157	.120	.177	.140	.197	.160	.257	.220
12.000–13.999	.126	.084	.136	.094	.146	.104	.172	.130	.192	.150	.212	.170	.272	.230
14.000–15.999	.141	.094	.151	.104	.161	.114	.187	.140	.207	.160	.227	.180	.287	.240
16.000–17.999	.156	.104	.166	.114	.176	.124	.202	.150	.222	.170	.242	.190	.302	.250

18.000–19.999	.171	.114	.181	.124	.191	.134	.217	.160	.237	.180	.257	.200	.317	.260
20.000–21.999	.186	.124	.196	.134	.206	.144	.232	.170	.252	.190	.272	.210	.332	.270
22.000–24.000	.201	.134	.211	.144	.221	.154	.247	.180	.267	.200	.287	.220	.347	.280

Closed-Space Dimensions

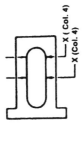

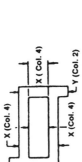

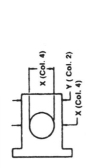

All dimensions designated "Y" are classed as "metal dimensions," and tolerances are determined from column 2.

Dimensions designated "X" are classed as "space dimensions through an enclosed void," and the tolerances applicable are determined from column 4 unless 75 percent or more of the dimension is metal, in which case column 2 applies.

① These Standard Tolerances are applicable to the average profile (shape); wider tolerances may be required for some profiles (shapes) and closer tolerances may be possible for others.

② The tolerance applicable to a dimension composed of two or more component dimensions is the sum of the tolerances of the component dimensions if all of the component dimensions are indicated.

③ When a dimension tolerance is specified other than as an equal bilateral tolerance, the value of the standard tolerance is that which applies to the mean of the maximum and minimum dimensions permissible under the tolerance for the dimension under consideration.

④ Where dimensions specified are outside and inside, rather than wall thickness itself, the allowable deviation (eccentricity) given in Column 3 applies to mean wall thickness. (Mean wall thickness is the average of two wall thickness measurements taken at opposite sides of the void.)

⑤ In the case of Class 1 Hollow Profiles (Shapes) the standard wall thickness tolerance for extruded round tube is applicable. (A Class 1 Hollow Profile (Shape) is one whose void is round and one inch or more in diameter and whose weight is equally distributed on opposite sides of two or more equally spaced axes.)

TABLE 3.4   Cross-Sectional Dimension Tolerances—Profiles①

Except for T3510, T4510, T6510, T73510, T76510 and T8510 Tempers⑦

Open-Space Dimensions

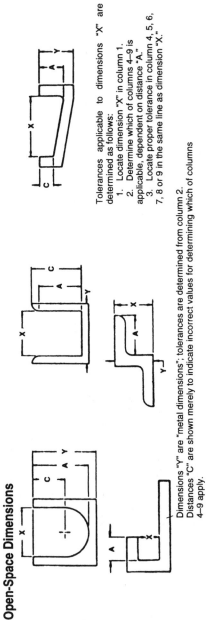

Dimensions "Y" are "metal dimensions"; tolerances are determined from column 2. Distances "C" are shown merely to indicate incorrect values for determining which of columns 4–9 apply.

Tolerances applicable to dimensions "X" are determined as follows:

1. Locate distance "B" in column 1.
2. Determine which of columns 4–9 is applicable, dependent on distance "A."
3. Locate proper tolerance in column 4, 5, 6, 7, 8 or 9 in the same line as value chosen in column 1.

Tolerances applicable to dimensions "X" are determined as follows:

1. Locate dimension "X" in column 1.
2. Determine which of columns 4–9 is applicable, dependent on distance "A."
3. Locate proper tolerance in column 4, 5, 6, 7, 8 or 9 in the same line as dimension "X."

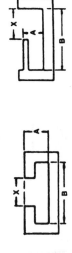

Tolerances applicable to dimensions "X" are not determined from Table 3.4; tolerances are determined by standard tolerances applicable to angles "A."

⑥ At points less than 0.250 inch from base of leg the tolerances in Col. 2 are applicable.

⑦ Tolerances for extruded profiles (shapes) in T3510, T4510, T6510, T73510, T76510 and T8510 tempers shall be as agreed upon between purchaser and vendor at the time the contract or order is entered.

⑧ The following tolerances apply where the space is completely enclosed (hollow profiles (shapes)):

For the width (A), the balance is the value shown in Col. 4 for the depth dimension (D).

For the depth (D), the tolerance is the value shown in Col. 4 for the width dimension (A).

In no case is the tolerance for either width or depth less than the metal dimensions (Col. 2) at the corners.

Example—Alloy 6061 hollow profile (shape) having 1 × 3 rectangular outside dimensions: width tolerance is ± 0.021 inch and depth tolerance ± .034 inch. (Tolerances at corners, Col. 2, metal dimensions, are ± 0.024 inch for the width and ± 0.012 inch for the depth.) Note that the Col. 4 tolerance of 0.021 inch must be adjusted to 0.024 inch so that it is not less than the Col. 2 tolerance.

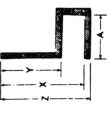

depth D

width A

⑨ These tolerances do not apply to space dimensions such as dimensions "X" and "Z" of the example (right), even when "Y" is 75 percent or more of "X." For the tolerance applicable to dimensions "X" and "Z," use Col. 4, 5, 6, 7, 8 or 9, dependent on distance "A."

3t or Greater

3t or Greater

⑩ The wall thickness tolerance for hollow or semihollow profiles (shapes) shall be as agreed upon between purchaser and vendor at the time the contract or order is entered when the nominal thickness of one wall is three times or greater than that of the opposite wall.

⑪ Limited to those alloys listed in *Aluminum Standards and Data*, Table 11.1.

55

3.1.2 Sheet and Plate

Introduction The rolled products with structural uses are addressed by ASTM B209, *Aluminum and Aluminum-Alloy Sheet and Plate* (49). *Sheet* is defined as rolled product with a thickness of at least 0.006 in. (anything thinner is called *foil*) and less than 0.25 in., although in structural applications the thinnest nominal sheet thickness of much practical use is 0.024 in. [0.6 mm] thick. *Plate* has a thickness of 0.25 in. or greater and is produced up to 8 in. [200 mm] thick in some alloys. The SI unit definitions are subtly different: sheet is rolled product thicker than 0.15 mm and up to 6.3 mm thick, and plate is thicker than 6.3 mm. (See Table 3.5.) Almost all aluminum alloys are produced in sheet or plate or both; they are the most common aluminum product forms.

What's Available Sheet and plate are available from producers in several forms. Common flat sheet widths are 24 in., 30 in., 36 in., 48 in., and 60 in. [600 mm, 750 mm, 900 mm, 1,200 mm, and 1,500 mm]; common lengths are 96 in., 120 in., and 144 in. [2,400 mm, 3,000 mm, and 3,600 mm]. Sheet is also sold rolled in *coils* (Figure 3.13) available in various alloys and thicknesses in widths of 30 in., 36 in., 48 in., 60 in., 72 in., 96 in., 102, and 108 in. [750 mm, 900 mm, 1200 mm, 1500 mm, 1800 mm, 2400 mm, 2600 mm, and 2700 mm], as well as others. The advantage of coils is that piece lengths are limited only by fabrication processes, such as brake press die lengths, rather than by the material stock size.

Common thicknesses are given in Tables 3.6 and 3.7. Thickness is also identified in terms of *gauge* number. *Aluminum sheet gauge thicknesses are different from steel.* For example, 16 gauge aluminum sheet is 0.0508 in. [1.29 mm] thick, while 16 gauge steel is 0.0625 in. [1.59 mm] thick. The *Aluminum Design Manual*, Part VI, Section Properties, Table 33, gives aluminum (in the first column) and steel gauge thicknesses. It is preferable when ordering sheet to identify thickness in inches rather than by gauge.

Pre-painted sheet is available from producers; Section 3.2 contains more information.

Tolerances *Tolerances* on thickness, width, length, *lateral bow*, squareness, and flatness for sheet and plate are given in *Aluminum Standards and Data*, Tables 7.7 through 7.18. Tolerances are larger for thicker and wider product.

TABLE 3.5   Flat-Rolled Products

Product	U.S. Units (in.)	SI Units (mm)
Foil	$t < 0.006$	$t \leq 0.15$
Sheet	$0.006 \leq t < 0.25$	$0.15 < t \leq 6.3$
Plate	$t \geq 0.25$	$t > 6.3$
t = thickness		

Figure 3.13 Aluminum sheet may be purchased in coil form. This allows pieces to be cut or nested to suit fabrication requirements, thereby reducing waste. (Courtesy of Temcor)

Also, aerospace alloys (2014, 2024, 2124, 2219, 2324, 2419, 7050, 7075, 7150, 7178, and 7475) have different tolerances on thickness than other alloys; in some cases the aerospace thickness tolerance is larger, and in others it's smaller than for other alloys. The width breakoffs for thickness tolerances are round numbers in metric units (in other words, they have been *hard-*

TABLE 3.6   Common Aluminum Sheet Thicknesses (U.S. Units)

Thickness (in.)	Gauge No.	Weight (lb/ft²)	Thickness (in.)	Gauge No.	Weight (lb/ft²)
0.025	22	0.360	0.080	12	1.15
0.032	20	0.461	0.090	11	1.30
0.040	18	0.576	0.100	10	1.44
0.050	16	0.720	0.125	8	1.80
0.063	14	0.907	0.160	6	2.30
0.071	13	1.02	0.190		2.74

Weight is based on a density of 0.1 lb/in.³.

TABLE 3.7 Common Aluminum Sheet and Plate Thicknesses (SI Units)

Thickness (mm)	Thickness (in.)	Mass (kg/m²)
0.50	0.020	1.35
0.60	0.024	1.62
0.80	0.031	2.16
1.0	0.039	2.70
1.2	0.047	3.24
1.6	0.063	4.32
2.0	0.079	5.40
2.5	0.098	6.75
3.0	0.118	8.10
3.5	0.138	9.45
4.0	0.157	10.8
4.5	0.177	12.15
5.0	0.197	13.5
6.0	0.236	16.2
7.0	0.276	18.9
8.0	0.315	21.6
10	0.394	27.0
12	0.472	32.4
16	0.630	43.2
20	0.787	54.0
25	0.984	67.5
30	1.18	81.0
35	1.38	94.5
40	1.57	108
50	1.97	135
60	2.36	162
80	3.15	216
100	3.94	270

Masses are based on a density of 2.70 kg/m³.

converted). Alclad alloys (see Section 3.2) have the same thickness tolerances as non-alclad alloys. It's customary in the tank and pressure vessel industry to limit the under run tolerance to 0.01 in. [0.25 mm], which is more stringent than standard aluminum mill tolerance for many commonly supplied widths and thicknesses. Rolling mills can generally meet the tighter tolerance if they are made aware of it beforehand.

Fabrication   Sheet and plate may be sheared, sawed, router-cut, nibbled by punching, plasma-arc-or laser-cut, or cut with an abrasive water jet. Shearing is generally limited to thicknesses less than $\frac{1}{2}$ in. [12 mm]. Coiled sheet may also be slit. Oxygen-cutting is not permitted because the heat is too great, causing melting or weakening of the aluminum. Plasma arc and laser cutting may produce edge cracks in heat-treatable alloys, so these edges must be planed or ground to sound metal.

Take care to avoid re-entrant cuts (Figure 3.14). To do this, first fabricate a hole near the intersection of cut lines. The cut then terminates at the hole, which serves to fillet the corner. Sometimes cuts can also be angled so any resulting notches occur in the drop. AWS D1.2-97, *Structural Welding Code—Aluminum*, (91) requires that corners in statically loaded structures be filleted to a radius of $\frac{1}{2}$ in. [13 mm] and $\frac{3}{4}$ in. [19 mm] in dynamically loaded structures.

Some cutting and forming operations involve the application of heat to the material. While this is not limited to sheet and plate, this issue most often arises with these products. The Aluminum *Specification* (in Section 6.3a) allows heating to a temperature not exceeding 400°F [200°C] for a period not exceeding 30 minutes before the effect on strength must be considered. Temperatures and periods greater than these produce partial annealing of the material, which reduces the strength toward the O temper of the alloy, and in certain alloys decreases corrosion resistance. (See Chapter 4.) Note that the effect of time at elevated temperature is cumulative, so you are not allowed to heat to 400°F for 30 minutes, let the metal cool, and then heat to 400°F again.

The length of time that an alloy can be held at an elevated temperature of less than 400°F [200°C] is not given in the Aluminum *Specification* since it's a function of the temperature and alloy. Aluminum may be held at temperatures lower than 400°F [200°C] for periods of time longer than 30 minutes before the loss in strength is significant. For example, 6061-T6 may be held at 375°F [191°C] for 2 hours, and 350°F [177°C] for 10 hours, without a loss of more than about 5% in strengths (1). The effect of heating on many alloy tempers can be quantified from *Properties of Aluminum Alloys: Tensile,*

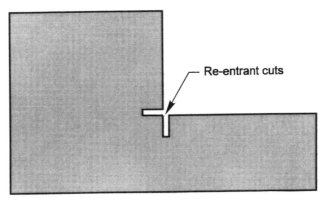

Figure 3.14   Re-entrant cuts in sheet material can precipitate tearing and should be avoided.

Creep, and Fatigue Data at High and Low Temperatures (118), which was used to prepare Table 8.23.

You'll want to carefully monitor the time and temperature of any heating above about 150°F [66°C]. This can be done by marking the metal with heat crayons with known melting points: when the marking starts to run, the metal has reached a benchmark temperature. Unlike steel, heating aluminum doesn't change its color while the metal is in a solid state, so you can't gauge the temperature of a piece of aluminum by looking at it, and you don't want to try by touching it. Most fabricators simply avoid the practice of heating aluminum for cutting or forming. Where aluminum must be heated above 400°F [200°C] in specific areas (such as at welds), the strength reduction is accounted for in structural calculations. Methods for addressing this effect from welding are discussed in Chapter 9.

Sheet and plate may be formed by various operations. *Roll forming* of sheet is used to make cross sections with full-length bends (Figure 3.15), such as for roofing and siding panels produced in large quantities, but it has high initial tooling costs. Press brakes are commonly used to form bends of various configurations. Brake capacity needed to form aluminum sheet and plate can be calculated from capacities given for steel by ratioing by the aluminum yield strength to the steel yield strength. Many commonly used aluminum sheet alloys require less brake capacity than steel.

Minimum bend radii for 90° cold bends for some alloy tempers and thicknesses are given in the *Aluminum Design Manual*, Part VII, Table 6–1. These radii are the smallest recommended in a standard press brake without fracturing and should be verified on trial pieces before production. Minimum bend radii for aluminum are generally greater than those for steel, so you should consider them carefully to avoid cracking (20). For heat-treatable alloys, bending at right angles to the direction that plate or sheet was rolled or extruded helps prevent cracks. For non-heat-treatable alloys, bends should be parallel to the direction of rolling to minimize cracking. Maximum sheet thickness for 180° cold bends (metal to metal) of some alloy tempers is given in the *Design Manual*, Part VII, Design Aids, Table 6-3, which is useful for designing sections with lockseams, such as that shown in Figure 3.16. For some alloy tempers and thicknesses, ASTM B209 gives the minimum diameter of a pin around which the material can be wrapped 180° without cracking.

The fact that a radius is usually required at bends affects cross-sectional dimensions. The sum of the outside dimensions of a bent cross section is different from the sum of the flat lengths (see Figure 3.17). There are two reasons for this difference: the material stretches slightly when bent, and the distance along a radius at the bends is less than the straight line distance to the point of intersection of the flat lengths. The *Aluminum Design Manual*, Part VI, Table 4, gives values for this difference, called the *developed length* of a bend. Another method for accounting for the effect of bends on dimensions is the following equation, which also applies to bend angles other than 90°:

Figure 3.15 Roll-forming aluminum. Flat sheet or coil material can be fed through a series of rollers to form bends and quickly produce a predetermined profile, such as this roofing. (Courtesy of Metform International)

Figure 3.16 Lockseam joints require bending sheet metal back on itself (180°).

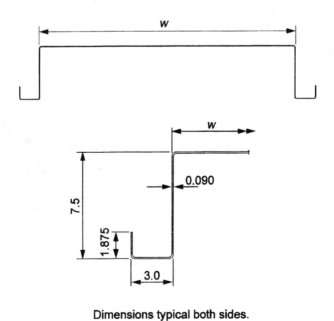

Dimensions typical both sides.

Figure 3.17 Cold-formed sheet profile. (See also Figure 10.8.)

$$L = (0.64t + 1.57R) \, A/90 \tag{3.1}$$

where:

L = length before bending
t = metal thickness
R = bend inside radius
A = bend angle in degrees

The material stretches at the outside of the bend and compresses on the inside of the bend, but because aluminum stretches slightly more readily than it compresses, the neutral axis is slightly closer to the inside surface than the outside surface. In this equation, the neutral axis is located approximately 40% of the material thickness from the inside surface. This location depends on several factors, including the material and the angle of bend, so for other materials the coefficient of t is not necessarily the same as that used in Equation 3.1.

Example 3.1: Determine the outside dimension (w) of the top flange of the cross section shown in Figure 3.17 for a section made from 60 in. wide 5052-H36 sheet 0.09 in. thick.

First, determine the radius to be used for making the six bends in the cross section. The *Aluminum Design Manual.* Part VII, Table 6-1, Recommended Minimum Bend Radii for 90 Degree Cold Bends, doesn't show the 0.09 in. thickness, but the recommended radius for the next thicker sheet ($\frac{1}{8}$ in.) is $2\frac{1}{2} t$. If we use this conservatively, the bend radius will be $2.5 \times 0.09 = 0.23$ in., which we'll round up to 0.25 in. for convenience.

At each bend the difference between the outside dimensions and the length of the formed sheet will be (see Figure 3.18):

$$2(t + R) - (0.64t + 1.57R) \, 90/90 =$$
$$2(0.09 + 0.25) - [0.64(0.09) + 1.57(0.25)] = 0.23 \text{ in.}$$

The sum of the outside dimensions of the cross section is:

$$2(1.875 + 3 + 7.5) + w = 24.75 + w$$

The material width of 60 in. will equal the sum of the outside dimensions less the difference between the outside dimensions and the formed sheet at each of the six bends calculated above:

$$60 = 24.75 + w - 6(0.23)$$
$$w = 36.63 \text{ in.}$$

When designing a section to be fabricated by bending, you must consider the

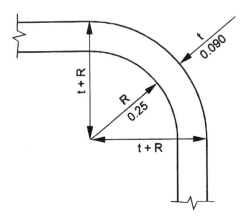

Figure 3.18   Detail of a corner bend in a cold-formed shape.

feasibility of forming bends. If the bends are too close together, it may not be possible to fabricate them. The proper distance between bends is a function of the bend radii and the dimensions of the brake and dies used. Also, to allow for variation in the actual developed length of bends versus the calculated length, the bend sequence may be planned so that the variance is accumulated in a non-critical dimension. Another consideration arises when the bend is longer than the clear distance between the supports of the press. In such instances, the distance from the edge of the piece to the bend cannot be too great or the piece cannot be inserted far enough into the press. When this occurs, bends nearest the edge of the piece may need to be made first.

Uncoiled sheet may experience *coil set*, a longitudinal bowing in the direction of coiling, especially near the end of the coil where the radius of curvature is smallest. To eliminate this, introduce small longitudinal stiffening bends. Large, flat expanses of sheet are undesirable even when made from flat sheet stock because they tend to show buckles or waves, action sometimes referred to as "oil-canning." Often these are not a structural concern but can adversely affect appearance. Small stiffening bends at sufficient intervals can overcome this.

Quality Assurance   Aluminum plate (and other products) may contain flaws (euphemistically called *discontinuities*) due to porosity, laminations, cracks, or foreign material. *Aluminum Standards and Data (ASD)* provides no requirements other than visual inspection except ultrasonic inspection of certain 2xxx and 7xxx series alloys (Table 6.3), usually referred to as "aerospace alloys." ASTM B548, *Ultrasonic Inspection of Aluminum-Alloy Plate for Pressure Vessels* (65), and ASTM B594, *Ultrasonic Inspection of Aluminum-Alloy Wrought Products for Aerospace Applications* (68), also provide specifications for ultrasonic inspection of aluminum products. Even if your application isn't a pressure vessel or aerospace vehicle, if structural demands warrant it, you might consider requiring similar inspection. The *ASD* test parameters are similar to the ASTM B594 aerospace inspection, but *ASD* establishes the maximum acceptable discontinuities. The lowest acceptable *ASD* quality class is B, for which discontinuities may not exceed $\frac{1}{8}$ in. [3 mm] in size. The ASTM B548 pressure vessel criteria limit discontinuities to 1.0 in. [25 mm]. AWS D1.2-97 section 3.2.4 provides procedures to address discontinuities larger than 1 in. [25 mm].

Costs   Sheet of non-heat-treatable alloys (the 1xxx, 3xxx, and 5xxx series) bought directly from a mill costs as little as about $1.30/lb [$3.00/kg] in 2001, which is similar to the costs for common extrusions summarized in Section 3.1.1. On the other hand, heat-treated rolled products (2xxx, 6xxx, and 7xxx) can cost more than $2.00/lb [$4.40/kg] due to the cost of the heat treatment. Costs vary widely by alloy, size, and quantity, so these figures are only a general guide. In 6xxx series alloys, extrusions are usually more economical than plate when widths are small enough to be extruded, which can be as large as 18 in. [457 mm] and even more in certain thicknesses and alloys.

Water Stains   The final note in this section is a warning that applies to all aluminum products, but most especially to sheet and plate. *Water stains* occur when moisture is trapped between aluminum surfaces in contact, causing stains varying in color from white to dark gray depending on the alloy and the conditions (Figure 3.19). (High-magnesium-content alloys, common in the 5xxx series, seem most susceptible.) The phenomenon is not unlike white rust, (which is also called *wet storage stain*), that occurs on galvanized parts in similar situations. When it occurs, the surface discoloration is often so pronounced that laypersons will question the structural integrity of the material. Actually, water staining only looks bad; it does not weaken the product or spread once the water has evaporated. However, restoring the original appearance is almost impossible, and for these reasons you'll want to avoid water staining. Furthermore, it must be removed from surfaces to be welded. The good news is that at least it isn't rust.

The Aluminum Association offers advice in its publication *AA TR3, Guidelines for Minimizing Water Staining of Aluminum* (17). Good practice is to keep aluminum that is closely packed together under roof, dry, and out of contact with the ground. Avoid bringing cold aluminum into heated areas to prevent condensation from forming on the material. If aluminum is received wet, dry it.

Water vapor present in the air will condense on any surface that is colder than the *dew point*. You've probably observed this phenomenon on a glass with a cold beverage in the summer. The dew point is a temperature that is a function of the relative humidity and the actual temperature. For example, suppose the air temperature in a fabrication shop is 70°F [21°C] and the

Figure 3.19 Water-stained beams in an aluminum cover. The beams will be covered by roofing panels, so the water stains pose no detriment to the structure, aesthetic or otherwise.

relative humidity is 50%. The dew point, which can be determined from a psychrometric chart, is 65°F [18°C] for these conditions. If cold metal is brought in from outside where the temperature is, say, 40°F [4°C], water will condense on the metal. When the relative humidity is 100%, the dew point temperature and the air temperature are the same. Since the dew point temperature can never exceed the air temperature, condensation won't occur as long as the metal is as warm as the air. For this reason, as well to avoid precipitation, it's advisable to store aluminum products indoors, especially products on which water can stand or become trapped between aluminum parts in contact. Moisture and water stains also wreak havoc with aluminum welding, so similar cautions apply to parts to be welded.

Protective techniques are also used to prevent water stains. The Aluminum Association provides specifications for protective oil to be applied to aluminum in *Aluminum Standards and Data*. Also, producers apply plastic films to aluminum sheet stock; they can be removed after the material is ready to be incorporated in the work in the field. This film should be removed before it is allowed to weather, however, or it degrades and becomes difficult to strip.

3.1.3 Forgings

Introduction *Forgings* are produced by pounding hot metal into a cavity. If you think this process sounds vaguely primitive, you're right—forging is one of the oldest fabrication methods. Modern forging processes, though, can produce high-strength parts to very accurate dimensions, and are heavily relied on for complex, critical members such as aircraft frames (8).

Two general categories of forgings exist: *hand forgings* (sometimes called *open die forgings*), and *die forgings* (also called *closed die forgings* or *impression die forgings*). Hand forgings are produced without lateral confinement of the material during the forging operation. Die forgings are more common and are produced by pressing the forging stock (made of ingot, plate, or extrusion) between a pair of dies and, thus, are also called closed die forgings. The complex shapes of parts suited to forgings are similar to those of castings. Forgings, however, have more uniform properties and generally better ductility, at least in the direction parallel to the grain, and unlike castings, they can be strain hardened by working to improve strength. Closed die forgings can be produced weighing up to 15,000 lb [6,800 kg]. Minimum thicknesses are a function of width, plan area, and forging method, but they may be as small as 0.09 in. [2 mm] for a 3 in. [75 mm] width.

Automobile wheels and aircraft structural framing members are good examples of forging applications, but connection plates for some building structures also fall in this category. Forgings are more expensive than castings, but they may be feasible if enough parts are needed and strength and ductility are important.

Die forgings are divided into four categories described below, from the least intricate, lowest quantity, and lowest cost to the most sharply detailed, highest quantity, and highest cost (see Table 3.8). Less intricate forgings are used when quantities are small because it is more economical to machine a few pieces than to incur higher one-time die costs. The most economical forging for a particular application depends on the dimensional tolerances and quantities required.

Blocker-type forgings have large fillet and corner radii and thick webs and ribs, so that only one set of dies is needed; generally, two squeezes of the dies are applied to the stock. Fillets are about 2 times the radius of conventional forgings, and corner radii about $1\frac{1}{2}$ times that of conventional forgings. Usually, all surfaces must be machined after forging. Blocker-type forgings may be selected if tolerances are so tight that machining would be required

TABLE 3.8 Die Forgings

	Blocker-Type	Finish-Only	Conventional	Precision
Typical Production Quantity	<200	500	>500	
Number of Sets of Dies	1	1	2 to 4	
Fillet Radius	2	1.5	1.0	<1.0
Corner Radius	1.5	1.0	1.0	<1.0

in any event, or if the quantity to be produced is small (typically up to 200 units). Blocker-type forgings can be of any size.

Finish-only forgings also use only one set of dies, like blocker-type forgings, but typically one more squeeze is applied to the part. Because of this, the die experiences more wear than for other forging types, but the part can be forged with tighter tolerances and reduced fillet and corner radii and web thicknesses. Fillets are about $1\frac{1}{2}$ times the radius of conventional forgings, and corner radii about the same as that of conventional forgings. The average production quantity for finish-only forgings is 500 units.

Conventional forgings are the most common of all die forging types. Conventional forgings require two to four sets of dies; the first set produces a blocker forging that is subsequently forged in finishing dies. Fillet and corner radii and web and rib thicknesses are the smaller than for blocker-type or finish-only forgings. Average production quantities are 500 or more.

Precision forgings, as the name implies, are made to closer than standard tolerances, and they include forgings with smaller fillet and corner radii and thinner webs and ribs.

You can categorize forgings in other ways. Can and tube forgings are cylindrical shapes that are open at one or both ends; these are also called *extruded forgings.* The walls may have longitudinal ribs or be flanged at one open end. No-draft forgings require no slope on vertical walls and are the most difficult to make. Rolled ring forgings are short cylinders circumferentially rolled from a hollow section.

The ASTM specification for aluminum forgings is B247 *Die Forgings, Hand Forgings, and Rolled Ring Forgings* (54).

Minimum Mechanical Properties and Allowable Stresses Of the 19 die forging alloys that the Aluminum Association has identified, alloys 2014, 2219, 2618, 5083, 6061, 7050, 7075, and 7175 are the most popular. Minimum mechanical properties (tensile yield strength, tensile ultimate strength, and elongation) are given in *Aluminum Standards and Data* (11). Because of the fabrication process, some forging alloys have slightly higher strengths and elongation in a direction parallel to grain flow compared to a direction perpendicular to grain flow, and, along with some plate alloys, are unique among aluminum products in this regard. Forgings produced to the 6061-T6 temper, a good candidate for construction applications, have the same minimum strengths in both directions, and the same strength as extrusions of this temper. In complex shapes, the grain flow direction may not be obvious, and microstructural analysis may be needed.

While tensile yield and ultimate strengths are listed in *Aluminum Standards and Data*, compressive yield and shear ultimate and yield strengths are not, nor are they given in the Aluminum *Specification*. If these properties are needed to perform structural design calculations, you should obtain them from the producer.

Quality Assurance for Die Forgings   *Aluminum Standards and Data* requires that compliance with minimum mechanical property requirements for die forgings be demonstrated by testing the following number of samples: (1) for parts weighing up to and including 5 lb [2.5 kg], one sample is taken for each 2,000 lb [1,000 kg], or part thereof, in a *lot*; and (2) for parts weighing more than 5 lb, one sample is taken for each 6,000 lb [3,000 kg], or part thereof, in a lot. Testing is applicable to a direction only when the corresponding dimension is more than 2.000 in. [50 mm] thick. ASTM B247 requires ultrasonic inspection for discontinuities for certain 2xxx and 7xxx alloys, forging weights, and thicknesses only, so if you want it otherwise you must specify it. Inspection methods for forgings are the same as for plate, as discussed in Section 3.1.2.

Dimensional *tolerances* are suggested in the *Aluminum Forging Design Manual* (7). Tolerances for forgings are subject to the same considerations as those for castings, as discussed in Section 3.1.4.

Identification markings are provided only as required by the forging drawing, so call for them there if you want them. As a minimum, it's a good idea to require identification, including part number, vendor identification, and alloy, and to do so with integrally forged, raised letters on the part.

3.1.4   Castings

Introduction   *Castings* are made by pouring molten metal into a mold. This process is useful for forming complex three-dimensional shapes, but it suffers some limitations for structural uses. Most castings have more variation in mechanical properties and less ductility than wrought (that is, worked, such as forged, extruded, or rolled) aluminum alloys. This is due to defects that are inherent to the casting process, such as porosity and oxide inclusions, and the variation in cooling rates within a casting. Casting alloys contain larger proportions of alloying elements than wrought alloys, resulting in a heterogeneous structure that is generally less ductile than the more homogeneous structure of wrought alloys. Cast alloys also contain more silicon than wrought alloys to provide the fluidity necessary to make a casting. On the other hand, with proper inspection and design, castings have been used in critical aerospace parts.

Another point to keep in mind is that for two reasons, castings are rarely hardened by *cold working* and so are almost always *heat-treated*. This is because castings are, by definition, already produced in essentially their final shape and they're brittle. Where some additional shaping is needed, it's usually in the form of machining, which doesn't affect the *temper*. Consequently, almost all castings are heat-treated and bear the T temper.

Cast Alloy Designation System   Cast aluminum alloys and wrought aluminum alloys have different designation systems. While both systems use four digits, most similarities end there. The cast alloy designation system has three digits, followed by a decimal point, followed by another digit (xxx.x). The first digit indicates the primary alloying element. The second two digits designate the alloy, or in the case of commercially pure aluminum cast alloys, the level of purity. The last digit indicates the product form: 1 or 2 for ingot (depending on impurity levels) and 0 for castings. (So unless you're a producer, the only cast designations you'll be concerned with will end in ".0.") Often, the last digit is dropped (for example, A356-T6); this generally implies that you're talking about a casting and simply didn't bother to append the ".0." A modification of the original alloy is designated by a letter prefix (A, B, C, etc.) to the alloy number. The series and their primary alloying elements are:

1xx.x - Commercially Pure Aluminum: These alloys have low strength. An application is cast motor rotors.

2xx.x - Copper: These are the strongest cast alloys and are used for machine tools, aircraft, and engine parts. Alloy 203.0 has the highest strength at elevated temperatures and is suitable for service at 400°F [200°C].

3xx.x - Silicon with Copper and/or Magnesium: The 3xx.x alloys have excellent fluidity and strength and are the most widely used aluminum cast alloys. Alloy 356.0 and its modifications are very popular and used in many applications. High silicon alloys have good wear resistance and are used for automotive engine blocks and pistons.

4xx.x - Silicon: Silicon provides excellent fluidity as it does for wrought alloys, so these alloys are well suited to intricate castings, such as typewriter frames, and they have good general corrosion resistance. Alloy A444.0 has modest strength but good ductility.

5xx.x - Magnesium: Cast alloys with magnesium have good corrosion resistance, especially in marine environments, for example, 514.0; good machinability; and have good finishing characteristics. They are more difficult to cast than the 2xx, 3xx, and 4xx series, however.

6xx.x - Unused.

7xx.x - Zinc: This series is difficult to cast and so is used where finishing or machinability is important. These alloys have moderate or better strengths and good general corrosion resistance, but they are not suitable for elevated temperatures.

8xx.x - Tin: This series is alloyed with about 6% tin and primarily used for bearings, being superior to most other materials for this purpose. These alloys are used for large rolling mill bearings and connecting rods and crankcase bearings for diesel engines.

9xx.x - Others: This series is reserved for castings alloyed with elements other than those used in series 1xx.x through 8.xx.

Before the Aluminum Association's promulgation of this standard designation system (also called the *ANSI system,* since the Association is the ANSI registrar for it), the aluminum industry used other designations. Also, other organizations such as the federal government, SAE, the U.S. military, and ASTM, assigned still other designations to cast alloys. Even though ASTM finally gave up in 1974, its designations and others are still used by some, so don't be alarmed if you occasionally encounter some strange cast alloy designations. (For example, 535.0 is still sometimes called Almag 35.) A cross-reference to former aluminum industry, federal, ASTM, SAE, and military designations is provided in *Standards for Aluminum Sand and Permanent Mold Castings* (AA-CS-M1-85) (23). ASTM B275, *Standard Practice for Codification of Certain Nonferrous Metals and Alloys, Cast and Wrought* (55), provides old ASTM designations.

Casting Processes   Two kinds of molds are commonly used for castings with construction applications, *sand* and *permanent mold.* Sand castings are addressed by ASTM B26, *Standard Specification for Aluminum-Alloy Sand Castings* (47), and use sand for the mold, which is usable only once. This method is used for larger castings without intricate shapes and produced in small quantities. Permanent mold castings (ASTM B108, *Standard Specification for Aluminum-Alloy Permanent Mold Castings* (48)) are made by introducing molten metal by gravity or low pressure into a mold made of some durable material, such as iron or steel. Permanent mold castings are more expensive but are held to tighter dimensional tolerances than sand castings, and can produce parts with wall thicknesses as little as about 0.09 in. [2 mm]. Sand castings as large as 7,000 lb [3,000 kg] and permanent mold castings as large as 400 lb [180 kg] have been produced in aluminum.

Some casting alloys can be made by only one process (sand or permanent mold), while others can be produced by both methods. Generally, minimum mechanical properties for an alloy that is sand cast are less than those of the same alloy when produced by permanent mold. 356.0 and its higher purity version A356.0 are by far the most commonly used cast alloys in nonaerospace structural applications. The cost premium for A356.0 is usually justifiable in light of its better properties (strength, ductility, and fluidity) compared to 356.0.

There are a number of other types of castings. Die castings (ASTM B85) are produced by the introduction of molten metal under substantial pressure into a metal die, producing castings with a high degree of fidelity to the die. Die castings are used for castings produced in large volumes because they can be produced rapidly within relatively tight dimensional tolerances. Investment castings (ASTM B618) are produced by surrounding, or investing, an expendable pattern (usually wax or plastic) with a refractory slurry that sets at room temperature, after which the pattern is removed by heating and the resulting cavity is filled with molten metal. ASTM also provides a specification for high-strength aluminum alloy castings (ASTM B686) used in airframe, missile, or other critical applications requiring high strength, ductility, and quality.

Dimensional Tolerances Suggested standards for dimensional *tolerances* for castings are given in the Aluminum Association's publication *Standards for Aluminum Sand and Permanent Mold Castings*. Tolerances are held on castings only when purchasers specify them with orders. This is unlike the case for rolled and extruded products, for which standard dimensional tolerances are held by the mill even when purchasers do not specify them. The dimensions of castings can be hard to control because of difficulty in predicting shrinkage during solidification and warping due to uneven cooling of thick and thin regions. Without requirements, castings may be produced with dimensions that deviate more widely than you anticipated.

Minimum Mechanical Properties *Minimum mechanical properties* for some cast alloys taken from ASTM B26 and B108 are shown in Table 3.9. Unless shown otherwise in the ASTM specifications and Table 3.9, the values given are the minimum values for separately cast test bars, not for production castings. Minimum strengths for *coupons* cut from actual production castings must be no less than 75% of the values given in Table 3.9. Elongation values of coupons from castings must be no less than 25% of the values in Table 3.9; in other words, they may be considerably less than the values in the table. Minimum mechanical properties (tensile ultimate and yield strengths and percent elongation in 2 in. [50 mm]) for separately cast test bars of cast aluminum alloys are also given in *Standards for Aluminum Sand and Permanent Mold Castings* and the *Aluminum Design Manual*, Part V, Material Properties. All strengths in these tables must be multiplied by 0.75 and all elongations by 0.25 to obtain values for specimens cut from castings. The ASTM B26 and B108 specifications are updated more frequently than the Aluminum Association publications and so may better reflect current industry standards.

TABLE 3.9 Minimum Mechanical Properties of Some Cast Aluminum Alloys

Alloy Temper	Specimen Location	Sand (S) or Permanent Mold (P)	Minimum Tensile Ultimate Strength		Minimum Tensile Yield Strength		Minimum Elongation (% in 2 in. or 4d)
			(ksi)	(MPa)	(ksi)	(MPa)	
356.0-T6	SC	S	30.0	205	20.0	140	3.0
A356.0-T6	SC	S	34.0	235	24.0	165	3.5
356.0-T6	SC	P	33.0	228	22.0	152	3.0
A356.0-T61	SC	P	38.0	262	26.0	179	5.0
A356.0-T61	DL	P	33.0	228	26.0	179	5.0
A356.0-T61	NL	P	28.0	193	26.0	179	5.0

Notes: Except for A356.0-T61, where properties are for separately cast test bars, the tensile ultimate strengths and tensile yield strengths of specimens cut from castings shall not be less than 75% of the values given in this table, and the elongation of specimens cut from castings shall not be less than 25% of the values given in this table.

SC = separately cast test specimen
DL = specimen cut from designated location in casting
NL = specimen cut from casting; no location specified

Allowable Stresses The Aluminum *Specification* gives allowable stresses for castings used in *bridge type structures* only (Allowable Stress Design Aluminum *Specification* Table 3.4-4), probably because of their well-known structural uses as bases for highway light poles and bridge rail posts. Ironically, though, these applications are *building type structures*. Castings have also been used, however, in actual building structures, especially as gussets and in connections. Allowable stresses for castings in building structures can be derived from those given in Table 3.4-4 of the Aluminum *Specification* by multiplying the values found there by the ratio of bridge to building safety factors (1.12). Allowable stresses for castings subject to fatigue, however, are not provided in the Aluminum *Specification*, and it requires that they be established by testing. No design stresses are given in the Load and Resistance Factor Design (LRFD) version of the Aluminum *Specification*.

Quality Assurance The most commonly used inspection techniques are *radiography*, liquid *penetrant,* and ultrasonic; all are non-destructive.

Radiography is done by x-raying the part to show discontinuities, such as gas holes, cracks, shrinkage, and foreign material. These discontinuities are rated by comparing them to reference radiographs (x-ray pictures) shown in ASTM E 155. The ratings are then compared to inspection criteria established beforehand by the customer and the foundry. The inspection criteria for quality and frequency of inspection can be selected and then specified from the Aluminum Association's casting quality standard AA-CS-M5-85, which is published in *Standards for Aluminum Sand and Permanent Mold Castings*. You can choose from seven quality levels and four frequency levels (the number of castings to be inspected). Quality level IV and frequency level 3 are considered reasonable requirements for building structures. Frequency level 3 is "foundry control" without a specified number of inspections, so the additional stipulation of how many radiographs must be taken per *lot* is required. Two radiographs per lot is typical for building structure parts.

ASTM B26 and B108 also provide radiographic inspection quality levels and corresponding acceptance criteria. Rather than the seven quality levels given in the Aluminum Association's casting quality standard, B26 and B108 provide only four levels (from highest quality to lowest: A, B, C, and D). Level A should be specified only if you need perfection since it allows no defects; conversely, if grade D is specified, not only are the radiograph-revealed discontinuity criteria very liberal, but also no tensile tests are required from coupons cut from castings. Regardless of the quality level you specify, you must also specify the number of radiographs required. No standard inspection frequency levels are provided in B26 or B108, so you have to make this up yourself.

The liquid penetrant inspection method is suitable only for detecting surface defects. Two techniques are available. The fluorescent penetrant procedure is to apply penetrating oil to the part, remove the oil, apply developer to absorbed oil bleeding out of surface discontinuities, and then inspect the casting under ultraviolet light. The dye penetrant method uses a color penetrant, enabling inspection in normal light; because of this method's simplicity, it tends to be used more often than the fluorescent penetrant. No interpretative standards exist for penetrant methods, but a penetrant-revealed defect is like pornography: it's hard to define, but you know it when you see it. Frequency levels are given in AA-CS-M5-85 for penetrant testing also. For building parts carrying calculated stresses, 100% penetrant testing is often considered appropriate since penetrant inspection is relatively inexpensive.

Ultrasonic inspection may be performed, but no provisions exist in ASTM B26 or B108 or standards for such inspection. This reflects the fact that ultrasonic inspection is probably as much art as science.

A test bar cast with each heat is also useful. It can be tested and results compared directly to minimum mechanical properties listed for the alloy in the relevant ASTM specification or *Standards for Aluminum Sand and Permanent Mold Castings*.

3.1.5 Prefabricated Products

A number of prefabricated aluminum structural products are available. Some of the most common are discussed here.

Roofing and Siding Commercial *roofing* and *siding* (Figure 3.20) is produced in various configurations, including corrugated, V-beam, and ribbed (Figure 3.21). For these old stand-by styles, dimensions and weights are given

Figure 3.20 Column-supported aluminum roof with preformed sheathing, covering a water storage reservoir. (Courtesy of Temcor)

in the *Aluminum Design Manual*, Part VI, Section Properties, Table 31, and section properties in Table 32. Tolerances on sheet thickness, depth of corrugations, length, parallelness of corrugations, and squareness are given in *Aluminum Standards and Data* Tables 7.26 through 7.30.

Maximum spans are given in the *Aluminum Design Manual*, Part VII, Design Aids, Tables 4-4 and 4-5, but the *ADM* 2000 and earlier versions of these tables contain errors. Correct versions are given here in Table 3.10 for corrugated and V-beam roofing and siding and in Table 3.11 for ribbed siding.

The maximum span tables are based on limiting stresses to the allowable stresses for Alclad 3004-H151,-H261, and -H361 (Section 3.2.5 explains alclad) and limiting deflections to $1/60$ of the span. The $L/60$ deflection limit matches that stated in the Aluminum *Specification,* Section 8.4.4. Because non-alclad versions of these alloys have slightly higher strengths, using the

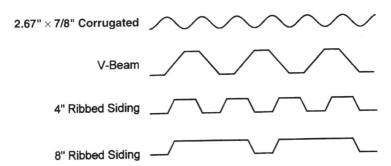

Figure 3.21 Examples of typical preformed roofing and siding profiles.

TABLE 3.10 Commercial Corrugated and V-Beam Roofing and Siding Maximum Recommended Spans (in.)

Design Load (psf)	One Strength	One Deflection	Two Strength	Two Deflection	Three Strength	Three Deflection	
Corrugated Roofing and Siding—0.024 in. Thick							
20	79	61	79	82	88	76	$I =$ 0.0307 in.^{4}/ft
25	70	57	70	76	79	70	$DL =$ 0.442 psf
30	64	54	64	72	72	66	$S =$ 0.0708 in.^{3}/ft
35	60	51	60	68	67	63	$F_b =$ 18.6 ksi
40	56	49	56	65	63	60	
45	53	47	53	63	59	58	
50	50	45	50	61	56	56	
Corrugated Roofing and Siding—0.032 in. Thick							
20	92	67	92	90	102	83	$I =$ 0.0409 in.^{4}/ft
25	82	63	82	84	92	77	$DL =$ 0.589 psf
30	75	59	75	79	84	73	$S =$ 0.0936 in.^{3}/ft
35	70	56	70	75	78	69	$F_b =$ 19.2 ksi
40	65	54	65	72	73	66	
45	62	52	62	69	69	64	
50	58	50	58	67	65	62	

Number of Equal Spans: One, Two, Three

V-Beam Roofing and Siding—0.032 in. Thick, $4\frac{7}{8}$ in. Pitch

Span							Properties
20	128	110	128	147	144	136	$I =$ 0.179 in.^{4}/ft
25	115	102	115	137	129	127	$DL =$ 0.623 psf
30	105	97	105	129	118	119	$S =$ 0.205 in.^{3}/ft
35	98	92	98	123	109	113	$F_b =$ 17.3 ksi
40	92	88	92	118	102	108	
45	86	85	86	113	97	104	
50	82	82	82	109	92	101	
55	78	79	78	106	87	98	
60	75	77	75	103	84	95	

V-Beam Roofing and Siding—0.040 in. Thick, $4\frac{7}{8}$ in. Pitch

Span							Properties
20	150	118	150	158	167	146	$I =$ 0.223 in.^{4}/ft
25	134	110	134	147	150	136	$DL =$ 0.779 psf
30	123	104	123	139	137	128	$S =$ 0.255 in.^{3}/ft
35	114	99	114	132	127	122	$F_b =$ 19.0 ksi
40	107	94	107	126	119	117	

TABLE 3.10 Commercial Corrugated and V-Beam Roofing and Siding Maximum Recommended Spans (in.) (continued)

Design Load (psf)	One		Two		Three			
	Strength	Deflection	Strength	Deflection	Strength	Deflection		
V-Beam Roofing and Siding—0.040 in. Thick, $4\frac{7}{8}$ in. Pitch (continued)								
45	101	91	101	122	113	112	$I =$	0.279 in.^{4}/ft
50	96	88	96	118	107	108	$DL =$	0.974 psf
55	91	85	91	114	102	105	$S =$	0.317 in.^{3}/ft
60	87	83	87	111	98	102	$F_b =$	20.2 ksi
V-Beam Roofing and Siding—0.050 in. Thick, $4\frac{7}{8}$ in. Pitch								
20	171	127	171	170	191	157		
25	154	118	154	158	172	146		
30	141	111	141	149	158	138		
35	131	106	131	142	146	131		
40	122	102	122	136	137	125		
45	116	98	116	131	129	121		
50	110	94	110	126	123	117		
55	105	92	105	123	117	113		
60	100	89	100	119	112	110		

V-Beam Roofing and Siding—0.032 in. Thick, $5\frac{1}{3}$ in. Pitch

Span							Properties
20	128	114	128	153	143	141	$I =$ 0.199 in.⁴/ft
25	115	106	115	142	129	131	$DL =$ 0.613 psf
30	105	100	105	134	118	124	$S =$ 0.229 in.³/ft
35	98	95	98	127	109	117	$F_b =$ 15.4 ksi
40	91	91	91	122	102	112	
45	86	88	86	117	96	108	
50	82	85	82	113	91	104	
55	78	82	78	110	87	101	
60	75	80	75	107	84	98	

V-Beam Roofing and Siding—0.040 in. Thick, $5\frac{1}{3}$ in. Pitch

Span							Properties
20	153	123	153	164	171	151	$I =$ 0.249 in.⁴/ft
25	137	114	137	153	154	141	$DL =$ 0.766 psf
30	126	108	126	144	141	133	$S =$ 0.285 in.³/ft
35	117	102	117	137	130	126	$F_b =$ 17.8 ksi
40	109	98	109	131	122	121	
45	103	94	103	126	115	116	
50	98	91	98	122	110	112	
55	93	88	93	118	104	109	
60	90	86	90	115	100	106	

TABLE 3.10 Commercial Corrugated and V-Beam Roofing and Siding Maximum Recommended Spans (in.) (continued)

Design Load (psf)	One		Two		Three			
	Strength	Deflection	Strength	Deflection	Strength	Deflection		
V-Beam Roofing and Siding—0.050 in. Thick, $5\frac{1}{3}$ in. Pitch								
20	176	132	176	176	197	163		$I =$ 0.311 in.^{4}/ft
25	158	123	158	164	177	151		$DL =$ 0.957 psf
30	145	116	145	155	162	143		$S =$ 0.354 in.^{3}/ft
35	134	110	134	147	150	136		$F_b =$ 19.1 ksi
40	126	105	126	141	141	130		
45	119	101	119	136	133	125		
50	113	98	113	131	126	121		
55	108	95	108	127	120	117		
60	103	92	103	124	115	114		

1. Maximum recommended spans are calculated in accordance with the *Specification for Aluminum Structures*, Allowable Stress Design, for building type structures.

2. Material is Alclad 3004-H151, -H261, or -H361 (which are stucco embossed tempers) or Alclad 3004-H16. Dimensions are given in the *Aluminum Design Manual*, Part VI, Table 31, and section properties are given in Part VI, Table 32.

3. The deflection limit is 1/60 of the span.

4. Shaded entries denote that the span is limited by strength considerations rather than deflection.

TABLE 3.11 Commercial Ribbed Siding Maximum Recommended Spans (in.)

Ribbed Siding—0.032 in. Thick, 4 in. Pitch

Design Load (psf)	One Strength[1]	One Strength[2]	One Deflection	Two Strength[1]	Two Strength[2]	Two Deflection	Three Strength[1]	Three Strength[2]	Three Deflection
20	98	101	85	101	98	114	113	110	106
25	88	91	79	91	88	106	101	98	98
30	80	83	75	83	80	100	93	90	93
35	75	77	71	77	75	95	86	83	88
40	70	72	68	72	70	91	80	78	84
45	66	68	66	68	66	88	76	74	81
50	63	64	63	64	63	85	72	70	78

Properties:
- l = 0.0836 in.^{4}/ft
- DL = 0.609 psf *
- S = 0.175 in.^{3}/ft **
- S = 0.16 in.^{3}/ft *
- F_b = 11.8 ksi **
- F_b = 13.7 ksi *
- c = 0.478 in. **
- c = 0.522 in. *

Ribbed Siding—0.040 in. Thick, 4 in. Pitch

Design Load (psf)	One Strength[1]	One Strength[2]	One Deflection	Two Strength[1]	Two Strength[2]	Two Deflection	Three Strength[1]	Three Strength[2]	Three Deflection
20	118	120	92	120	118	123	134	132	113
25	106	107	85	107	106	114	120	118	105
30	97	98	80	98	97	108	110	108	99
35	90	91	76	91	90	102	102	100	94
40	84	85	73	85	84	98	95	94	90
45	79	80	70	80	79	94	90	89	87
50	75	76	68	76	75	91	85	84	84

Properties:
- l = 0.104 in.^{4}/ft
- DL = 0.76 psf *
- S = 0.217 in.^{3}/ft **
- S = 0.198 in.^{3}/ft *
- F_b = 13.8 ksi **
- F_b = 15.6 ksi *
- c = 0.478 in. **
- c = 0.522 in. *

Ribbed Siding—0.032 in. Thick, 8 in. Pitch

Design Load (psf)	One Strength[1]	One Strength[2]	One Deflection[1]	One Deflection[2]	Two Strength[1]	Two Strength[2]	Two Deflection[1]	Two Deflection[2]	Three Strength[1]	Three Strength[2]	Three Deflection[1]	Three Deflection[2]
20	62	77	74	79	77	62	100	105	78	69	92	97
25	56	69	69	73	69	56	93	98	70	63	85	90
30	51	63	65	69	63	51	87	92	64	57	81	85

Properties:
- l = 0.648 in.^{4}/ft
- DL = 0.539 psf *
- S = 0.235 in.^{3}/ft *
- S = 0.0895 in.^{3}/ft **

TABLE 3.11 Commercial Ribbed Siding Maximum Recommended Spans (in.) (continued)

Design Load (psf)	Number of Equal Spans													
	One				Two				Three					
	Strength[1]	Strength[2]	Deflection[1]	Deflection[2]	Strength[1]	Strength[2]	Deflection[1]	Deflection[2]	Strength[1]	Strength[2]	Deflection[1]	Deflection[2]		
Ribbed Siding—0.032 in. Thick, 8 in. Pitch (continued)														
35	47	59	62	65	59	47	83	88	59	51	77	81	F_b = 3.5 ksi	*
40	44	55	59	63	55	44	79	84	55	43	73	77	F_b = 14.3 ksi	**
45	42	52	57	60	52	42	76	81	52	47	70	74	c = 0.276 in.	*
50	40	49	55	58	49	40	74	78	49	44	68	72	c = 0.724 in.	**
Ribbed Siding—0.040 in Thick, 8 in. Pitch													l = 0.081 in.^{4}/ft	
20	75	91	80	84	91	75	107	113	94	84	99	104	DL = 0.674 psf	
25	67	82	74	79	82	68	100	105	85	76	92	97	S = 0.289 in.^{3}/ft	*
30	62	75	70	74	75	62	94	99	78	70	87	91	S = 0.111 in.^{3}/ft	**
35	57	69	67	70	66	55	89	94	68	61	82	87	F_b = 4.2 ksi	*
40	54	65	64	67	55	46	85	90	57	51	79	83	F_b = 16.1 ksi	**
45	51	61	61	65	62	51	82	87	63	57	76	80	c = 0.276 in.	*
50	48	58	59	63	58	48	79	84	60	54	73	77	c = 0.724 in.	**

[1] Wide flat on loaded side; load is toward neutral axis.
[2] Narrow flat on loaded side; load is toward neutral axis.
[3] Maximum recommended spans are calculated in accordance with the *Specification for Aluminum Structures, Allowable Stress Design*, for building type structures.
[4] Material is Alclad 3004-H151, -H261, or -H361 (which are stucco embossed tempers) or Alclad 3004-H16. Dimensions are given in *Aluminum Design Manual*, Part VI, Table 31, and section properties are given in Part VI, Table 32.
[5] The deflection limit is 1/60 of the span.
[6] Shaded entries denote that the span is limited by strength considerations rather than deflection.
* Wide flat in compression.
** Narrow flat in compression.

maximum span tables is conservative for non-alclad versions of these alloys. The *Aluminum Design Manual*, Part VIII, Illustrative Example 29 shows how section moduli, load carrying capacity, and deflections are calculated for 8 in. ribbed siding. Ribbed roofing is not addressed in the *Manual*. Ribbed siding is not symmetric about its neutral axis, so its ability to span depends on how it's oriented with respect to the direction of the load.

Standard finishes available for roofing and siding as purchased from the manufacturer are given in *Aluminum Standards and Data*, Table 7.20. These include mill finish, stucco embossed, painted, and low reflectance. Stucco embossed sheet is an economical way to reduce the effect on appearance of smudges and other surface marring. Chances are the manufacturer of your refrigerator used this trick. The low reflectance finish is measured in accordance with ASTM Method D 523. More information on finishes is given in Section 3.2.

Roofing and siding can be produced in almost any length because they are made from coil. However, shipping considerations and allowance for thermal expansion may limit length. Thermal expansion and contraction is discussed further in Section 10.1, Cold-Formed Construction.

Some manufacturers produce corrugated, V-beam, and ribbed sheet in thicknesses not given in the Aluminum Association publications. Other styles that various manufacturers produce include *standing seam* types and those with attachment fasteners that are concealed from view. Dimensions for proprietary styles should be obtained from the manufacturer; allowable spans may be provided by the manufacturer or calculated using the method from Example 29. Obtaining load capacities from manufacturers is easier said than done. A practical guide is: If the manufacturer doesn't readily provide the information necessary to check load capacity (i.e., section modulus per unit width), the section probably doesn't carry enough load to bother calculating it. Many standing seam sections fall into this category because they are intended to be used with a continuous substrate providing support. Some are capable of spanning a worthwhile distance between purlins, though. A method for calculating the bending deflection of a cold-formed sheet member is given in Section 10.1.

Tread Plate   *Tread plate*, which is also called *checkered plate, diamond plate,* and *floor plate,* has a raised diamond pattern providing a slip-resistant surface on one side. The ASTM specification is B632, *Aluminum-Alloy Rolled Tread Plate* (69). A load-span table is given in the *Aluminum Design Manual*, Part VII, Design Aids, Table 4-3. This table is based on 6061-T6 plate with simply supported edges and a limiting deflection of $L/150$, (where L is the shorter of the two dimensions spanned by the plate) for tread plate thicknesses of 0.188 in., 0.25 in., 0.375 in., and 0.5 in. [5 mm, 6 mm, 10 mm, 12 mm]. (Tread plate is also available in 0.100 in., 0.125 in., 0.156 in., and 0.625 in. [2.5 mm, 3.2 mm, 4 mm, 8 mm, and 16 mm thicknesses]). Tread plate is commonly sold in 48 in. [1,220 mm] widths; 60 in. [1,520 mm] and even 72 in. [1,830 mm] widths are available in certain thicknesses. Tolerances for tread plate dimensions (thickness, width, length, height of pattern, camber of pattern line, lateral bow, and squareness) are given in *Aluminum Standards and Data*, Tables 7.37 through 7.43. Load-span tables for other support conditions, thicknesses, and spans can be calculated from cases given in Roark's *Formulas for Stress and Strain*, Table 26 (140).

The most common alloy for tread plate is 6061, but it is also produced in 3003-H231 with a bright finish and 5086-H34 mill finish.

Punches for holes in tread plate should enter the material from the raised pattern side.

Bar Grating   Aluminum grating is sometimes used as a substitute for plate. It is available in a number of configurations, such as rectangular bar and I-beam bar grating, with various strengths. The grating may be either serrated for skid-resistance or plain (smooth) and is usually made of 6061-T6, 6063-T5, or 6063-T6. Information, including dimensions and load-carrying capacity, is available from manufacturers and the National Association of Architectural Metal Manufacturers (NAAMM).

Corrugated Pipe   Two types of corrugated aluminum pipe are available:

Factory-made corrugated aluminum pipe made of alclad 3004-H34 sheet is used for storm-water drainage and culverts. ASTM B745, *Corrugated Aluminum Pipe for Sewers and Drains* (72), addresses corrugated pipe, which is available in two cross-sectional shapes: Type I, which is circular, and Type II, which is circular on top and approximately flat on bottom and called *pipe-arch*. Both shapes are also available with an inner liner of smooth (uncorrugated) sheet. Circular cross sections are also available with perforations (of several styles) to permit water to flow in or out. Two types of seams are used: *helical lock seams,* which are mechanically staked (indented) and *riveted seams.* The nominal sheet thicknesses range from 0.036 in. [0.91 mm] to 0.164 in. [4.17 mm], and circular pipes' nominal inside diameters range from 4 in. [100 mm] to 120 in. [3,000 mm]. Installation practices are given in ASTM B788, *Installing Factory-Made Corrugated Aluminum Culverts and Storm Sewer Pipe* (74); structural design practices are given in ASTM B790, *Structural Design of Corrugated Aluminum Pipe, Pipe Arches, and Arches for Culverts, Storm Sewers, and Other Buried Conduits* (76).

Corrugated aluminum structural plate pipe (ASTM B746, *Corrugated Aluminum Alloy Structural Plate for Field-Bolted Pipe, Pipe-Arches, and Arches* (73) is suitable for more structurally demanding applications. It's available in circular and pipe-arch shapes and made with 5052-H141 sheet and plate and 6061-T6 or 6063-T6 extruded stiffeners. Seams are bolted and stiffening ribs are used; corrugations are 9 in. × 2.5 in. [230 mm × 64 mm]. Nominal thicknesses range from 0.100 in. [2.54 mm] to 0.250 in. [6.35 mm] in 0.025 in. [0.64 mm] increments. Nominal diameters for pipe range from 60 in. [1,525 mm] to 312 in. [7,925 mm]. Installation practices are given in ASTM

B789, *Installing Corrugated Aluminum Structural Plate Pipe for Culverts and Sewers* (75); structural design practices are given in ASTM B790.

Large, aluminum box culverts used as small bridges and grade-separation structures are addressed in ASTM B864 *Corrugated Aluminum Box Culverts* (77). They use the corrugated plate described in ASTM B746 to make a conduit that has a long-radius crown segment, short-radius haunch segments, and straight-side segments supported by a foundation and stiffened with 6061-T6 ribs.

Honeycomb Panels Honeycomb panels are made by adhesively bonding a honeycomb aluminum foil core to two thin aluminum face sheets. The honeycomb core can be made of perforated foil so that there are no isolated voids in the core. Aerospace and naval applications use 2024, 5052, and 5056 foil cores; general commercial applications include architectural panels, storage tank covers, walls, doors, and tables, and utilize 3003 cores. Commercial grade honeycomb is manufactured in several cell sizes with densities from about 1 lb/ft^3 to 5 lb/ft^3 [16 kg/m^3 to 80 kg/m^3] (compared to 170 lb/ft^3 [2,700 kg/m^3] for solid aluminum) and in overall thicknesses between 0.125 in. and 20 in. [3 mm and 500 mm]. Face sheets are made of an aluminum alloy compatible with the core.

Aluminum Foam Closed-cell aluminum foam is made by bubbling gas or air through aluminum alloys or aluminum metal matrix composites to create a strong but lightweight product. The foam density is 2% to 20% that of solid aluminum. Foamed aluminum's advantages include a high strength-to-weight ratio and good rigidity and energy absorbency. Current applications include sound insulation panels. Standard size blocks up to 6 in. [150 mm] thick, as well as parts with complex shapes, can be cast.

Aluminum Composite Material Aluminum composite material (ACM), developed in the 1970s, is made of aluminum face sheets adhesively bonded to plastic cores to make sandwich panels used for building cladding, walls, and signs. One of the earliest and most dramatic installations was the cladding for the EPCOT sphere in Disney World (Figure 3.22). Because the first manufacturers were not American, standard thicknesses are in round metric dimensions such as 3 mm, 4 mm, 5 mm, and 6 mm [0.118 in., 0.157 in., 0.197 in., and 0.236 in.], with 4 mm the most common. The aluminum face sheets are usually 0.020 in. thick [0.51 mm] and come in a wide variety of colors (painted or anodized), while the cores are made of either polyethylene or proprietary fire-resistant plastic, the latter being useful in building applications. ACM has come into its own upon inclusion in the *International Building Code* (Section 1407) (114).

Load-deflection tables are available from manufacturers. The *L*/60 deflection limit (see roofing and siding above) is customary, and wind load is the primary consideration. The Aluminum *Specification* does not offer specific

Figure 3.22 Aluminum-clad panels cover the sphere at the entrance to Walt Disney World's EPCOT Center in Florida.

design provisions for sandwich panels, but they are discussed in the *Aluminum Design Manual*, Part III, Section 6.0.

ACM is easy to work with: panels can be routed, punched, drilled, bent, and curved with common fabrication equipment, and manufacturers provide concealed fastener panel attachment systems. Installed costs naturally depend on volume, but for more than a few thousand square feet are under $20/ft² [$200/m²], about 20% of which is for installation.

3.2 COATINGS AND FINISHES

A variety of finishes are available for aluminum. Factors in the selection of the type of finish include the usual suspects: cost, appearance, and durability. The primary purpose of this book is to provide guidance in structural design, but sometimes design engineers inherit the challenge of specifying a finish, so we'll give you a little orientation.

Common finishes include *mill finish* (generally, a fancy way to say "bare aluminum"), *anodizing*, painting, and mechanical finishes. Some considerations in their use are discussed in this section.

An identification system for aluminum finishes is given in the Aluminum Association publication *Designation System for Aluminum Finishes* (14). This system identifies the finish by its mechanical, chemical, and coating characteristics, and examples of its use are given below.

3.2.1 Mill Finish

The Aluminum Association publication *Specifications for Aluminum Sheet Metal Work in Building Construction* (21) states:

> "One of aluminum's most useful characteristics is its tendency to develop an extremely thin, tough, invisible oxide coating on its surface immediately on exposure to air. This oxide film, although only 2 to 4 ten-millionths of an inch thick on first forming, is almost completely impermeable and highly resistant to attack by corroding atmospheres."

This aluminum-oxide film does not weather away or stain adjacent materials. Whereas most steel must be coated for protection, aluminum is generally coated only when a certain appearance is desired. When aluminum is left in its natural state, its initial bright appearance will weather through shades of gray.

When a coating is desired, aluminum may be anodized or painted. Coatings, however, may impose a maintenance concern that would not be present with mill-finish aluminum since coatings must be maintained. Coatings also affect the price and delivery schedule of aluminum components. Anodizing is limited to pieces that will fit in anodizing tanks and cannot be done in the field. Tank sizes vary by anodizer, with lengths up to 30 ft [10 m] commonplace. Few tanks, however, are deeper than 8 ft [2.5 m], so anodizing limits shop-assembly sizes. Painting of aluminum is also typically performed in the factory to take advantage of controlled conditions. This limits the size of an assembly to that which fits in paint booths and ovens. Simply using aluminum in the unadorned state in which it leaves the mill (i.e., mill finish) eliminates these limitations on assembly sizes.

Mill finish may be specified under the Aluminum Association system by the designation "Mill finish AA-M10 as fabricated."

3.2.2 Anodized Finishes

Anodizing is a process that accelerates the formation of the oxide coating on the surface of aluminum, producing a thicker oxide layer than would occur naturally. An anodized coating is actually part of the metal, and so has excellent durability. One of the primary advantages of anodizing is its resistance to abrasion. The color, however, depends on the alloying elements and, thus, varies among different alloys and even among parts of the same alloy due to nonuniform mixing of alloying elements.

Anodized color tends to be more consistent in some alloys than others. When aluminum is to be anodized, the color match between components of varying alloys and fabrications should be considered, but it may result in a trade-off with other considerations, such as price, strength, and corrosion resistance. Alloy 3003 is often used for wide sheet, for example, but alloy 5005 AQ (anodizing-quality) may be substituted (at a premium price) for improved color consistency. Extrusions of alloys 6463 and 6063 are generally regarded as providing better anodized appearance than 6061, but they are not as strong. When their use is not constrained by price, strength, or availability, 5005 sheet material and 6063 extrusion alloy are preferred for color consistency. 6463 is used when a bright anodized finish is desired.

Architectural coatings are sometimes specified to tone down the glare from mill finish, as well as to minimize the appearance of handling marks, surface imperfections, and minor scratches. A dull gray initial appearance may be cost effectively achieved with a Class II clear anodizing of 3003 alloy, following an appropriate chemical etch (e.g., C22 medium matte). Over time Class II clear anodizing might weather unevenly, but it will retain variations of a dull gray, low-sheen appearance.

Anodizing can also affect at least one structural issue: the selection of weld filler alloy. As mentioned above, when anodized by the same process, different alloys will anodize to different colors. Because filler alloys are typically different than the alloy of the base metal, filler alloys should be selected for good color match if weldments are to be anodized. An example is weldments of 6061, for which the Aluminum *Specification* (Table 7.2-1) recommend the use of 4043 filler alloy. However, 4043 anodizes much darker than 6061, resulting in the welds being a distinctly different color than the rest of the anodized assembly. To avoid this two-tone appearance, 5356 filler alloy should be used. Since anodizing must be removed before welding, assemblies requiring both welding and anodizing should be welded before anodizing. The size of the completed assembly should, therefore, be limited to that which will fit in an anodizing tank.

Specifying an Anodized Finish   Designating an anodized finish requires the selection of color, process, and architectural class. The architectural class is an Aluminum Association designation indicating the thickness of the oxide layer. Class I is the thicker, and more expensive, designation. Some anodizers recommend Class II as the best value, whereas others believe that Class I will retain a more even appearance over time and is the more suitable choice. Anything thinner than Class II is not recommended for exterior use. Thicknesses are given in Table 3.12.

Anodizing may be colored or clear. Clear anodizing retains the gray aluminum color, but it results in a more even appearance than mill finish. The

TABLE 3.12   Anodizing Thicknesses

Description	Thickness (mils)	Thickness (μm)	AA Designation
Protective and Decorative	$t < 0.4$	$t < 10$	A2×
Architectural Class II	$0.4 \le t \le 0.7$	$10 \le t < 18$	A3×
Architectural Cass I	$t \ge 0.7$	$t > 18$	A4×

Note:
   1 mil = 0.001 in.

availability of other colors depends on the anodizing process used. These processes, and the resulting typical colors, are summarized below:

Integral Coloring: The color is inherent in the oxide layer itself in this process and ranges from pale champagne to bronze to black. This process has been largely replaced by electrolytic coloring.

Electrolytic (two-step) Coloring: The first step in this process is to clear-anodize the surface, and the second is to deposit another metal oxide in the pores of the aluminum by means of an electric current. Electrolytic deposition of tin is commonly used as a more efficient method of producing the bronze shades that have been associated with integral coloring. Shades of burgundy and blue, which are achieved by modulating power-supply wave form or by depositing copper salts, are also commercially available.

Impregnated Coloring: In this process, color is produced by the use of dyes. While this allows a greater variety of color to be achieved, you should take extreme care in selecting a dye that is suitable for exterior use. Only a handful are colorfast, including a gold that has been used with good results and certain reds and blues that have been reported to have superior performance. All dyed, anodized coatings for exterior use should be processed to a Class I coating thickness.

Anodizing may be specified in accordance with the Aluminum Association's *Designation System for Aluminum Finishes* (14). The nomenclature is illustrated as follows:

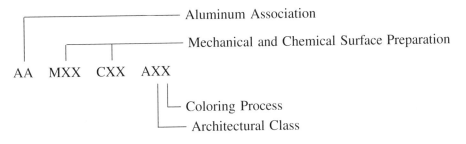

where AA indicates Aluminum Association

M = prefix for mechanical finishes
C = prefix for chemical etches
A = prefix for the architectural class and coloring process to be used in the anodizing.

Common mechanical and chemical preparations for anodizing are designated M12 and C22, respectively. The architectural class designations are A3X for Class II and A4X for Class I. The coloring process designations are:

AX1—Clear
AX2—Integral (no longer in common use)
AX3—Impregnated (dyed)
AX4—Electrolytic (two-step).

Examples:

AA M12C22 A31:   The 3 following the A indicates that this is a Class II finish, and the trailing 1 indicates that it is clear. This is a relatively low-cost finish to reduce sheen.

AA M12C22 A44:   The first 4 after the A indicates that this is the more durable Class I finish, and the second 4 signifies that it is produced by the electrolytic process. This designation can be used to specify a durable, moderate-cost bronze or black finish. The shade in the bronze-to-black spectrum would also have to be specified.

AA M12C22 A43:   This also has a 4 after the A for a Class I finish, but the 3 indicates that the color is to be imparted by the impregnated (dyeing) process. A highly decorative metallic gold is one of the colors that can be achieved with this designation.

The American Architectural Manufacturers Association (AAMA) provides two voluntary specifications concerning anodizing:

- *Cleaning and Maintenance of Architectural Anodized Aluminum* (AAMA 609-93) (28), which recommends methods of cleaning anodized aluminum
- *Anodized Architectural Aluminum* (AAMA 611-98) (29), which provides requirements for inspection of material prior to anodizing, appearance, color and gloss uniformity, coating thickness, abrasion resistance, corrosion resistance, color durability, ability to seal the surface, craze resistance, and sealant compatibility.

Anodizing is a cost-effective way to enhance the initial appearance of aluminum. Even clear anodizing will minimize the handling marks and dull the glare inherent in a mill finish. When color selection and chemical resistance are more important than abrasion resistance, however, paint is generally recommended.

3.2.3 Painted Finishes

Coatings based on polyvinylidene fluoride (PVDF) resins (commonly known by the Elf Atochem North America tradename of Kynar 500 but also available under other tradenames from other sources) have become the standard for architectural applications of painted aluminum. These products are chemically inert and exhibit excellent weatherability. The many licensed applicators are continually introducing new products based on this resin. A wide variety of standard colors are available, and custom colors can generally be matched on large orders. Premium metallic colors are also available, including products that are intended to give the appearance of anodizing with the consistency of paint.

These products vary in availability and cost, but are typically more expensive than anodizing when applied to fabricated assemblies. If you're working from a budget price for a generic Kynar coating, be aware that the cost probably assumes a "standard" finish and a "standard" color. Should you then decide to specify a custom primer, such as urethane, under a custom color with a metallic finish, and then cover it all with a clear topcoat, you will have approximately tripled the dry-film thickness and caused a corresponding escalation of the price.

As with any coating application, preparation of the substrate is critical. Once again invoking the Aluminum Association's system, a sample paint designation could be "AA C12C42 R1X." The C12 and C42 refer to an inhibited chemical cleaning and a chemical conversion coating, respectively. No mechanical finish is specified, and the R1X simply indicates that it is an organic coating. The coating itself, including primer and topcoat, would have to be further specified in terms of material, dry-film thickness, color, and gloss.

A practical caution when specifying baked-on organic coatings is to remember that they are generally factory-applied in an electrostatic process, which tends to draw some overspray to the back side of a part being painted. If only one side is to be painted and overspray is unacceptable on the back side, then it should be masked for protection.

Pre-painted sheet and coil are also available. These materials significantly reduce the cost of the coating as compared to painting post-fabricated assemblies, but there are limitations to the fabrication that the preapplied coatings can withstand. Certain fabrication methods, including bending, may be employed when proper procedures are followed. Recommended minimum bend radii for selected coating types, alloys, and thicknesses are given in *Aluminum Standards and Data,* Table 7.19 (11). These bend radii are generally larger than those required for mill-finish material to avoid damaging the coating. The recommended bend radii should be taken as a starting point for trial bends for specific applications.

Generally, painting reduces minimum strengths from those of the bare product because the paint is baked on at temperatures that tend to anneal the aluminum. (See Section 3.1.2 for more on aluminum at elevated temperatures.) Minimum mechanical properties of painted sheet must, therefore, be obtained from the manufacturer.

The American Architectural Manufacturers Association (AAMA) provides voluntary specifications concerning painting:

- *Pigmented Organic Coatings on Aluminum Extrusions and Panels* (AAMA 2603-98) (30)
- *High Performance Organic Coatings on Aluminum Extrusions and Panels* (AAMA 2604-98) (31)
- *Superior Performing Organic Coatings on Aluminum Extrusions and Panels* (AAMA 2605-98) (32)

These specifications provide requirements for: material pretreatment, color uniformity, gloss, film hardness, film adhesion, impact resistance, abrasion resistance, chemical resistance (to muriatic acid, mortar, nitric acid, detergent, and window cleaner), corrosion resistance (to humidity and salt spray), and weathering (color retention, chalk resistance, gloss retention, and erosion resistance). The order of quality, from lowest to highest, is 2603, 2604, and 2605. The 2603 spec is primarily for interior or residential applications, 2604 is used for low-rise buildings, and 2605 is used for high rises. Although the 2605 spec is labeled a 10-year-specification, 20-year warranties are available when it's used. Finishing costs compare at about \$2/ft^2 for anodizing, \$3/ft^2 for 2604 spec paint, and \$4/ft^2 for 2605.

The less corrosion-resistant aluminum alloys are sometimes painted not for aesthetics, but to provide protection from corrosion. Typical of this group are the 2xxx series alloys and some of the 7xxx series group. In fact, these alloys are sometimes clad with a more corrosion-resistant aluminum alloy as a coating. (See the cladding paragraph below.) A general indication of the suitability of an aluminum alloy in a particular service may be found in the Aluminum Association's publication *Guidelines for the Use of Aluminum with Food and Chemicals* (18). There is no substitute, however, for testing in actual service conditions before selecting an alloy or coating.

Painting is sometimes considered to be a method of isolating aluminum from dissimilar materials. Guidance on painting for this purpose is found in Aluminum *Specification* Section 6.6.

3.2.4 Mechanical Finishes

Mechanical methods, such as abrasion blasting and buffing, may enhance the appearance of aluminum without coatings. These are sometimes called for as economical ways to reduce reflectance initially, rather than waiting for natural weathering to occur. Because abrasion blasting introduces compressive stress to the blasted surface, it tends to curl thin material, and so is limited by Aluminum *Specification* Section 6.7 to material thicker than $\frac{1}{8}$ in. [3 mm]. A floor buffer with a stainless steel pad can be run over flat aluminum sheet to produce an inexpensive finish and can be used on thinner material.

Both abrasion blasting and buffing have the potential to be performed unevenly. To achieve a more consistent mechanical finish, run sheet or coil material through a set of textured rollers, thereby producing an embossed finish. The most common embossed pattern is called *stucco*, but leather-grain and diamond patterns are also available. These textured patterns tend to hide smudges, scratches, and other minor surface blemishes. Embossing is readily available on material up to 48 in. [1,220 mm] wide, but finding a facility that can handle anything wider is difficult.

3.2.5 Cladding

Some wrought aluminum products (sheet and plate, tube, and wire) may receive a metallurgically bonded coating of high-purity aluminum (such as 1230), or corrosion-resistant aluminum alloy, (such as 7008 and 7072), to provide improved corrosion resistance or certain finish characteristics (like reflectivity), or to facilitate brazing. When such a coating is applied for this purpose, the product is referred to as *alclad* (sometimes abbreviated "alc"). The thickness of the coating is expressed as a percentage of the total thickness on a side for sheet and plate, the total wall thickness for tube, and the total cross-sectional area for wire. Tube is clad on either the inside or the outside, while plate and sheet may be clad on one side or both, but alclad sheet and plate, unless designated otherwise, is clad on both sides. The only tube commonly clad is of alloy 3003, and the only wire commonly clad is 5056. Among sheet and plate, 2014, 2024, 2219, 3003, 3004, 6061, 7050, 7075, 7178, and 7475 are clad. Nominal cladding thicknesses range from 1.5% to 10%. More details are given in *Aluminum Standards and Data*, Table 6.1 (11). Alloying elements are often added to the base metal to increase its strength, and the cladding typically has lower strength than the base alloy being clad. This is enough to affect the overall minimum mechanical properties, so alclad material has slightly lower design strengths than non-alclad material of the same alloy and temper. This is accounted for in the minimum mechanical properties for alclad sheet; some examples are given in Table 3.13. For all tempers of the 3003 and 3004 alloys, the alclad tensile ultimate strength is 1 ksi [5 MPa] less than the non-alclad strength.

TABLE 3.13 Alclad vs. Non-Alclad Minimum Strengths

Alloy Temper	Non-Alclad F_{tu} (ksi)	Alclad F_{tu} (ksi)	Non-Alclad F_{tu} (MPa)	Alclad F_{tu} (MPa)
3003-H12	17	16	120	115
3003-H14	20	19	140	135
3003-H16	24	23	165	160
3003-H18	27	26	185	180
3004-H32	28	27	190	185
3004-H34	32	31	220	215
3004-H36	35	34	240	235
3004-H38	38	37	260	255

The cost premium for alclad products is a function of the cladding thickness and the product. Alclad 3004 sheet for roofing and siding costs about 6% more than the non-clad 3004 sheet.

3.2.6 Roofing and Siding Finishes

Finishes available for roofing and siding as purchased from the manufacturer are given in *Aluminum Standards and Data*, Table 7.20 (11). These include mill finish, embossed, painted, and low reflectance. As mentioned previously, embossed sheet is an economical way to reduce the effect on appearance of smudges and other surface marring. The low reflectance finish is measured in accordance with ASTM Method D 523, and is specified in terms of a specular gloss number. Before you get too creative in specifying low reflectance or any other finish, it would be wise to check with suppliers to determine what's available here in the real world.

3.3 ERECTION

As with other topics peripheral to our central theme of providing guidance in the structural design of aluminum, we share only a few observations concerning the erection of aluminum structures. The properties of aluminum that differ from steel and most significantly affect erection are density and modulus of elasticity. An aluminum member of exactly the same size and shape as a steel member will weigh one-third as much and deflect about three times as far under live load. Aluminum members are often sized larger than steel members in the same application, however, in order to limit deflections, stresses, or fatigue effects. The net result is that aluminum structural components and assemblies weigh about one-half those of steel designed to support the same loads. This means they require less lifting capacity to handle and place and usually can be erected more quickly (Figure 3.23)

Figure 3.23 An aluminum dome being set in place with a crane. Aluminum domes as large as 190 ft [58 m] in diameter have been built alongside the structure to be covered, and then set in place in this manner. (Courtesy of Conservatek Industries, Inc.)

Field welding of aluminum should be performed only by companies that specialize in this type of construction, such as erectors of welded-aluminum storage tanks. Even then this work is subject to limitations, including frequent radiographic inspection. The old maxim, "Cut to fit, paint to match," should never be applied to aluminum. It cannot be cleanly cut with an oxygen-acetylene torch, and even if it could, the application of heat would affect its strength. It would be preferable to not have any equipment that can generate temperatures above 150°F [65°C], such as welding equipment and torches, on site. (The temptation to try to straighten out that beam they bent while unloading the truck is just too great for some field crews to resist.) Consequently, most aluminum structures are designed to be field assembled by bolting.

One fastener that can be installed very quickly, even under field conditions, is the *lockbolt*. This fastener system (see Figure 8.5) requires access to both sides of the joint and consists of a pin that has ribs and a narrow neck section, as well as a collar that is swaged onto the pin. When the tension force in the pin exceeds the strength of its neck, the neck breaks, resulting in a predictable clamping force between the parts joined. The lockbolt cannot be removed without destroying the fastener. This can be readily achieved, if necessary, by splitting the collar with bolt cutters. The need for involved installation procedures, such as those using torque wrenches or turn-of-the-nut methods, is eliminated, and visual inspection is sufficient. Lockbolts useful in aluminum structures are available in aluminum alloys and *austenitic stainless steel*. They do, however, require special installation tools, which can be bought or rented. Information on the design of fasteners for aluminum structures is given in Section 8.1.

When aluminum weldments are used, consider inspection after installation, even if the welding was performed in the shop. Shipping and erection stresses can damage weldments of any material, and aluminum is no exception. Information on welding and weld inspection can be found in Section 8.2.

Aluminum members and assemblies may be marked in the shop for field identification by stamping, scribing, or marking with paint or ink. Stamping should not be used on parts subject to fatigue and when used should be placed away from highly stressed areas. Because the location of marking may be difficult for the engineer to control, paint or ink marking may be preferred.

Take care storing aluminum on the job site. Aluminum that is allowed to stay wet or in contact with other objects will stain. Once stained, the original appearance cannot be restored. (See Section 3.1.2 regarding water staining.) No harm is done to the strength or life of the material by surface staining, so it is generally accepted on concealed components. Water staining of exposed components, however, can leave the structure "aesthetically challenged" (i.e., quite ugly). If sheet material that is to be used on an exposed surface is left lying flat, either on the ground or in contact with other sheets, you can expect the owner to develop an attitude. That attitude might be expressed as "Somebody's going to pay!" On the other hand, this entire scenario can be avoided by simply storing material properly.

Aluminum should also be protected from splatter from uncured concrete or mortar, as well as muriatic acid used to clean or prepare concrete and masonry. Once these are allowed to stand on aluminum surfaces, the resulting stains cannot be readily removed.

There is no aluminum equivalent to the AISC's *Code of Standard Practice for Steel Buildings and Bridges* (37), which addresses such erection issues as methods, conditions, safety, and tolerances. The most frequent question is: which tolerances apply to general aluminum construction? Unless a specific code, such as ASME B96.1 for welded-aluminum storage tanks (85), applies, the AISC *Code of Standard Practice* is a good starting point. Some differences apply, however, due to differences in properties of steel and aluminum. For example, thermal expansion and contraction of aluminum is about twice that of steel, so expect about $\frac{1}{4}$ in. of movement per 100 ft for each 15°F change in an aluminum structure (versus about $\frac{1}{8}$ in. in steel), or 2.4 mm per 10,000 mm for each 10°C change. Also, don't expect field tolerances to be any tighter than the cumulative effect of aluminum material and fabrication tolerances.

A final caution: because of to its relatively high scrap value, aluminum may need to be secured against what might be charitably termed "premature salvage."

PART II
Structural Behavior of Aluminum

Aluminum barrel vault covers. (Courtesy of Koch Refining Co.)

4 Material Properties for Design

Before we discuss the properties of aluminum alloys, it's useful to review how alloys are commonly identified in the aluminum industry. Unlike steel, which is usually identified by its ASTM specification and grade, (for example, A709 grade 50 steel), aluminum alloys are identified by their Aluminum Association alloy and temper, (for example, 3003-H16). Since the Aluminum Association designation system is used throughout most of the world, this method of identification is more portable than the one used in the U.S. for steel. On the other hand, just identifying the alloy and temper can, in a few instances, still leave some doubt as to the material's properties since different product forms of the same aluminum alloy may have different properties. A notable example is 6061-T6; 6061-T6 sheet has a minimum tensile ultimate strength of 42 ksi [290 MPa], whereas 6061-T6 extrusions have a minimum tensile ultimate strength of 38 ksi [260 MPa]. But if this were simple, you wouldn't be paid the big bucks.

4.1 MINIMUM AND TYPICAL PROPERTIES

Because of small but inevitable variations in production, two specimens of the same alloy, temper, and product form are likely to have slightly different strengths. Because strengths are a statistical distribution about an average, as shown in Figure 4.1, the minimum strength is arbitrary. The aluminum industry has historically used a very rigorous definition of a minimum strength for mill products—it's the strength that will be exceeded by 99% of the parts 95% of the time. This definition is not used by the steel industry and may have its roots in the use of aluminum in critical parts, such as aircraft components, which predates aluminum's use in construction. The U.S. military calls such minimum values "A" values and defines "B" values as those that 90% of samples will equal or exceed with a probability of 95%, a slightly less stringent criterion that yields higher minimum values. For 6061-T6 extrusions, for example, the A-basis minimum tensile yield strength is 35 ksi [240 MPa], while the B-basis minimum tensile yield strength is 38 ksi [260 MPa]. As for other structural materials, minimum strengths are used for aluminum structural design, and since the A-basis minimum strengths are the ones given in ASTM aluminum product specifications, they are the ones used for general structural design.

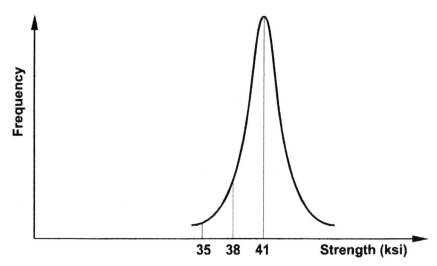

Figure 4.1   Distribution of material strengths.

Typical (average) strengths are given in the *Aluminum Design Manual*, Part V, but they are not used in structural design. Typical strength values are used instead of minimum strengths, however, to calculate the necessary capacity of fabrication equipment, such as punches and shears. Typical strengths are about 15% higher than minimum strengths.

4.2   STRENGTHS

The only minimum strengths that are required in product specifications for most mill products, and, thus, are said to be guaranteed are tensile yield and ultimate strengths. (*Wire* and *rod* to be made into fasteners must also meet minimum shear ultimate strength requirements.) The tensile test procedure is given in ASTM B557, *Tension Testing Wrought and Cast Aluminum and Magnesium Alloy Products* (66), and the shear test procedure is given in ASTM B565, *Shear Testing of Aluminum and Aluminum Alloy Rivets and Cold Heading Wire and Rods* (67).

Yield strength is determined by the 0.2% offset method because the stress-strain curves for aluminum alloys are not precisely linear up to yield and don't exhibit a sharply defined yield point. By this method, a line is drawn parallel to the initial linear portion of the curve; the stress at which this line intersects the actual stress-strain curve is defined as the yield strength. This definition is used worldwide.

The Aluminum *Specification* Table 3.3-1 provides minimum strengths for 22 alloys by temper and product form. The strengths given are:

- tensile yield strength (F_{ty})
- tensile ultimate strength (F_{tu})
- compression yield strength (F_{cy})
- shear yield strength (F_{sy})
- shear ultimate strength (F_{su})

The tensile yield and tensile ultimate strengths are the minimum specified strengths from *Aluminum Standards and Data* (11) (a subset of those in the corresponding ASTM specifications). Shear ultimate strengths (for material other than fasteners) and compression yield strengths given in the table are based on test data since they are not given in product specifications. The shear yield strengths given were derived by dividing the tensile yield strengths by $\sqrt{3}$ (equivalent to multiplying by approximately 0.58 and called the Von Mises yield criterion) and rounding the results.

Compressive ultimate strength is not a measured mechanical property for aluminum alloys. For material as ductile as aluminum, compressive ultimate strength has no meaning; in compressive tests, the metal swells but does not exhibit brittle fracture.

We've provided the minimum mechanical properties for the alloys listed in the Aluminum *Specification* Table 3.3-1, as well as for some additional alloys, in Appendix C. For other alloys, you must obtain these properties from the producer, other publications, or approximately derived from their relationships with tensile strengths. Approximate relationships between tensile yield and ultimate strengths and other strengths are:

$$F_{cy} = 0.9 \, F_{ty} \quad \text{For strain-hardened } (-\text{H}) \text{ tempers}$$

$$F_{cy} = F_{ty} \quad \text{For all other tempers}$$

$$F_{su} = 0.6 \, F_{tu}$$

$$F_{sy} = 0.6 \, F_{ty}$$

The strength of thick material tends to be less than for thin material, as shown in Table 4.1 for 5086-H112 plate. The thickness that should be used to determine the strength of a product is the original thickness of the material, not the final fabricated thickness. For example, if a part is machined to a thinner dimension, the mechanical properties for the original thickness are the proper ones to be used in structural calculations.

4.3   MODULUS OF ELASTICITY (E), SHEAR MODULUS (G), AND POISSON'S RATIO (ν)

A material's modulus of elasticity E (also called *Young's modulus*) is the slope of the elastic portion of the stress-strain curve and is a measure of the ma-

TABLE 4.1 Tensile Strengths of 5086-H112 Plate as a Function of Thickness

Thickness (in.)	Minimum Tensile Ultimate Strength (ksi)	Minimum Tensile Yield Strength (ksi)	Thickness (mm)	Minimum Tensile Ultimate Strength (MPa)	Minimum Tensile Yield Strength (MPa)
0.250 to 0.499	36	18	4.00 to 12.50	250	125
0.500 to 1.000	35	16	12.50 to 40.00	240	105
1.001 to 2.000	35	14	40.00 to 80.00	235	95
2.001 to 3.000	34	14			

terial's stiffness and buckling strength. The modulus of elasticity of aluminum alloys is roughly the weighted average of the moduli of its constituent alloying elements, so it tends to be about the same for alloys in a given series. (For example, for 6xxx series alloys, E is 10,000 ksi [69,000 MPa].) Temper does not significantly affect the modulus of elasticity. Steel's modulus (29,000 ksi [200,000 MPa]) is about three times that of aluminum, making steel considerably stiffer.

Compressive moduli of elasticity are given in Aluminum *Specification* (Table 3.3-1 [3.3-1M]) because the modulus of elasticity's effect on strength is limited to compression members, as we'll discuss in Section 5.2.1. The modulus varies from 10,100 ksi [69,600 MPa] to 10,900 ksi [75,200 MPa] (about 10%) for the alloys listed in the Aluminum *Specification*. Aluminum's tensile modulus of elasticity is about 2% less than its compressive modulus. Since bending involves both tensile and compressive stress, bending deflection calculations use an average modulus, found by subtracting 100 ksi [700 MPa] from the compressive modulus.

At strains beyond yield, the slope of the stress-strain curve is called the *tangent modulus* and is a function of stress, decreasing as the stress increases. The Ramberg-Osgood parameter n defines the shape of the stress-strain curve in this inelastic region and is given in the U.S. Military *Handbook on Metallic Materials and Elements for Aerospace Structures* (MIL HDBK 5) (137) for many aluminum alloys. The Ramberg-Osgood equation is:

$$\varepsilon = \sigma/E + 0.002 \, (\sigma/F_y)^n$$

where:

ε = strain
σ = stress
F_y = yield strength

For example, for 6061-T6 extrusions in longitudinal compression, the typical Ramberg-Osgood parameter n is 38. The parameter varies widely by alloy, temper, product, type of stress, and direction of stress with respect to product dimensions.

The shear modulus of aluminum (also called the *modulus of rigidity*) is generally calculated from its relationship with the modulus of elasticity and *Poisson's ratio* (v) for aluminum. Poisson's ratio (v) is the negative of the ratio of transverse strain that accompanies longitudinal strain caused by axial load in the elastic range. Poisson's ratio is approximately $\frac{1}{3}$ for aluminum alloys, similar to the ratio for steel. While the ratio varies slightly by alloy and decreases slightly as temperature decreases, such variations are insignificant for most applications. For a modulus of elasticity of 10,000 ksi [70,000 MPa], the shear modulus G is:

$$G = E/[2(1 + v)] = E/[2(1 + \tfrac{1}{3})] = 3E/8 = 3,800 \text{ ksi} \qquad [26,000 \text{ MPa}]$$

4.4   FRACTURE PROPERTIES

Ductility is the ability of a material to withstand plastic strain before fracture. Material that is not ductile may fracture at a lower tensile stress than its minimum ultimate tensile strength because it is unable to deform plastically at local stress concentrations that inevitably arise in the real world, even if they don't in machined laboratory-test specimens. Instead, brittle fracture may occur at a stress raiser. Since this is premature failure of the part, it's important to have a way to measure the ductility of structural alloys. Unfortunately, that's not so easy.

Fracture toughness is a measure of a material's resistance to the extension of a crack. Aluminum has a face-centered cubic crystal structure, so unlike steel, it does not exhibit a transition temperature below which the material suffers a significant loss in fracture toughness. Furthermore, alloys of the 1xxx, 3xxx, 4xxx, 5xxx, and 6xxx series are so tough that their fracture toughness cannot be readily measured by the methods commonly used for less tough materials and is rarely of concern. Alloys of the 2xxx and 7xxx series are less tough, and when they are used in such fracture critical applications as aircraft, their fracture toughness is of interest to designers. However, the notched bar impact tests that were developed for steel to determine a transition temperature, such as Charpy and Izod tests, don't work on aluminum alloys because they don't have a transition temperature to determine (26).

Instead, the plane-strain fracture toughness (K_{Ic}) can be measured by ASTM B645 (70). For those products whose fracture toughness cannot be measured by this method—such as sheet, which is too thin to apply B645—nonplane-strain fracture toughness (K_c) may be measured by ASTM B646. Fracture toughness is measured in units of ksi $\sqrt{\text{in}}$. [MPa $\sqrt{\text{m}}$] and is a function of thickness (thicker material has lower fracture toughness). Fracture toughness limits established by the Aluminum Association are given in *Aluminum Standards and Data* (11) for certain tempers of several plate and sheet alloys (2124, 7050, and 7475), but none of these alloys are included in the Aluminum *Specification*. Fracture toughness is a function of the orientation of the specimen and the notch relative to the part, so toughness is identified by three letters: L for the length direction, T for the width (long transverse) direction, and S for the thickness (short transverse) direction. The first letter denotes the specimen direction perpendicular to the crack; the second letter, the direction of the notch.

Elongation is defined as the percentage increase in the distance between two gauge marks of a specimen tested to tensile failure. Typical elongation values for some aluminum alloys are given in the *Aluminum Design Manual*, Part V, Table 5. Minimum elongation values for most alloys and product forms are given in *Aluminum Standards and Data* and ASTM specifications. They are dependent on the original gauge length (2 in. [50 mm] is standard for wrought products) and original dimensions, such as thickness, of the specimen. Elongation values are greater for thicker specimens. For example, typical elongation values for 1100-O material are 35% for a $\frac{1}{16}$ in. [1.6 mm] thick specimen, and 45% for a $\frac{1}{2}$ in. [12.5 mm] diameter specimen. Elongation is also very dependent on temperature, being lowest at room temperature and increasing at both lower and higher temperatures.

The elongation of aluminum alloys tends to be less than mild carbon steels. For example, while A36 steel has a minimum elongation of 20%, the comparable aluminum alloy, 6061-T6, has a minimum elongation requirement of 8% or 10%, depending on the product form. The elongation of annealed tempers is greater than that of strain-hardened or heat-treated tempers, while the strength of annealed tempers is less. Therefore, annealed material is more workable and has the capacity to undergo more severe forming operations without cracking.

All other things being equal, the greater the elongation, the greater the ductility. Unfortunately, other things usually aren't equal for different alloys, so elongation isn't a very useful indicator of ductility. The percentage reduction in area caused by necking in a tensile specimen tested to failure is similar to elongation in this regard. Unlike elongation, however, the reduction in area is not usually reported for aluminum tensile tests and is not generally available to designers.

Probably the most accurate measure of ductility (we saved the best for last) is the ratio of the tensile strength of a notched specimen to the tensile yield strength, called the *notch-yield ratio* (119, 120). Various standard notched-specimen specifications are available, but 60° notches with a sharp (0.0005 in. [0.01 mm] radius) tip seem to provide good discrimination between various alloys. The notched-specimen strength alone is not very useful since stress raisers in actual structures aren't likely to match the test specimen's notch. But if the notched strength is less than the yield strength of the material, then not much yielding can take place before fracture, a condition to be avoided. Of the alloys in the Aluminum *Specification*, only 2014-T6 and 6070-T6 suffer this fate. Therefore, these alloys (and 6066-T6, which is similar to 6070-T6) are considered to be notch-sensitive. The effect this has on design is discussed in Section 5.1.1.

4.5   THE EFFECT OF WELDING ON MECHANICAL PROPERTIES

Although alloying aluminum with elements like copper or magnesium increases the strength of aluminum, much of aluminum alloys' strength comes from precipitation-heat treatments (all tempers beginning with T5, T6, T7, T8, T9, or T10) or strain-hardening (all tempers beginning with H1, H2, H3, or H4). Heating the material tends to erase the effect of precipitation-heat treatments and strain-hardening, so arc welding, which brings a localized portion of the metal to the melting point, decreases the material's strength at the weld. For non-heat-treatable alloys (the ones that are strengthened by strain-hardening, that is, the 1xxx, 3xxx, and 5xxx series), the heat-affected zone's strength is reduced to the annealed strength. For heat-treatable alloys (the 2xxx, 6xxx, and 7xxx series), the heat-affected zone's strength is reduced to slightly below the solution-heat-treated strength (T4 temper), wiping out the effect of precipitation-heat treatment. Since the strength of mild carbon steel is not reduced by welding, welding's effect on aluminum can be an unpleasant surprise to some structural engineers. Welding does not, however, affect the modulus of elasticity (E) of the metal.

The effect of welding on strength varies by alloy. The reduction in strength is smallest for some of the 5xxx series alloys designed to be especially suitable for welding. For example, the unwelded tensile ultimate strength of 5086-H112 plate under 0.500 in. [12.5 mm] thick is 36 ksi [250 MPa]; its welded strength is 35 ksi [240 MPa]. At the other extreme, most alloys containing appreciable amounts of copper (these tend to be the highest strength aluminum alloys and include all of the 2xxx series and many alloys of the 7xxx series) are difficult to weld at all.

The effect of welding depends on many factors, including how much heat is applied during welding. This can vary by welding process (gas tungsten or gas metal arc welding), thickness of the parts welded, welding speed, and preheating. It also depends on the filler alloy used. All these effects are averaged to establish the minimum welded strengths given in Aluminum *Specification* Table 3.3-2, and are distinguished from the unwelded strengths by adding w to the subscript. (For example, the minimum welded tensile yield strength is F_{tyw}.) Two warnings should be heeded in using this table:

1) The welded tensile ultimate strengths (F_{tuw}) must be multiplied by 0.9 before they may be used in formulas for member strength in the *Specification*.

This factor is intended to account for the possibility that welds might receive only visual inspection, so their strengths might not be as reliably attained as that of the AWS qualification welds from which welded tensile ultimate strengths are established.

2) The welded yield strengths (F_{tyw} and F_{cyw}) for tempers other than annealed material (−O) must be multiplied by 0.75 before they may be used in formulas for member strength in the *Specification*. This multiplication has been performed, and the results tabulated in Appendix C. This factor effectively adjusts the yield strength from a 10 in. [250 mm] to a 2 in. [50 mm] gauge length, which is more representative of weld-affected material. This is because the 10 in. gauge length yield strengths were obtained from specimens with a transverse groove weld that affected only approximately 2 in. [50 mm] of the base metal (1 in. [25 mm] to either side of the weld centerline).

The ultimate tensile strengths for welded aluminum alloys given in Aluminum *Specification* Table 3.3-2 are defined there as the AWS D1.2 *Structural Welding Code–Aluminum* (91) weld qualification test values. This means that these are the strengths that must be attained in a tensile test of a groove weld in order for the weld procedure to be qualified according to AWS requirements. These same values are also required by the American Society of Mechanical Engineers (ASME) *Boiler and Pressure Vessel Code* to qualify a weld procedure. Table 4.2 of AWS D1.2-97 gives values of minimum tensile ultimate strength for thicknesses not included in Aluminum *Specification* Table 3.3-2, as well as for additional alloys, including several cast alloys.

Section 9.1 explains the structural design of welded members.

4.6 THE EFFECT OF TEMPERATURE ON ALUMINUM PROPERTIES

As temperature decreases from room temperature, the tensile strengths and elongation values of aluminum alloys increase, and unlike steel, aluminum suffers no transition temperature below which the risk of brittle fracture increases markedly. This makes aluminum an excellent choice for low-temperature structural applications (2). (This low-temperature ductility is a pleasant surprise to engineers familiar with steel.)

On the other hand, aluminum has a considerably lower melting point than steel (about 1,220°F [660°C]), and the decrease in aluminum strengths is fairly significant for most alloys above 200°F [95°C]. Figure 4.2 shows the variation in typical tensile ultimate strength with respect to temperature for various aluminum alloys. (See *Aluminum Design Manual*, Part V, for typical tensile strength and elongation values of alloys at temperatures from −320 to 700°F [−195 to 370°C].) Generally, the longer the alloy is held at an elevated temperature, the lower the strength. Under a constant stress, the deformation of an aluminum part may increase over time; this behavior is known as *creep*.

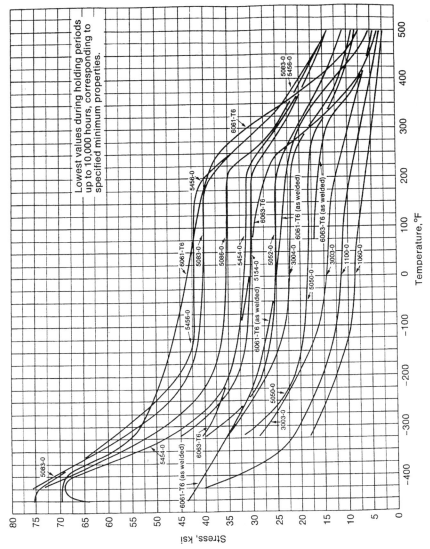

Figure 4.2 Minimum tensile ultimate strengths at various temperatures. (Courtesy of the Aluminum Association).

107

Creep effects increase as the temperature increases. At room temperature, very little creep occurs unless stresses are near the tensile strength. Creep usually is not a factor unless stresses are sustained at temperatures more than about 200°F [95°C]. Properties of aluminum alloys at high and low temperatures are given in Kaufman's *Properties of Aluminum Alloys—Tensile, Creep, and Fatigue Data at High and Low Temperatures* (118).

For most alloys, the minimum mechanical properties given in the Aluminum *Specification* are used for service temperatures up to 200°F [95°C], but above this are decreased to account for reduced strength. An example of how allowable stresses at elevated temperatures are decreased is given in ASME B96.1, *Welded Aluminum-Alloy Storage Tanks* (85). Aluminum alloys are used in structural applications to 400°F [200°C] with appropriately reduced design strengths.

Alloys that have more than 3% magnesium content (such as 5083, 5086, 5154, and 5456, as well as filler alloys 5183, 5356, 5556, and 5654) should not be held above 150°F [65°C] because these elevated temperatures make them sensitive to *exfoliation corrosion*. Exfoliation is a delamination or peeling of layers of the metal in planes approximately parallel to the metal surface. Unfortunately, magnesium content is directly related to strength, so this leaves only modest-strength 5xxx alloys available for elevated-temperature applications.

4.7   FIRE RESISTANCE

When it comes to fire resistance, aluminum can be confusing (16). On one hand, finely divided aluminum powder burns so explosively that it's used as rocket fuel; on the other hand, aluminum is routinely melted (every time it's welded) without any concern regarding combustion. Because of the relatively low melting point of aluminum alloys (around 1,100°F [600°C]), however, aluminum is not suitable when exposed to high temperatures, such as in fireplaces, where it quickly turns to mush. In fires, thin-gauge aluminum sheet can melt so fast that it appears to be burning. But in reality, semi-fabricated aluminum products, such as extrusions, sheet, plate, castings, and forgings, are not combustible under ordinary conditions.

Building codes classify a material as noncombustible if it passes ASTM E136. In this test, solid pieces of material $1\frac{1}{2}$ in. × $1\frac{1}{2}$ in. × 2 in. [38 mm × 38 mm × 50 mm] are placed in an oven and heated to 1,382°F [750°C]; unless flames are produced, the material is called noncombustible. Aluminum alloys subjected to this test have been classified noncombustible since no flames result even though aluminum's melting point is lower than the test temperature. Other fire performance tests, such as ASTM E108 for exterior fire resistance and ASTM E119 for interior fire endurance, exist. These test a building assembly (for example, an insulated metal roofing assembly) and must be evaluated for specific designs because they are dependent on the dimensions and combination of materials of the assembly being tested.

4.8 HARDNESS

The hardness of aluminum alloys isn't used directly in structural design, but since it's relatively easy to measure, it's frequently reported. Hardness can be measured by several methods, including Webster hardness (ASTM B647), Barcol hardness (ASTM B648), Newage hardness (ASTM B724), Vickers hardness, and Rockwell hardness (ASTM E18). The Brinnell hardness (ASTM E10) for a 500 kg load on a 10 mm ball is used most often and is given in the *Aluminum Design Manual*, Part V, Table 5 [5M]. Hardness measurements are sometimes used for quality-assurance purposes on temper. The Brinnell hardness number (BHN) multiplied by 0.56 is approximately equal to the ultimate tensile strength (F_{tu}) in ksi; this relationship can be useful to help identify material or estimate its strength based on a simple hardness test. The relationship between hardness and strength, however, is not as dependable for aluminum as for steel. Hardness measurements are sometimes used to measure the effect on strength at various points in the heat-affected zone across a weld.

Typical hardness values for some common alloy tempers are given in Table 4.2.

4.9 PHYSICAL PROPERTIES

Properties other than mechanical properties are usually referred to under the catch-all heading of "physical properties." The significant ones for structural engineers include density and coefficient of thermal expansion. These properties vary among aluminum alloys; they're given to three significant figure accuracy in the *Aluminum Design Manual*, Part V. Often, however, the variations between alloys aren't large enough to matter for structural engineering purposes.

TABLE 4.2 Typical Hardness Values

Alloy Temper	Typical Brinnell Hardness (500 kg load 10 mm ball)
1100-H14	32
2014-T6	135
3003-H16	47
5005-H34	41
5052-H32	60
5454-H32	73
6061-T6	95
6063-T5	60
6063-T6	73
6101-T6	71

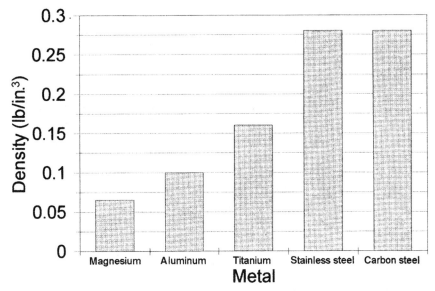

Figure 4.3 Densities of several metals.

Density doesn't vary much by alloy because alloying elements make up such a small portion of the composition. For aluminum alloys, density averages about 0.1 lb/in³ [2,700 kg/m³]; most alloys are within 5% of this number. This is about one-third the density of carbon and stainless steels, which weigh about 0.283 lb/in³ [7,640 kg/m³]. Since density is the weighted average of the densities of the elements that constitute the alloy, the 5xxx and 6xxx series alloys are the lightest because magnesium is the lightest of the main alloying elements. The actual density of wrought alloys is usually close to the nominal value, but the density of castings is about 95% to 100% of the theoretical density of the cast alloy due to porosity. Nominal densities for wrought aluminum alloys are listed in Table 4.3—if you need to be that precise.

The *coefficient of thermal expansion*, which is the rate at which material expands as its temperature increases, is itself a function of temperature, increasing with increasing temperature. Average values are used for a range of temperatures, usually from room temperature (68°F [20°C]) to water's boiling temperature (212°F [100°C]). The coefficient also varies by alloy. The average coefficient is approximately 13 × 10⁻⁶/°F [23 × 10⁻⁶/°C]. Table 4.4 compares aluminum's coefficient to some other materials.

4.10 ALUMINUM MATERIAL SPECIFICATIONS

A number of sources of specifications for aluminum mill products exist, including the American Society for Testing and Materials (ASTM), Military

TABLE 4.3 Nominal Densities of Aluminum Alloys in the Aluminum *Specification*

Alloy	Density (lb/in.^{3})	Density (kg/m^3)
1100	0.098	2,710
2014	0.101	2,800
2024	0.100	2,780
3003	0.099	2,730
3004	0.098	2,720
3005	0.098	2,730
3105	0.098	2,720
4043	0.097	2,690
5005	0.098	2,700
5050	0.097	2,690
5052	0.097	2,680
5083	0.096	2,660
5086	0.096	2,660
5154	0.096	2,660
5183	0.096	2,660
5356	0.096	2,640
5454	0.097	2,690
5456	0.096	2,660
5554	0.097	2,690
5556	0.096	2,660
5654	0.096	2,660
6005	0.097	2,700
6061	0.098	2,700
6063	0.097	2,700
6066	0.098	2,720
6070	0.098	2,710
6105	0.097	2,690
6351	0.098	2,710
6463	0.097	2,690
7075	0.101	2,810

TABLE 4.4 Coefficients of Thermal Expansion

Material	Coefficient of Thermal Expansion	
	$\times 10^{-6}/°F$	$\times 10^{-6}/°C$
Copper	18	32
Magnesium	15	27
Aluminum	13	23
304 Stainless steel	9.6	17
Carbon Steel	6.5	12
Concrete	6.0	11

(MIL), federal, Aerospace Materials Specifications (AMS) published by the Society of Automotive Engineers (SAE), American Society of Mechanical Engineers (ASME), and American Welding Society (AWS) (for welding electrodes and rods). The MIL and federal specifications are being phased out—mercifully, since they often duplicated private-sector specifications—but in many cases, AMS has merely replaced the federal spec prefix with "AMS" and continues to publish it.

Of these, the ASTM specifications are probably the most widely used. Most ASTM aluminum-material specifications include a number of different alloys and tempers. What these alloys have in common in a given ASTM specification is that each is used to make the product covered by the specification. For example, ASTM B210 (50) is for aluminum drawn seamless tubes and includes alloys 1100, 2014, 3003, 5005, 5050, and 6061. The ASTM specifications and alloys listed in the Aluminum *Specification* are tabulated in Table 4.5.

TABLE 4.5   ASTM Specifications and Alloys Listed in the Aluminum *Specification*

ASTM Specification	Product	Alloys
B209	Sheet and Plate	1100, 2014, 3003, Alclad 3003, 3004, Alclad 3004, 3005, 3105, 5005, 5050, 5052, 5083, 5086, 5154, 5454, 5456, 6061
B210	Drawn Seamless Tubes	1100, 2014, 3003, Alclad 3003, 5005, 5050, 5052, 5083, 5086, 5154, 5456, 6061, 6063
B211	Bar, Rod, and Wire	1100, 2014, 3003, 5052, 5154, 6061
B221	Extruded Bars, Rods, Wire, Profiles, and Tubes	1100, 2014, 3003, Alclad 3003, 3004, 5052, 5083, 5086, 5154, 5454, 5456, 6005, 6061, 6063, 6066, 6070, 6105, 6351, 6463
B241	Seamless Pipe and Seamless Extruded Tube	1100, 2014, 3003, Alclad 3003, 5052, 5083, 5086, 5454, 5456, 6061, 6063, 6066, 6351
B247	Die Forgings, Hand Forgings, and Rolled Ring Forgings	1100, 2014, 3003, 5083, 6061, 6066
B308	Standard Structural Profiles	6061
B316	Rivet and Cold Heading Wire and Rods	1100, 2017, 2117, 5056, 6053, 6061, 7050, 7075
B429	Extruded Structural Pipe and Tube	6061, 6063

The ASTM aluminum specifications are maintained by ASTM's Committee B-7 on Light Metals and Alloys. This committee is composed mostly of representatives of aluminum producers, and many of these representatives are also members of the Aluminum Association's Technical Committee on Product Standards (TCPS), which maintains the Aluminum Association's *Aluminum Standards and Data* (11) publication. Hence, it isn't too surprising that the ASTM and *Aluminum Standards and Data* requirements are almost identical.

ASME aluminum specifications are nearly identical to ASTM specifications of the same name except that ASME specifications include only those alloy tempers included in the ASME *Boiler and Pressure Vessel Code* and the ASME specification numbers are prefixed with an "S" (for example, ASME SB-209 is similar to ASTM B209). The ASME aluminum welding electrode and rod specifications SFA-5.3 and SFA-5.10 are identical to AWS specifications A5.3 and A5.10, respectively (87)(88).

4.11 ALLOY IDENTIFICATION

A color code has been assigned to some common wrought aluminum alloys for identification purposes (11). The color may be used on tags attached to the material, or in the case of extrusions, rods, and bars, may be painted on the end of the part. Table 4.6 lists the color assignments for alloys listed in the Aluminum *Specification*. While the color code identifies the alloy, it does not reveal anything about the material's temper.

4.12 CERTIFICATION DOCUMENTATION

Frequently, owners will require that materials incorporated into the work be accompanied by "mill certification." Often, the specific properties to be certified are not identified. Various forms of documentation are available to purchasers to certify that suppliers have inspected and tested the material and that it meets requirements, typically those established by the Aluminum Association (11) or the appropriate ASTM specification for the material. Customers buying aluminum for structural applications may obtain one of the following types of documentation:

Certificate of Compliance: Issued to cover shipments made over a period of time (usually one year), rather than a specific shipment. No test results are included.

Certificate of Inspection: Similar to the certificate of compliance, but for a specific shipment of material. This certificate is what ASTM specifications require manufacturers to provide when purchasers require material certification in their ordering information.

TABLE 4.6 Color Code for Aluminum
Specification **Alloys**

Alloy	Color
1100	White
2014	Gray
2024	Red
3003	Green
3004	—
3005	—
3105	—
5005	—
5050	—
5052	Purple
5083	Red and Gray
5086	Red and Orange
5154	Blue and Green
5454	—
5456	Gray and Purple
6005	—
6061	Blue
6063	Yellow and Green
6066	Red and Green
6070	Blue and Gray
6105	—
6351	Purple and Orange
7075	Black

Certificate of Inspection and Test Results: Gives minimum and maximum mechanical properties obtained from testing each lot and lists the applicable chemical-composition limits (not the actual chemical-composition).

Certificate of Inspection and Test Results including Chemical Analysis: Gives minimum and maximum mechanical properties and chemical analysis obtained from testing each lot.

These certificates are listed in ascending order of cost. For most routine structural applications, a certificate of compliance or inspection is usually sufficient.

5 Explanation of the Aluminum *Specification*

This chapter could also have been titled "Structural Design with Metals" because it covers a range of metal behavior beyond that covered by the metal specification we all learned in school (the AISC *Specification* [38, 39]). In a number of ways, hot-rolled steel design is just a special case of metal design, while the Aluminum *Specification* offers a more panoramic view. Ironically, learning about aluminum can actually give you better insight into steel design; in college, most of us studiously avoided any metal but hot-rolled steel.

Design with hot-rolled steel is typically limited to the so-called "*compact sections.*" These shapes have sufficiently stocky cross-sectional *elements* so *local buckling* is not a concern. In the real world, many metal structures are made of lighter gauge components instead, whose capacities may be influenced by local buckling. We'll lead you into the land of *postbuckling strength,* where no compact section has ever gone. If you master this concept, you'll not only be able to deal with the wide range of shapes available in aluminum, you will also gain some insight into the design of light-gauge steel structures. While the Aluminum *Specification* addresses the full spectrum of *element slenderness*, in steel you have to leave the comfort of the *Steel Manual* and wade into the complex provisions of the AISI cold-formed steel specification (40) to design members that have very slender elements. Although the Aluminum *Specification* and the cold-formed steel specification use entirely different methods to address slender elements, they both deal with the same fundamental behavior.

We'll discuss the Aluminum *Specification*'s approach to metal design in this chapter, giving examples and highlighting sections of the *Specification* as we cover them. For the plug-and-chug oriented, you can skip to Chapter 7, where we'll keep the discussion to a minimum and provide the step-by-step procedures for applying the Aluminum *Specification*. Finally, for a checklist of the applicable provisions of the *Specification* for each mode of behavior, refer to Appendix G.

5.1 TENSION MEMBERS

If you're like most engineers, you appreciate the simplicity of tensile stresses. After all, tensile stresses and strengths are so much easier to calculate than other types of stress. Tensile stresses can cause fracture, however, which is a failure mode that's sudden and, therefore, usually catastrophic. This is just about the worst thing that can happen to your structure.

Tensile stresses may also cause yielding, but the consequences are, perhaps, not quite as disastrous as fracture. Some yielding occurs in many structures, whether in a bolt torqued to installation requirements or in a member at a stress riser, without robbing the structure of its usefulness. It's only when the yielding is excessive that failure has occurred.

We'll keep these *limit states* in mind as we discuss the Aluminum *Specification*'s requirements. Identifying the limit of a structure's usefulness and then providing an adequate safeguard against reaching that limit are what design specifications are all about.

5.1.1 Tensile Strength

Aluminum *Specification* Section 3.4.1 effectively defines axial tensile strength as:

1) Yield strength on the gross area
2) Ultimate strength on the *net area.*

Each of these strengths is factored by its respective safety factor in allowable stress design. (The strengths are factored by a resistance factor in load and resistance factor design [LRFD]. We'll stick to allowable stress design for now.) The resulting allowable strength of a member is:

$$F_t = \min \left(\frac{F_{ty} A_g}{n_y}, \frac{F_{tu} A_n}{k_t n_u} \right) \tag{5.1}$$

where:

A_g = gross area of the member
A_n = net area of the member
F_{ty} = tensile yield strength
F_{tu} = tensile ultimate strength
k_t = coefficient for tension members (see Aluminum *Specification* Table 3.4-2)
n_y = factor of safety on yield
n_u = factor of safety on ultimate strength

This *Specification* section is highlighted in Aluminum *Specification* Table 3.4-3 reproduced here as Figure 5.1. The general equations are evaluated for various alloys in the *Aluminum Design Manual*, Part VII, Tables 2-2 through 2-23, with 6061-T6 being found on Table 2-21 (reproduced here as Figure 5.2).

GENERAL FORMULAS FOR DETERMINING ALLOWABLE STRESSES FROM SECTION 3.4

Type of Stress	Type of Member or Component	Sub-Sec.	Allowable Stress F_{ty}/n_y or $F_{tu}/(k_t n_u)$
TENSION, axial	Any tension member	1	F_{ty}/n_y or $F_{tu}/(k_t n_u)$
TENSION IN BEAMS, extreme fiber, net section	Rectangular tubes, structural shapes bent around strong axis	2	F_{ty}/n_y or $F_{tu}/(k_t n_u)$
	Round or oval tubes	3	$1.17 F_{ty}/n_y$ or $1.24 F_{tu}/(k_t n_u)$
	Shapes bent about weak axis, bars, plates	4	$1.30 F_{ty}/n_y$ or $1.42 F_{tu}/(k_t n_u)$
BEARING	On rivets and bolts	5	$2F_{tu}/n_u$
	On flat surfaces and pins and on bolts in slotted holes	6	$2F_{tu}/(1.5 n_u)$

Type of Stress	Type of Member or Component	Sub-Sec.	Allowable Stress Slenderness ≤ S_1	Slenderness Limit S_1	Allowable Stress Slenderness Between S_1 and S_2	Slenderness Limit S_2	Allowable Stress Slenderness ≥ S_2
COMPRESSION IN COLUMNS, axial, gross section	All columns	7	$\dfrac{F_{cy}}{n_y}$	$\dfrac{kL}{r} = \dfrac{B_c - \frac{n_u F_{cy}}{n_y}}{D_c}$	$\dfrac{1}{n_u}\left(B_c - D_c\,\dfrac{kL}{r}\right)$	$\dfrac{kL}{r} = C_c$	$\dfrac{\pi^2 E}{n_u\left(\frac{kL}{r}\right)^2}$
COMPRESSION IN COMPONENTS OF COLUMNS gross section	Flat plates supported along one edge - columns buckling about a symmetry axis	8	$\dfrac{F_{cy}}{n_y}$	$\dfrac{b}{t} = \dfrac{B_p - \frac{n_u F_{cy}}{n_y}}{5.1 D_p}$	$\dfrac{1}{n_u}\left(B_p - 5.1 D_p\,\dfrac{b}{t}\right)$	$\dfrac{b}{t} = \dfrac{k_1 B_p}{5.1 D_p}$	$\dfrac{k_2\sqrt{B_p E}}{n_u(5.1 b/t)}$
	Flat plates supported along one edge - columns not buckling about a symmetry axis	8.1	$\dfrac{F_{cy}}{n_y}$	$\dfrac{b}{t} = \dfrac{B_p - \frac{n_u F_{cy}}{n_y}}{5.1 D_p}$	$\dfrac{1}{n_u}\left(B_p - 5.1 D_p\,\dfrac{b}{t}\right)$	$\dfrac{b}{t} = \dfrac{C_p}{5.1}$	$\dfrac{\pi^2 E}{n_u(5.1 b/t)^2}$
	Flat plates with both edges supported	9	$\dfrac{F_{cy}}{n_y}$	$\dfrac{b}{t} = \dfrac{B_p - \frac{n_u F_{cy}}{n_y}}{1.6 D_p}$	$\dfrac{1}{n_u}\left(B_p - 1.6 D_p\,\dfrac{b}{t}\right)$	$\dfrac{b}{t} = \dfrac{k_1 B_p}{1.6 D_p}$	$\dfrac{k_2\sqrt{B_p E}}{n_u(1.6 b/t)}$
	Flat plates with one edge supported and other edge with stiffener	9.1			See Section 3.4.9.1		
	Flat plates with both edges supported and with an intermediate stiffener	9.2			See Section 3.4.9.2		
	Curved plates supported on both edges, walls of round or oval tubes	10*	$\dfrac{F_{cy}}{n_y}$	$\dfrac{R_b}{t} = \left(\dfrac{B_t - \frac{n_u F_{cy}}{n_y}}{D_t}\right)^2$	$\dfrac{1}{n_u}\left(B_t - D_t\sqrt{\dfrac{R_b}{t}}\right)$	$\dfrac{R_b}{t} = C_t$	$\dfrac{\pi^2 E}{16 n_u\left(\dfrac{R_b}{t}\right)\left(1+\dfrac{\sqrt{R_b/t}}{35}\right)^2}$

*For tubes with circumferential welds, equations of Sections 3.4.10, 3.4.12, and 3.4.16.1 apply for $R/t \le 20$.

Figure 5.1 General requirements of the Aluminum *Specification* for axial tension.

Type of Stress	Type of Member or Component	Sub-Sec.	Allowable Stress Slenderness ≤ S_1	Slenderness Limit S_1	Allowable Stress Slenderness Between S_1 and S_2	Slenderness Limit S_2	Allowable Stress Slenderness ≥ S_2
COMPRESSION IN BEAMS, extreme fiber gross section	Single web beams bent about strong axis	11	$\dfrac{F_{cy}}{n_y}$	$\dfrac{L_b}{r_y} = \dfrac{1.2(B_c - F_{cy})}{D_c}$	$\dfrac{1}{n_y}\left(B_c - D_c\dfrac{L_b}{1.2 r_y}\right)$	$\dfrac{L_b}{r_y} = \dfrac{1.2 C_c}{D_c}$	$\dfrac{\pi^2 E}{n_y\left(\dfrac{L_b}{1.2 r_y}\right)^2}$
	Round or oval tubes	12*	$\dfrac{1.17 F_{cy}}{n_y}$	$\dfrac{R_b}{t} = \left(\dfrac{B_{tb} - 1.17 F_{cy}}{D_{tb}}\right)^2$	$\dfrac{1}{n_y}\left(B_{tb} - D_{tb}\sqrt{\dfrac{R_b}{t}}\right)$	$\dfrac{R_b}{t} = \left(\dfrac{n_s B_{tb} - B_t}{n_s D_{tb} - D_t}\right)^2$	Same as Section 3.4.10
	Solid rectangular and round section beams	13	$\dfrac{1.3 F_{cy}}{n_y}$	$\dfrac{d}{t}\sqrt{\dfrac{L_b}{d}} = \dfrac{B_{tb} - 1.3 F_{cy}}{2.3 D_{br}}$	$\dfrac{1}{n_y}\left(B_{tb} - 2.3 D_{br}\dfrac{d}{t}\sqrt{\dfrac{L_b}{d}}\right)$	$\dfrac{d}{t}\sqrt{\dfrac{L_b}{d}} = \dfrac{C_{br}}{2.3}$	$\dfrac{\pi^2 E}{5.29 n_y\left(\dfrac{d}{t}\right)^2 \dfrac{L_b}{d}}$
	Rectangular tubes and box sections	14	$\dfrac{F_{cy}}{n_y}$	$\dfrac{L_b S_c}{0.5\sqrt{I_y J}} = \left(\dfrac{B_c - F_{cy}}{1.6 D_c}\right)^2$	$\dfrac{1}{n_y}\left(B_c - 1.6 D_c\sqrt{\dfrac{L_b S_c}{0.5\sqrt{I_y J}}}\right)$	$\dfrac{L_b S_c}{0.5\sqrt{I_y J}} = \left(\dfrac{C_c}{1.6}\right)^2$	$\dfrac{\pi^2 E}{2.56 n_y\left(\dfrac{L_b S_c}{0.5\sqrt{I_y J}}\right)}$
COMPRESSION IN COMPONENTS OF BEAMS, (component under uniform compression), gross section	Flat plates supported on one edge	15	$\dfrac{F_{cy}}{n_y}$	$\dfrac{b}{t} = \dfrac{B_p - F_{cy}}{5.1 D_p}$	$\dfrac{1}{n_y}\left(B_p - 5.1 D_p\dfrac{b}{t}\right)$	$\dfrac{b}{t} = \dfrac{k_1 B_p}{5.1 D_p}$	$\dfrac{k_2\sqrt{B_p E}}{n_y(5.1 b/t)}$
	Flat plates with both edges supported	16	$\dfrac{F_{cy}}{n_y}$	$\dfrac{b}{t} = \dfrac{B_p - F_{cy}}{1.6 D_p}$	$\dfrac{1}{n_y}\left(B_p - 1.6 D_p\dfrac{b}{t}\right)$	$\dfrac{b}{t} = \dfrac{k_1 B_p}{1.6 D_p}$	$\dfrac{k_2\sqrt{B_p E}}{n_y(1.6 b/t)}$
	Curved plates supported on both edges	16.1*	$\dfrac{1.17 F_{cy}}{n_y}$	$\dfrac{R_b}{t} = \left(\dfrac{B_t - 1.17 F_{cy}}{D_t}\right)^2$	$\dfrac{1}{n_y}\left(B_t - D_t\sqrt{\dfrac{R_b}{t}}\right)$	$\dfrac{R_b}{t} = C_t$	$\dfrac{\pi^2 E}{16 n_y\left(\dfrac{R_b}{t}\right)\left(1 - \sqrt{\dfrac{R_b/t}{35}}\right)^2}$
	Flat plates with one edge supported and the other edge with stiffener	16.2	See Section 3.4.16.2				
	Flat plates with both edges supported and with an intermediate stiffener	16.3	See Section 3.4.16.3				
COMPRESSION IN COMPONENTS OF BEAMS, (component under bending in own plane) gross section	Flat plates with compression edge free, tension edge supported	17	$\dfrac{1.3 F_{cy}}{n_y}$	$\dfrac{b}{t} = \dfrac{B_{br} - 1.3 F_{cy}}{3.5 D_{br}}$	$\dfrac{1}{n_y}\left[B_{br} - 3.5 D_{br}\left(\dfrac{b}{t}\right)\right]$	$\dfrac{b}{t} = \dfrac{C_{br}}{3.5}$	$\dfrac{\pi^2 E}{n_y(3.5 b/t)^2}$
	Flat plate with both edges supported	18	$\dfrac{1.3 F_{cy}}{n_y}$	$\dfrac{h}{t} = \dfrac{B_{br} - 1.3 F_{cy}}{0.67 D_{br}}$	$\dfrac{1}{n_y}\left[B_{br} - 0.67 D_{br}\left(\dfrac{h}{t}\right)\right]$	$\dfrac{h}{t} = \dfrac{k_1 B_{br}}{0.67 D_{br}}$	$\dfrac{k_2\sqrt{B_{br} E}}{n_y(0.67 h/t)}$
	Flat plate with horizontal stiffener, both edges supported	19	$\dfrac{1.3 F_{cy}}{n_y}$	$\dfrac{h}{t} = \dfrac{B_{br} - 1.3 F_{cy}}{0.29 D_{br}}$	$\dfrac{1}{n_y}\left[B_{br} - 0.29 D_{br}\left(\dfrac{h}{t}\right)\right]$	$\dfrac{h}{t} = \dfrac{k_1 B_{br}}{0.29 D_{br}}$	$\dfrac{k_2\sqrt{B_{br} E}}{n_y(0.29 h/t)}$
SHEAR IN WEBS, gross section	Unstiffened flat webs	20	$\dfrac{F_{sy}}{n_y}$	$\dfrac{h}{t} = \dfrac{B_s - F_{sy}}{1.25 D_s}$	$\dfrac{1}{n_y}\left[B_s - 1.25 D_s\left(\dfrac{h}{t}\right)\right]$	$\dfrac{h}{t} = \dfrac{C_s}{1.25}$	$\dfrac{\pi^2 E}{n_y(1.25 h/t)^2}$
	Stiffened flat webs $a_e = a_1/\sqrt{1 + 0.7(a_1/a_2)^2}$	21	$\dfrac{F_{sy}}{n_y}$	$\dfrac{a_e}{t} = \dfrac{B_s - (n_a F_{sy}/n_y)}{1.25 D_s}$	$\dfrac{1}{n_a}\left[B_s - 1.25 D_s\left(\dfrac{a_e}{t}\right)\right]$	$\dfrac{a_e}{t} = \dfrac{C_s}{1.25}$	$\dfrac{\pi^2 E}{n_a(1.25 a_e/t)^2}$

Figure 5.1 (*continued*)

118

Table 2-21
Allowable Stresses for BUILDING and Similar Type Structures

6061-T6, -T651, -T6510, -T6511

Extrusions up thru 1 in., Sheet & Plate, Pipe, Standard Structural Shapes, Drawn Tube, Rolled Rod and Bar, 6351-T5 Extrusions

WHITE BARS apply to nonwelded members and to welded members at locations farther than 1.0 in. from a weld.

SHADED BARS apply within 1.0 in. of a weld.

Equations that straddle the shaded and unshaded areas apply to both.

*For tubes with circumferential welds, equations of Sections 3.4.10, 3.4.12, and 3.4.16.1 apply for $R_b/t < 20$.

Type of Stress	Type of Member or Component	Sec. 3.4.	Allowable Stress
TENSION, axial	Any tension member	1	19 / 11t
TENSION IN BEAMS, extreme fiber, net section	Rectangular tubes, structural shapes bent around strong axis	2	19 / 11t
	Round or oval tubes	3	24 / 13.5t
	Shapes bent about weak axis, bars, plates	4	28 / 16t
BEARING	On rivets and bolts	5	39 / 23t
	On flat surfaces and pins and on bolts in slotted holes	6	26 / 16t

Type of Stress	Type of Member or Component	Sec. 3.4.	Allowable Stress Slenderness $\le S_1$	Slenderness Limit S_1	Allowable Stress Slenderness Between S_1 and S_2	Slenderness Limit S_2	Allowable Stress Slenderness $\ge S_2$
COMPRESSION IN COLUMNS, axial, gross section	All columns	7	— / 12t	$kL/r = 0$ / —	$20.2 - 0.126\,(kL/r)$ / 12t	$kL/r = 66$ / 65t	$51{,}000/(kL/r)^2$ / $51{,}000/(kL/r)^2$
COMPRESSION IN COMPONENTS OF COLUMNS, gross section	Flat plates supported along one edge - columns buckling about a symmetry axis	8	21 / 12t	$b/t = 2.7$ / —	$23.1 - 0.79\,(b/t)$ / 12t	$b/t = 10$ / 13t	$154(b/t)$ / $154(b/t)$
	Flat plates supported along one edge - columns not buckling about a symmetry axis	8.1	21 / 12t	$b/t = 2.7$ / —	$23.1 - 0.79\,(b/t)$ / 12t	$b/t = 12$ / 13t	$1970(b/t)^2$ / $1970(b/t)^2$
	Flat plates with both edges supported	9	21 / 12t	$b/t = 8.4$ / —	$23.1 - 0.25\,(b/t)$ / 12t	$b/t = 33$ / 41t	$490(b/t)$ / $490(b/t)$
	Flat plates with one edge supported and other edge with stiffener	9.1	See Section 3.4.9.1				
	Flat plates with both edges supported and with an intermediate stiffener	9.2	See Section 3.4.9.2				
	Curved plates supported on both edges, walls of round or oval tubes	10*	21 / 12t	$R_b/t = 2.2$ / $R_b/t = 9.0$	$22.2 - 0.80\sqrt{R_b/t}$ / $13.5 - 0.50\sqrt{R_b/t}$	$R_b/t = 141$ / $R_b/t = 290$	$\dfrac{3200}{(R_b/t)(1 + \sqrt{R_b/t}/35)^2}$

Figure 5.2 Requirements of the Aluminum *Specification* for axial tension, for the specific case of alloy 6061-T6.

Type of Stress	Type of Member or Component	Sec. 3.4.	Allowable Stress Slenderness ≤ S_1	Slenderness Limit S_1	Allowable Stress Slenderness Between S_1 and S_2	Slenderness Limit S_2	Allowable Stress Slenderness ≥ S_2
COMPRESSION IN BEAMS, extreme fiber gross section	Single web beams bent about strong axis	11	21 / 10.5	$L_b/r_y = 23$ / —	$23.9 - 0.124 L_b/r_y$ / 10.5	$L_b/r_y = 79$ / $L_b/r_y = 91$	$\dfrac{87,000}{(L_b/r_y)^2}$
	Round or oval tubes	12*	25 / 12	$R_b/t = 28$ / $R_b/t = 53$	$39.3 - 2.70\sqrt{R_b/t}$ / $18.6 - 1.00\sqrt{R_b/t}$	$R_b/t = 81$ / $R_b/t = 134$	Same as Section 3.4.10
	Solid rectangular and round section beams	13	28 / 13.5	$d/t\sqrt{L_b/d} = 13$ / —	$40.5 - 0.93\,d/t\sqrt{L_b/d}$ / 13.5	$d/t\sqrt{L_b/d} = 29$ / $d/t\sqrt{L_b/d} = 29$	$\dfrac{11,400}{(d/t)^2(L_b/d)}$
	Rectangular tubes and box sections	14	21 / 10.5	$L_b S_c/.5\sqrt{I_y J} = 146$	$23.9 - 0.24\sqrt{L_b S_c/.5\sqrt{I_y J}}$ / 10.5	$L_b S_c/.5\sqrt{I_y J} = 1700$ / $L_b S_c/.5\sqrt{I_y J} = 2290$	$\dfrac{24,000}{L_b S_c/.5\sqrt{I_y J}}$
COMPRESSION IN COMPONENTS OF BEAMS, (component under uniform compression), gross section	Flat plates supported on one edge	15	21 / 10.5	$b/t = 6.8$ / —	$27.3 - 0.93\,b/t$ / 10.5	$b/t = 10$ / $b/t = 17$	$183(b/t)$
	Flat plates with both edges supported	16	21 / 13.5	$b/t = 22$ / —	$27.3 - 0.29\,b/t$ / 10.5	$b/t = 33$ / $b/t = 55$	$580(b/t)$
	Curved plates supported on both edges	16.1*	25 / 12	$R_b/t = 1.6$ / $R_b/t = 1.3$	$26.2 - 0.94\sqrt{R_b/t}$ / $12.4 - 0.35\sqrt{R_b/t}$	$R_b/t = 141$ / $R_b/t = 340$	$\dfrac{3800}{(R_b/t)(1+\sqrt{R_b/t/35})^2}$
	Flat plates with one edge supported and the other edge with stiffener	16.2	See Section 3.4.16.2				
	Flat plates with both edges supported and with an intermediate stiffener	16.3	See Section 3.4.16.3				
COMPRESSION IN COMPONENTS OF BEAMS, (component under bending in own plane) gross section	Flat plates with compression edge free, tension edge supported	17	28 / 13.5	$b/t = 8.9$ / —	$40.5 - 1.41\,b/t$ / 13.5	$b/t = 19$ / $b/t = 19$	$4,900/(b/t)^2$
	Flat plate with both edges supported	18	28 / 13.5	$h/t = 46$ / —	$40.5 - 0.27\,h/t$ / 13.5	$h/t = 75$ / $h/t = 113$	$1520/(h/t)$
	Flat plate with horizontal stiffener, both edges supported	19	28 / 13.5	$h/t = 107$ / —	$40.5 - 0.117\,h/t$ / 13.5	$h/t = 173$ / $h/t = 260$	$3500/(h/t)$
SHEAR IN WEBS, gross section	Unstiffened flat webs	20	12 / 6	$h/t = 36$ / —	$15.6 - 0.099\,h/t$ / 6	$h/t = 65$ / $h/t = 81$	$39,000/(h/t)^2$
$a_e = a_1/\sqrt{1 + 0.7(a_1/a_2)^2}$	Stiffened flat webs	21	12 / 6	—	12 / 6	$a_e/t = 66$ / $a_e/t = 94$	$53,000/(a_e/t)^2$

Figure 5.2 (*continued*)

Substituting for the factors of safety n_y and n_u for building structures (from the Aluminum *Specification* Table 3.4-1), the tensile stress checks become:

$$F_t = \min \left(\frac{F_{ty} A_g}{1.65}, \frac{F_{tu} A_n}{1.95 \, k_t} \right) \tag{5.2}$$

Yielding on the net section is not considered a limit state because it's assumed that the net section exists over only a small portion of the total length of the member. Therefore, yielding on the net section will result in very little elongation of the member in relation to its overall length. Since the gross area usually equals or exceeds the net area and k_t is 1 in most cases, fracture will govern the allowable tensile strength if $F_{ty} > 0.85 \, F_{tu}$. An example is 6061-T6 where $F_{ty}/1.65 = (35 \text{ ksi})/1.65 = 21.2$ ksi and $F_{tu}/[(k_t)1.95] = (38 \text{ ksi})$ $/[(1.0)(1.95)] = 19.5$ ksi. Since the stress on the gross area is always less than or equal to the stress on the net area and the allowable stress on the gross area is higher than on the net area, the fracture strength on the net area always governs. The *Aluminum Design Manual,* Part VII, allowable stress design aid tables are of limited usefulness because they give only the lesser of the allowable stress based on yield and that based on fracture, even though the stresses these allowable stresses are compared to differ.

The tension coefficient k_t is 1.0 for all alloys that appear in the Aluminum *Specification* except 2014-T6, for which k_t is 1.25, and 6066-T6 and 6070-T6, for which it is 1.1. These alloys are *notch-sensitive* and, thus, require, in effect, an additional factor of safety against fracture. Notch-sensitive means that the ultimate stress on the net section attained in standard tests of specimens with sharp notches is less than the specified minimum tensile yield strength. For alloys not listed in the Aluminum *Specification,* users must determine the k_t factor. (See Section 4.4 of this book for more on this.)

The Aluminum *Specification*'s provisions for allowable stresses for tension members are actually very similar to those in the Steel *Specification.* The similarities may be obscured, however, because the aluminum allowable stresses are determined by dividing aluminum strengths by a factor of safety, while the Steel *Specification* requires multiplying steel strengths by the reciprocal of the safety factor. The steel allowable tensile strength is:

$$0.60 F_{ty} \; (= F_{ty}/1.67) \text{ on the gross area} \tag{5.3}$$

$$0.50 F_{tu} \; (= F_{tu}/2.00) \text{ on the effective net area} \tag{5.4}$$

The steel safety factors, then, are 1.67 on yield strength (versus 1.65 for aluminum) and 2.0 on ultimate strength (versus 1.95 for aluminum).

At room temperature, not all aluminum alloys exhibit the ability to sustain stresses well above yield, a property some structural designers may take for granted with A36 steel. This is demonstrated by examining their stress-strain curves up to failure (see Figure 5.3). For example, at a minimum, A36 steel

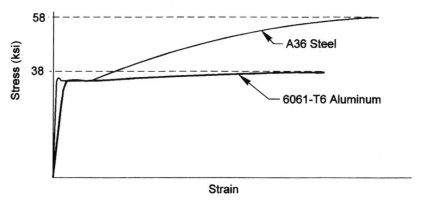

Figure 5.3 Comparison of the stress-strain curves of A36 steel and 6061-T6 aluminum.

yields at 36 ksi [250 MPa] and breaks at 58 ksi [400 MPa], while 6061-T6 extrusion, its aluminum counterpart, yields at 35 ksi [240 MPa] but breaks at 38 ksi [260 MPa]. Consequently, the allowable strength of 6061-T6 members is governed by its ultimate strength, unlike the allowable strength of typical A36 steel members, which is usually governed by its yield strength.

Unlike that of steel members, the strength of aluminum members is affected by welding. We'll discuss this effect in Section 9.1, Welded Members. Repeated application of tensile stresses can also affect aluminum member strength, as covered in Section 9.2, Fatigue. Axial tension in aluminum bolts is addressed in Section 8.1.9.

5.1.2   Net Area

The net area of aluminum members is calculated in a manner similar to that for steel. Aluminum *Specification* Section 5.1.6 gives requirements for determining the net section, which is the product of the thickness and the least net width. The net width for any chain of holes (as shown in Figure 5.4) is obtained by deducting from the gross width the sum of the diameters of all holes in the chain and adding the quantity $s^2/4g$ for each gauge space in the chain. The center-to-center spacing of the holes in the direction parallel to the direction of the load is the pitch s and the center-to-center spacing of the holes in the direction perpendicular to the force is the gauge g.

Many designers and fabricators prefer to use a standard pitch and gauge formula for fasteners to standardize detailing and tooling. If the pitch and gauge are standardized as 2.0D, where D is the diameter of the fastener, fastener spacing is 2.82D, exceeding the minimum spacing requirement of 2.5D in the Aluminum *Specification* Section 5.1.9.

For purposes of calculating net width, the diameter (D_h) of punched holes is taken as $\frac{1}{32}$ in. [0.8 mm] greater than the nominal hole diameter (*Specifi-*

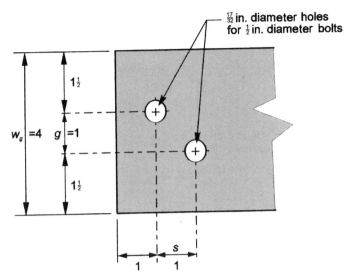

Figure 5.4 Net width for a chain of holes.

cation Section 5.1.6). The nominal hole diameter is the fabrication dimension, which is typically slightly larger than the fastener diameter. (See Section 8.1). The diameter (D_h) of drilled or reamed holes is taken as the nominal hole diameter. This differs from the Steel *Specification*, where the diameter of all holes is taken as $\frac{1}{16}$ in. [1.6 mm] greater than the nominal dimension of the hole (Steel *Specification* Section B2).

Example 5.1: Consider the $\frac{1}{4}$ in. $\times$ 4 in. 5005-H32 aluminum plate member shown in Figure 5.4. Holes for two $\frac{1}{2}$-in. diameter fasteners are to be drilled $\frac{1}{32}$ in. oversize. Calculate the allowable tensile force for the member.

The tensile ultimate strength and tensile yield strength of 5005-H32 plate are taken from Table 3.3-1 of the Aluminum *Specification:*

$$F_{tu} = 17 \text{ ksi, } F_{ty} = 12 \text{ ksi} \tag{5.5}$$

By *Specification* Table 3.4-2, $k_t = 1.0$.
The gross area of the member is:

$$A_g = (\tfrac{1}{4} \text{ in.})(4 \text{ in.}) = 1.00 \text{ in.}^2 \tag{5.6}$$

To find the net area of the member, first determine the least net width. The net width for a failure path through one hole is:

$$w_n = 4 \text{ in.} - (\tfrac{1}{2} + \tfrac{1}{32}) \text{ in.} = 3.47 \text{ in.}$$

(The actual hole diameter is used since the hole is drilled.)
The net width for a diagonal failure path through two holes is:

$$w_n = 4 \text{ in.} - 2(\tfrac{1}{2} + \tfrac{1}{32}) \text{ in.} + 1^2/(4)(1) = 3.19 \text{ in.}$$

Using the lesser of these widths, the net area is calculated:

$$A_n = (3.19 \text{ in.})(\tfrac{1}{4} \text{ in.}) = 0.797 \text{ in.}^2$$

The allowable tensile force is the lesser of:

$$T_y = F_{ty} A_g/n_y = (12 \text{ k/in.}^2)(1.00 \text{ in.}^2)/1.65 = 7.27 \text{ k}$$

and:

$$T_u = F_{tu} A_n/(k_t n_u) = (17 \text{ k/in.}^2)(0.797 \text{ in.}^2)/[(1.0)(1.95)] = 6.95 \text{ k}$$

So, the allowable tensile force is 6.95 k.

5.1.3   Effective Net Area

Depending on how a member is connected at its ends, the entire net section may not be effective in tension. Figure 5.5 shows an angle splice that is connected through only one leg. The leg that is connected is effectively continuous through the joint, but the other leg is not. How much, if any, of an unconnected cross-sectional *element* may be considered effective in tension? (See Section 5.2.2 for a definition of *element*).

   The aluminum and steel specifications offer different guidelines to determine *effective net area*. The Aluminum *Specification* doesn't provide effective net area criteria for all the cases covered by Steel *Specification* Section B3. The effective net area cases for aluminum are given in Aluminum *Specification* Section 5.1.7 and cover single and double angles only. A comparison with steel is made in Table 5.1. The effectiveness of unconnected elements is a function of the proportions of the elements and the length of the connection. While the 8th edition of the Steel *Specification* introduced criteria for the effective net area of I-beams and tees that are not connected through their webs, the Aluminum *Specification* requires only a reduction on the outstanding legs of angles.

   When all elements of the cross section are not connected, the resultant tensile force and the centroid of the section may not coincide. Examples of this include single and double angles that are not connected through both legs and tees that are not connected through both the flange and web. In these

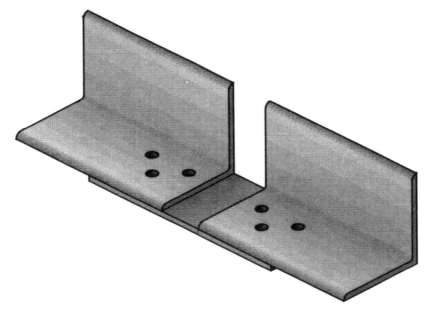

Figure 5.5 Angle splice connected through only one leg.

situations, you must also consider combined bending and axial stress (see Section 5.5).

5.1.4 Maximum Slenderness Ratios for Tension Members

The Aluminum *Specification* doesn't establish maximum slenderness ratios. However, the practice for steel buildings is to limit slenderness ratios for

TABLE 5.1 Comparison of Aluminum and Steel Specifications for Effective Net Area

Case	Aluminum *Specification*	Steel *Specification*
Equal leg angle with 1 leg connected, 2 bolts in line	$0.5 + (\frac{1}{3}).5 = 0.67$	0.75
Double equal leg angles, 2 bolts in line	$0.5 + (\frac{2}{3}).5 = 0.83$	0.75
Equal leg angle with 1 leg connected, 3 bolts in line	0.67	0.85
Double equal leg angles,3 bolts in line	0.83	0.85
Tee with flange connected	no provision	0.75 or 0.85

tension members to 240 for main members and to 300 for secondary members or bracing, except rods. This is to limit vibrations and protect against incidental loads, such as the weight of workers for which stress calculations may not be computed. Including such considerations result in a more robust structure; this can be applied to aluminum also. Such a structure is also capable of resisting loads that have the audacity to be slightly different from the loads chosen for structural checks.

5.2   COMPRESSION MEMBERS

In this section, we address members in axial compression. The shorthand term in the Aluminum *Specification* for such members is *columns*. This term is not limited to members actually serving as vertical columns in a building, but rather applies to any member subject to axial compression regardless of its location or orientation in a structure.

Wherever compression occurs, *buckling* lurks as a consideration. Buckling is strongly influenced by a member's dimensions, especially cross-sectional widths, thicknesses, and configuration. The configuration of aluminum *shapes* that can readily be fabricated by such methods as cold-forming and extruding is almost unlimited, and the Aluminum *Specification* places no limits on the proportions of members. Consequently, no buckling modes can be dismissed without consideration. It's not that aluminum buckles in ways no self-respecting piece of steel would; any metal produced in the variety of shapes that aluminum is would require similar buckling checks. This discussion is, thus, an explanation of the buckling behavior of metal columns in general.

First, let's deal with a matter of terminology. Each structural shape is considered to consist of one or more *elements*. (In the Aluminum *Specification*, elements of shapes are also called *components* of shapes.) Elements are defined as plates either rectangular or curved in cross section and connected only along their edges to other elements (Figure 5.6). For example, an I-beam consists of five elements: two rectangular plate elements in each of the two flanges, and one rectangular plate element called the *web*. Examples of shapes, called *profiles* in the aluminum industry, and their component elements are shown in Figure 5.7. An angle is an example of a shape with only two elements; one element for each leg. Other shapes may be considerably more complex, but all are treated as assemblies of elements, each of which are approximated by a rectangular plate or by a curved plate of a single radius.

It is helpful to first list all the ways that buckling can occur in a member, called *buckling modes,* and then deal with them one at a time. Buckling may be divided into two types: *overall buckling*, which occurs over the length of the member (the way a yardstick buckles), and *local buckling,* which is confined to an element of a cross section over a length about equal to the width of the element. Figure 5.8 illustrates an example of overall buckling, which

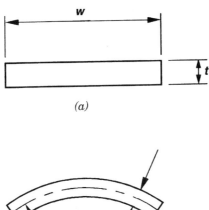

(a)

(b)

Figure 5.6 Elements of shapes.

is discussed below in Section 5.2.1. Local buckling is shown in Figure 5.9 and treated in Section 5.2.2.

Finally, we'll put it all together: overall and local buckling considerations and member design. An interaction can occur between overall buckling and local buckling, causing overall buckling to occur at a lower load than it would if local buckling were not present. We'll deal with this insidious effect also.

Local buckling is more likely to limit member strength where slender elements make up the cross section. Designers of hot-rolled steel shapes usually deal with what is called *compact sections*. These are composed of elements deliberately proportioned, so that overall buckling occurs before local buckling. This limits concern to overall buckling. While this limitation simplifies the design process, material can often be used more efficiently when greater variation in the proportions of the profile is allowed. When the restriction to compact shapes is removed, however, and more slender shapes are considered, both overall and local buckling need attention.

Slender is not necessarily synonymous with thin gage; a plate element of any thickness can buckle locally before overall buckling occurs if it is wide enough. The parameter that divides the two camps is the *slenderness ratio* of the elements of the shape. For rectangular plate elements, this is the ratio of the element's width to its thickness. Even a thick element can be relatively

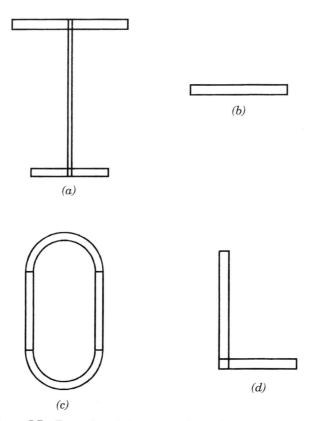

Figure 5.7   Examples of shapes and their component elements.

wide and, thus, have a high slenderness ratio, forcing us to consider the element to be slender.

Finally, a basic ground rule: To calculate compressive stresses, divide the axial load by the *gross area* of the cross section, which includes the area of holes filled by fasteners. (This is one of the reasons for the Aluminum *Specification* requirement that holes be no more than $\frac{1}{16}$ in. [1.6 mm] larger than the fastener diameter.) Unfilled holes, sometimes used to reduce weight or for convenience in mass fabrication, should be accounted for in computing the gross area. The cold-formed steel *Specification* (40) allows unfilled holes to be ignored if they exist over less than 1.5% of the length of the column (AISI Section C4). They also offer more complicated methods of calculating the effective width of compression elements supported along both edges and with circular holes (AISI Section B2.2). The Aluminum *Specification* does not address unfilled holes in columns.

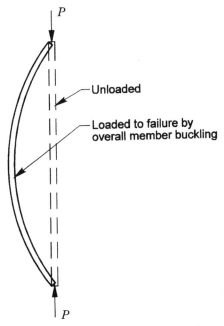

Figure 5.8 Overall buckling of a column.

5.2.1 Overall Buckling (Columns)

Overall member buckling, the action a yardstick exhibits when compressed at its ends, is the subject of this section. This is the behavior that Aluminum *Specification* Section 3.4.7 covers and applies to members in axial compression (columns) of any cross section or length.

A Qualitative View   Let's examine the general performance of a column as a function of its slenderness. Imagine a very short metal ruler with a rectangular cross section under an axial compressive load. As the load increases, the ruler does not buckle, but it is eventually shortened by yielding of its cross section, as shown in Figure 5.10a.

Now imagine a longer ruler of the same cross section, again loaded as a column. The load is increased until the ruler buckles laterally. Once buckled, the member cannot support any additional load. When the load is removed, the column retains its deflected shape (see Figure 5.10b). Because of this, the column is said to have undergone *inelastic buckling*. At the buckle, the metal has been stressed beyond its yield strength, so the buckled shape is permanent.

Finally, imagine a very long ruler under an axial compression load. As the load increases, this long ruler buckles much sooner than the intermediate-

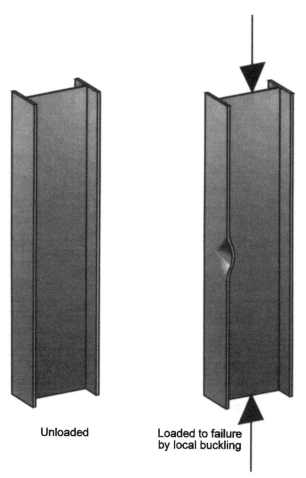

Unloaded

**Loaded to failure
by local buckling**

Figure 5.9 Local buckling of a column.

length ruler, displacing laterally by a distance that varies along its height. The buckled shape is shown in Figure 5.10c. At the midheight, the lateral displacement is greatest. Once buckled, the member cannot support any additional load. If the ruler is long enough, at no point on the column is the yield strength exceeded when the buckling occurs, so when the load is removed, the ruler springs back to its original straight shape. Because the deflection is not permanent, this behavior is called *elastic buckling.*

What's This Have to Do With the Aluminum Specification? The three modes of behavior described above correspond to the three cases listed in Aluminum *Specification* Section 3.4.7 and Table 3.4-3 (highlighted and reproduced here as Figure 5.11 and evaluated for 6061-T6 alloy in Figure 5.12):

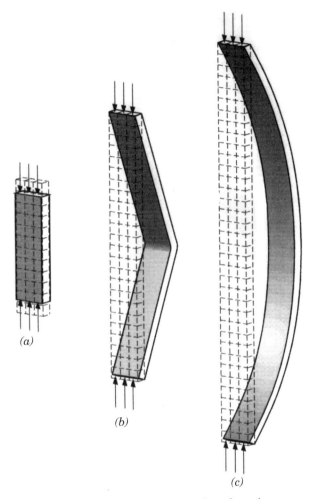

Figure 5.10 Overall failure modes of a column.

3.4.7(a), for stocky columns, which undergo yielding like the shortest ruler when loaded to failure

3.4.7(b), for columns of intermediate slenderness, which undergo inelastic buckling like the ruler of intermediate length when loaded to failure

3.4.7(c), for slender columns, which undergo elastic buckling like the long-est ruler when loaded to failure.

Column strength is determined in the Aluminum *Specification* by a different equation for each of these three regimes of slenderness. While we introduced these regimes in the preceding discussion in the order in which they appear

GENERAL FORMULAS FOR DETERMINING ALLOWABLE STRESSES FROM SECTION 3.4

Type of Stress	Type of Member or Component	Sub-Sec.	Allowable Stress
TENSION, axial	Any tension member	1	F_{ty}/n_y or $F_{tu}/(k_t n_u)$
TENSION IN BEAMS, extreme fiber, net section	Rectangular tubes, structural shapes bent around strong axis	2	F_{ty}/n_y or $F_{tu}/(k_t n_u)$
	Round or oval tubes	3	$1.17 F_{ty}/n_y$ or $1.24 F_{tu}/(k_t n_u)$
	Shapes bent about weak axis, bars, plates	4	$1.30 F_{ty}/n_y$ or $1.42 F_{tu}/(k_t n_u)$
BEARING	On rivets and bolts	5	$2F_{tu}/n_u$
	On flat surfaces and pins and on bolts in slotted holes	6	$2F_{tu}/(1.5 n_u)$

*For tubes with circumferential welds, equations of Sections 3.4.10, 3.4.12, and 3.4.16.1 apply for $R/t \leq 20$.

Type of Stress	Type of Member or Component	Sub-Sec.	Allowable Stress Slenderness $\leq S_1$	Slenderness Limit S_1	Allowable Stress Slenderness Between S_1 and S_2	Slenderness Limit S_2	Allowable Stress Slenderness $\geq S_2$
COMPRESSION IN COLUMNS, axial, gross section	All columns	7	$\dfrac{F_{cy}}{n_y}$	$\dfrac{kL}{r} = \dfrac{B_c - \dfrac{n_u F_{cy}}{n_y}}{D_c}$	$\dfrac{1}{n_u}\left(B_c - D_c\,\dfrac{kL}{r}\right)$	$\dfrac{kL}{r} = C_c$	$\dfrac{\pi^2 E}{n_u\left(\dfrac{kL}{r}\right)^2}$
COMPRESSION IN COMPONENTS OF COLUMNS gross section	Flat plates supported along one edge - columns buckling about a symmetry axis	8	$\dfrac{F_{cy}}{n_y}$	$\dfrac{b}{t} = \dfrac{B_p - \dfrac{n_u F_{cy}}{n_y}}{5.1 D_p}$	$\dfrac{1}{n_u}\left(B_p - 5.1 D_p\,\dfrac{b}{t}\right)$	$\dfrac{b}{t} = \dfrac{k_1 B_p}{5.1 D_p}$	$\dfrac{k_2\sqrt{B_p E}}{n_u(5.1 b/t)}$
	Flat plates supported along one edge - columns not buckling about a symmetry axis	8.1	$\dfrac{F_{cy}}{n_y}$	$\dfrac{b}{t} = \dfrac{B_p - \dfrac{n_u F_{cy}}{n_y}}{5.1 D_p}$	$\dfrac{1}{n_u}\left(B_p - 5.1 D_p\,\dfrac{b}{t}\right)$	$\dfrac{b}{t} = \dfrac{C_p}{5.1}$	$\dfrac{\pi^2 E}{n_u(5.1 b/t)^2}$
	Flat plates with both edges supported	9	$\dfrac{F_{cy}}{n_y}$	$\dfrac{b}{t} = \dfrac{B_p - \dfrac{n_u F_{cy}}{n_y}}{1.6 D_p}$	$\dfrac{1}{n_u}\left(B_p - 1.6 D_p\,\dfrac{b}{t}\right)$	$\dfrac{b}{t} = \dfrac{k_1 B_p}{1.6 D_p}$	$\dfrac{k_2\sqrt{B_p E}}{n_u(1.6 b/t)}$
	Flat plates with one edge supported and other edge with stiffener	9.1	See Section 3.4.9.1				
	Flat plates with both edges supported and with an intermediate stiffener	9.2	See Section 3.4.9.2				
	Curved plates supported on both edges, walls of round or oval tubes	10*	$\dfrac{F_{cy}}{n_y}$	$\dfrac{R_b}{t} = \left(\dfrac{B_t - \dfrac{n_u F_{cy}}{n_y}}{D_t}\right)^2$	$\dfrac{1}{n_u}\left(B_t - D_t\sqrt{\dfrac{R_b}{t}}\right)$	$\dfrac{R_b}{t} = C_t$	$\dfrac{\pi^2 E}{16 n_u\left(\dfrac{R_b}{t}\right)\left(1 + \dfrac{\sqrt{R_b/t}}{35}\right)^2}$

Figure 5.11 General requirements of the Aluminum *Specification* for overall buckling of columns.

Allowable Stresses for BUILDING and Similar Type Structures

6061-T6, -T651, -T6510, -T6511

Extrusions up thru 1 in., Sheet & Plate, Pipe, Standard Structural Shapes, Drawn Tube, Rolled Rod and Bar, 6351-T5 Extrusions

WHITE BARS apply to nonwelded members and to welded members at locations farther than 1.0 in. from a weld.

SHADED BARS apply within 1.0 in. of a weld.

Equations that straddle the shaded and unshaded areas apply to both.

*For tubes with circumferential welds, equations of Sections 3.4.10, 3.4.12, and 3.4.16.1 apply for $R_b/t \leq 20$.

Type of Stress	Type of Member or Component	Sec. 3.4.	Allowable Stress
TENSION, axial	Any tension member	1	19 / 11†
TENSION IN BEAMS, extreme fiber, net section	Rectangular tubes, structural shapes bent around strong axis	2	19 / 11†
	Round or oval tubes	3	24 / 13.5†
	Shapes bent about weak axis, bars, plates	4	28 / 16†
BEARING	On rivets and bolts	5	39 / 25†
	On flat surfaces and pins and on bolts in slotted holes	6	26 / 16†

Type of Stress	Type of Member or Component	Sec. 3.4.	Allowable Stress Slenderness ≤ S₁	Slenderness Limit S₁	Allowable Stress Slenderness Between S₁ and S₂	Slenderness Limit S₂	Allowable Stress Slenderness ≥ S₂
COMPRESSION IN COLUMNS, axial, gross section	All columns	7	— / 12†	$kL/r = 0$ / —	$20.2 - 0.126\,(kL/r)$ / 12†	$kL/r = 66$ / $kL/r = 65$†	$51{,}000/(kL/r)^2$ / $51{,}000/(kL/r)^2$†
COMPRESSION IN COMPONENTS OF COLUMNS gross section	Flat plates supported along one edge - columns buckling about a symmetry axis	8	21 / 12†	$b/t = 2.7$ / —	$23.1 - 0.79\,(b/t)$ / 12†	$b/t = 10$ / $b/t = 13$†	$154/(b/t)$ / $154/(b/t)$†
	Flat plates supported along one edge - columns not buckling about a symmetry axis	8.1	21 / 12†	$b/t = 2.7$ / —	$23.1 - 0.79\,(b/t)$ / 12†	$b/t = 12$ / $b/t = 13$†	$1970(b/t)^2$ / $1970(b/t)^2$†
	Flat plates with both edges supported	9	21 / 12†	$b/t = 8.4$ / —	$23.1 - 0.25\,(b/t)$ / 12†	$b/t = 33$ / $b/t = 41$†	$490/(b/t)$ / $490/(b/t)$†
	Flat plates with one edge supported and other edge with stiffener	9.1	See Section 3.4.9.1				
	Flat plates with both edges supported and with an intermediate stiffener	9.2	See Section 3.4.9.2				
	Curved plates supported on both edges, walls of round or oval tubes	10*	21 / 12†	$R_b/t = 2.2$ / $R_b/t = 9.0$	$22.2 - 0.80\sqrt{R_b/t}$ / $13.5 - 0.50\sqrt{R_b/t}$	$R_b/t = 141$ / $R_b/t = 290$†	$\dfrac{3200}{(R_b/t)(1 + \sqrt{R_b/t}/35)^2}$

Figure 5.12 Requirements of the Aluminum *Specification* for overall buckling of columns, for the specific case of alloy 6061-T6.

133

in the Aluminum *Specification* (i.e., compressive yield, inelastic buckling, and elastic buckling), we'll reverse this order in the following discussion.

The strength of slender columns (those that buckle elastically) was predicted by Leonhard Euler in 1757. His equation is still used today for both steel and aluminum, so the elastic buckling strength is also called the *Euler buckling* strength of the column:

$$\text{elastic buckling stress} = \frac{\pi^2 E}{\left(\dfrac{kL}{r}\right)^2} \tag{5.7}$$

This equation is graphed in Figure 5.13. Note that it is independent of the yield or ultimate strength properties of the material. The elastic buckling strength of a column made of 6061-T6 aluminum alloy, which has a compressive yield strength of 35 ksi [240 MPa], is exactly the same as that of a column of the same length and cross section of 3003-H12, which yields at 10 ksi [70 MPa]. The only mechanical property that influences the elastic buckling strength is the *modulus of elasticity*, E, the slope of the stress-strain curve. In the elastic range, E is constant, so the elastic buckling strength for a given alloy varies only with the slenderness of the column, expressed as the ratio kL/r, discussed further below.

Inelastic buckling occurs when the stress at the buckle is greater than the yield strength, which happens in columns of intermediate slenderness. We can also use the Euler buckling equation to predict buckling strength of such columns. However, at these higher stresses, the modulus of elasticity, or slope

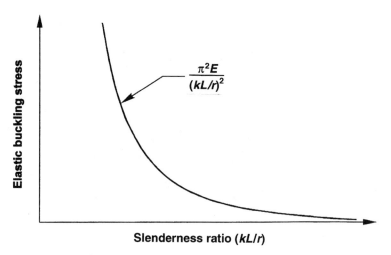

Figure 5.13   Euler column buckling stress.

of the stress-strain curve, varies depending on the strain, and is called the *tangent modulus of elasticity, E_t.* (See Figure 5.14.) Substituting E_t for E:

$$\text{inelastic buckling stress} = \frac{\pi^2 E_t}{\left(\dfrac{kL}{r}\right)^2} \qquad (5.8)$$

Fortunately, tests have shown that rather than using the tangent modulus, which varies with stress, this equation can be conveniently reduced to a linear function of the slenderness kL/r:

$$\text{inelastic buckling strength} = B_c - D_c(kL/r) \qquad (5.9)$$

In this equation, B_c is the stress at which the inelastic buckling strength line intersects the y-axis, and D_c is the slope of the inelastic buckling line (Figure 5.15). Formulas for B_c and D_c have been determined for aluminum alloys by testing.

Inelastic buckling strength is graphed with the strengths of the other slenderness regimes in Figure 5.16 with safety factors applied. The slenderness ratio at which the inelastic buckling curve (line) intersects the elastic buckling curve is called C_c (as it is for steel). This slenderness (C_c) is called the *slenderness limit S_2* because it's the upper limit of applicability of the inelastic buckling equation. When the slenderness ratio is above this limit, buckling will be elastic. Collectively, B_c, D_c, and C_c are called *buckling formula con-*

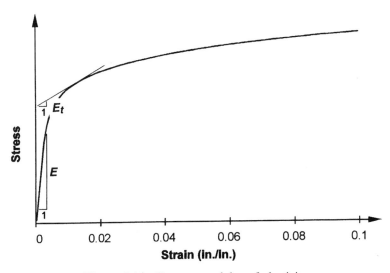

Figure 5.14   Tangent modulus of elasticity.

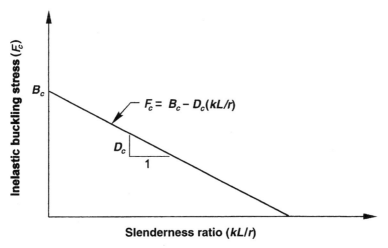

Figure 5.15   Inelastic column buckling stress.

stants, and they are based on the shape of the stress-strain curve above yield. As we mentioned in Section 2.4, aluminum alloys are divided into two groups: those that are not *artificially aged* (tempers -O, -H, -T1, -T2, -T3, -T4) and those that are (tempers -T5 through -T9). Now the reason is revealed: These two groups have stress-strain curves with different shapes, the latter having a

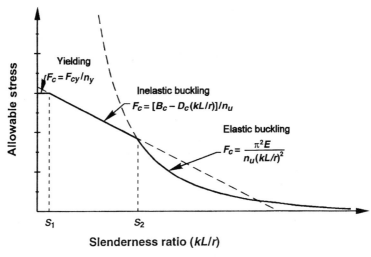

Figure 5.16   The three slenderness regimes for overall column buckling stress.

flatter curve beyond yield, as shown in Figure 5.17. This means that the variation of tangent modulus with stress is different for the two groups, hence, the different buckling formula constants for these groups.

Expressions for the buckling formula constants are given in Aluminum *Specification* Table 3.3-3 for alloys not artificially aged, and Table 3.3-4 for those that are. These expressions are a function of yield strength and modulus of elasticity only. Aluminum designers do not often use these expressions because they have been conveniently evaluated for most alloys, and values have been tabulated in the *Aluminum Design Manual*, Part VII, Table 2-1 and Appendix K of this book. It's useful to remember buckling formula constants more precisely as inelastic buckling formula constants because they're used for calculating inelastic buckling strength.

Finally, the strength of columns so stocky that they will not buckle is just the compressive yield strength (F_{cy}):

$$\text{compressive yield strength} = F_{cy} \tag{5.10}$$

This is sometimes referred to as the "squash load." (Prior to the 7th edition of the Aluminum *Specification*, the compressive yield strength was divided by a coefficient k_c which was greater than or equal to one and whose purpose was to increase the range of slenderness ratios over which the simpler compressive yield equation could be used. This conservative simplification has been dropped.)

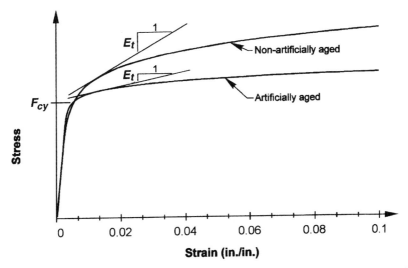

Figure 5.17 Comparison of stress-strain curves for artificially aged and non-artificially aged aluminum alloys.

Comparison to Steel   The AISC Steel *Specification* (39) doesn't distinguish compressive yield of the cross section from inelastic buckling and uses just two equations (inelastic and elastic buckling) to approximate the behavior of columns. The Aluminum *Specification* uses three equations (yielding, inelastic buckling, and elastic buckling), as shown in Figure 5.18. In steel, predicting inelastic buckling for any slenderness ratio below the intersection of the elastic and inelastic buckling equations requires a rather complex equation. The use of two equations to represent this range in the Aluminum *Specification* allows the use of simpler expressions, but it does require you to determine which equation to use or to evaluate both equations and use the lesser strength from the two.

Safety Factors   Until now we've discussed only the strength of columns in compression, leaving untouched the issue of how much of a margin there should be between working stress and strength for allowable stress design. The factors of safety used in the three regimes are shown in Figure 5.19 and contrasted with those used for steel. From Figure 5.19, we see that the factor of safety for aluminum columns in the compressive yield regime is the factor of safety on yield ($n_y = 1.65$) since this is the mode of failure. The

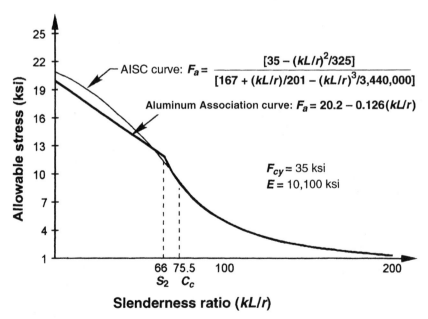

Figure 5.18   Comparison of the column buckling stress equations for steel and aluminum.

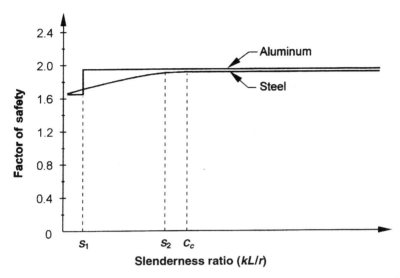

Figure 5.19 Comparison of the safety factors on overall buckling for steel and aluminum columns.

factor of safety on inelastic and elastic buckling is higher: the factor of safety on ultimate (n_u = 1.95) is used.

The most noticeable difference between steel and aluminum column factors of safety is in the realm of inelastic buckling slenderness ratios. The AISC Steel *Specification* uses a sliding scale for steel column safety factors, varying the factor from 1.67 (= 5/3) for the shortest columns to 1.92 (= 23/12) for columns at the upper limit of inelastic buckling. The reason is that initial imperfections are judged to affect the strength of short columns less significantly when the shapes are compact, as they are for hot-rolled steel. Aluminum shapes (profiles) may not necessarily be compact; they may also be composed of thin-gauge elements that are deemed to be connected less rigidly and more eccentrically at their ends. Thus, the Aluminum *Specification*, while using a factor of safety of 1.65 for the shortest columns, uses the higher factor of safety (1.95) throughout the whole range of buckling (inelastic and elastic). If this rationale seems rather nebulous, you may find solace by considering the relatively arbitrary nature of safety factors in the first place and that load and resistance factor design smooths out the rough spots. (See Chapter 11 for more on LRFD.)

Different Kinds of Overall Buckling and Their Slenderness Ratios So far, we've taken the slenderness ratio, (kL/r), for granted. This probably hasn't been too disturbing because kL/r is familiar to most structural engineers, as well as the designation:

$$\lambda = kL/r \qquad\qquad (5.11)$$

where:

k = effective length factor (more on this below)
L = length of the column between points of restraint against buckling (unbraced length)
r = radius of gyration of the column about the axis of buckling.

But what is "the axis of buckling" referred to in the definition of r? Here's a hint: It's a function of the kind of overall buckling.

Up to this point, we've described only one kind of overall buckling, the kind called *flexural buckling,* demonstrated by the ruler discussed above. Use of the term "flexural" when we're talking about columns can be confusing, so let's clarify this. Flexural buckling refers to the lateral bowing of the ruler as it buckles, taking a shape as if it were being bent, even though only an axial force is applied to the member. This kind of buckling is covered in Aluminum *Specification* Section 3.4.7.1. The axis of buckling is the cross-sectional bending axis about which the flexure occurs, which will be the axis with the greater slenderness ratio. Because the effective length factor (k) and the radius of gyration may be different for the two principal axes, designers must calculate the slenderness ratio for each axis to determine which is greater. Many engineers are familiar only with this kind of overall buckling, which is sufficient for typical closed shapes (i.e., round and rectangular tube).

How else can columns buckle? Another kind of overall buckling is *torsional buckling,* which is a twisting or corkscrewing of the column. Equal leg cruciform sections, which are *point symmetric,* tend to buckle this way. No lateral displacement takes place along the member length during buckling, only twisting. The axis of buckling is the longitudinal axis of the member for torsional buckling. Last, shapes that are not doubly symmetric may buckle in the overall buckling mode called *torsional-flexural buckling,* a combination of twisting and lateral deflection.

A way to treat these buckling modes involving torsion is to replace kL/r with an "effective" kL/r. A method for determining the effective slenderness ratio $(kL/r)_e$ for both torsional or torsional-flexural buckling for doubly or singly symmetric sections is given in Aluminum *Specification* Section 3.4.7.2. However, the *Specification* doesn't provide a method for determining the effective radius of gyration for unsymmetric shapes. Also, designers must already know if a shape is subject to torsional or torsional-flexural buckling in order to choose which Aluminum *Specification* Section (3.4.7.1 or 3.4.7.2) applies.

To clear these hurdles, consult Table 5.2, it shows which shapes are subject to which modes of overall buckling.

TABLE 5.2 Overall Column Buckling Modes

Shape	Flexural Buckling	Torsional Buckling	Torsional-Flexural Buckling
Closed	Yes	No	No
Open—doubly symmetric	Yes	Yes	No
Open—singly symmetric	Yes	No	Yes
Open—unsymmetric	No	No	Yes

Next, you must determine the slenderness ratio for each applicable buckling mode since you know that buckling will occur about the axis with the highest slenderness ratio. Here's how:

The equivalent slenderness ratio for torsional buckling (1) is:

$$\left(\frac{kL}{r}\right)_e = \lambda_\phi = \sqrt{\frac{Ar_o^2}{0.038\,J + \dfrac{C_w}{(k_\phi L)^2}}} \tag{5.12}$$

where:

λ_ϕ = equivalent slenderness ratio for torsional buckling
A = area of the cross section
r_o = polar radius of gyration of the cross section about the shear center:

$$r_o = \sqrt{r_x^2 + r_y^2 + x_o^2 + y_o^2} \tag{5.13}$$

x_o, y_o = distances between the centroid and shear center, parallel to the principal axes
r_x, r_y = radii of gyration of the cross section about the principal axes
J = torsion constant
C_w = warping constant
k_ϕ = effective length coefficient for torsional buckling
L = length of the column

The equivalent slenderness ratio for torsional-flexural buckling is λ_c, which may be solved for by trial and error from the following equation (133):

$$\left[1 - \left(\frac{\lambda_c}{\lambda_x}\right)^2\right]\left[1 - \left(\frac{\lambda_c}{\lambda_y}\right)^2\right]\left[1 - \left(\frac{\lambda_c}{\lambda_\phi}\right)^2\right] - \left(\frac{x_o}{r_o}\right)^2\left[1 - \left(\frac{\lambda_c}{\lambda_y}\right)^2\right]$$

$$- \left(\frac{y_o}{r_o}\right)^2\left[1 - \left(\frac{\lambda_c}{\lambda_x}\right)^2\right] = 0 \tag{5.14}$$

where:

λ_c = equivalent slenderness ratio for torsional-flexural buckling

λ_x, λ_y = effective slenderness ratios for flexural buckling about the x and y axes, respectively

λ_ϕ = equivalent slenderness ratio for torsional buckling

r_o = polar radius of gyration of the cross section about the shear center (see Equation 5.13)

x_o, y_o = distances between the centroid and shear center, parallel to the principal axes.

Formulas for the torsion constant (J) and the warping constant (C_w) for several kinds of shapes are given in the *Aluminum Design Manual,* Part VI, Table 29. Calculated values for the torsion and warping constants for the Aluminum Association standard I-beams and channels are not given in the *Aluminum Design Manual,* but we have calculated them. You can find them in Appendix B.

The Effective Length Factor (k)   The effective length factor (k) is multiplied by the actual unbraced column length (L) to obtain the length between points of inflection, or the length of Euler's pin-ended column. The k factor is a measure of the restraint against rotation and the resistance to lateral deflection at the ends of the unbraced length. Figure 5.20 gives k values for various cases. Aluminum *Specification* Section 3.2, Nomenclature, requires that k be taken as larger than or equal to 1 "unless rational analysis justifies a smaller value."

The effective length factor is determined for each kind of overall buckling (flexural or torsional), based on the restraint to rotation about the axis of buckling rotation. If the mode of buckling is flexural, k is based on the restraint placed on the column ends against bending rotation. If the mode of buckling is torsional, k is based on the restraint on the column ends against rotation about the member's longitudinal axis (twisting and warping).

Deviation from Straight   The tests (133) used to establish aluminum column buckling strengths limited the initial crookedness of the columns to less than $L/1000$. For this degree of crookedness, the column strength is 80% of that of a perfectly straight column. The Aluminum Association allows a mill tolerance of $L/960$ for most prismatic member products in *Aluminum Standards and Data* (11), and AWS D1.2 *Structural Welding Code—Aluminum* (91) has the same tolerance on straightness of welded members. This slight unconservatism was justified by recognizing that in practice columns idealized as pin-ended have an effective k of less than 1 due to partial restraint against rotation at end connections, even though columns idealized as having ends fixed against rotation are not completely fixed in practice. Beware that *Aluminum*

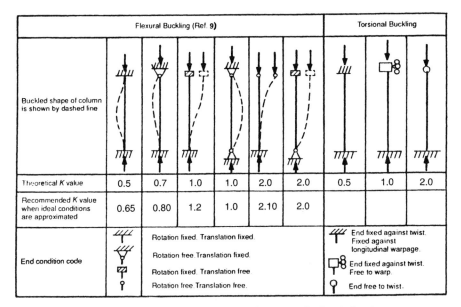

	Flexural Buckling (Ref. 9)						Torsional Buckling		
Buckled shape of column is shown by dashed line									
Theoretical K value	0.5	0.7	1.0	1.0	2.0	2.0	0.5	1.0	2.0
Recommended K value when ideal conditions are approximated	0.65	0.80	1.2	1.0	2.10	2.0			
End condition code		Rotation fixed. Translation fixed.						End fixed against twist. Fixed against longitudinal warpage.	
		Rotation free.Translation fixed.							
		Rotation fixed. Translation free.						End fixed against twist. Free to warp.	
		Rotation free.Translation free.						End free to twist.	

Figure 5.20   Effective length factors for centrally loaded columns.

Standards and Data does not require some extrusions, such as T6511 tempers with wall thicknesses less than 0.095 in. [2.4 mm], to meet the *L*/960 straightness requirement, so they may require special straightening if used as compression members.

How to Check a Column for Overall Buckling

Example 5.2: What is the overall buckling allowable stress of a 6061-T6 Aluminum Association standard I-beam I 12 × 11.7 (shown in Figure 5.21) that is 66 in. tall and is braced at midheight against minor axis translation? Assume the column ends are free to rotate about the major and minor axes of the section but are not free to twist or warp.

Properties for this shape are tabulated in Appendix B. The first step is to determine the axis of buckling; this is the axis about which the slenderness ratio is largest. Torsional-flexural buckling will not occur because the section is doubly symmetric. You have three axes of buckling to consider:

The major axis slenderness ratio is:

$$\frac{kL}{r} = \frac{(1)(66 \text{ in.})}{(5.07 \text{ in.})} = 13$$

The minor axis slenderness ratio is:

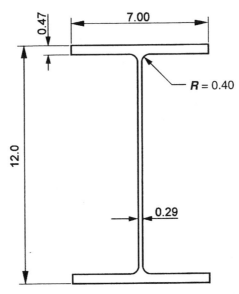

Figure 5.21   Cross section of an Aluminum Association standard I-beam I 12 × 11.7.

$$\frac{kL}{r} = \frac{(1)[(66 \text{ in.}/2)]}{(1.65 \text{ in.})} = 20$$

Since the shape is doubly symmetric, $Ar_o^2 = I_x + I_y$. The longitudinal axis slenderness ratio (equivalent slenderness ratio for torsional buckling) is:

$$\left(\frac{kL}{r}\right)_e = \sqrt{\frac{I_x + I_y}{0.038J + \frac{C_w}{(k_\phi L)^2}}} = \sqrt{\frac{256 + 26.9}{0.038(0.621) + \frac{894}{[(0.5)(66)]^2}}} = 18.3 \text{ in.}$$

So, flexural buckling about the minor axis will occur first because the slenderness ratio associated with it, 20, is the largest. From the *Aluminum Design Manual*, Part VII, Table 2-21 (see Figure 5.12), where allowable stresses have been tabulated for 6061-T6, we see that this slenderness ratio is between slenderness limits S_1 (0) and S_2 (66). Thus, the overall buckling allowable stress is:

$$20.2 - 0.126 \ (kL/r) = 20.2 - 0.126 \ (20) = 17.7 \text{ ksi}$$

Keep in mind that this allowable stress might have to be reduced due to the effect of local buckling, the "insidious" effect we mentioned in Section 5.2

You won't know for sure until you read the paragraph on the potential inter-action between local buckling and overall buckling in Section 5.2.2.

Some Final Observations An understanding of the column buckling pro-visions of the Aluminum *Specification* is very useful in getting the gist of the specification provisions that follow (Aluminum *Specification* Sections 3.4.8 through 3.4.21), all of which deal with stresses that involve compression. The form of these other provisions is very similar to those for columns in that three regimes of slenderness are identified, and an equation for strength is given for each of the three.

There's no magic to the column slenderness limits S_1 and S_2. The formulas for them are simply the result of solving for the points of intersection of the yielding and inelastic buckling equations (S_1) and the inelastic and elastic buckling equations (S_2), respectively, as shown in Figure 5.16. In fact, the Aluminum *Specification* places no limits on slenderness ratios, so it might be better to call S_1 and S_2 *slenderness intersections.*

The situation with slender columns is mildly disconcerting since their strength is only a function of the modulus of elasticity. Aluminum's modulus is only one-third that of steel and doesn't vary much by alloy. Made of aero-space alloy or recycled beer cans, a slender aluminum column is going to buckle at the same stress, which is about one-third that of a steel column of the same dimensions. Looking at the bright side, you will see that pining (or paying) for an exotic alloy is pointless if the column is so slender that a stronger metal won't make a stronger column.

As mentioned, the Aluminum *Specification* places no limit on the slender-ness ratio for a column. The Steel *Specification* recommends a maximum of 200, which may be reasonable for aluminum columns also in order to avoid buckling from unforeseen loads. Allowable stresses for artificially aged alu-minum alloys with a compressive yield strength of 35 ksi and a compressive modulus of elasticity of 10,100 ksi (like 6061-T6) for slenderness ratios from 0 to 160 are given in Appendix F.

5.2.2 Local Buckling (Components of Columns)

In Section 5.2, Compression Members, we introduced and defined *elements* of *shapes* and showed examples in Figures 5.6 and 5.7. *Local buckling* is all about the buckling of these elements. First let's talk briefly about individual elements, and then we'll discuss how they buckle.

Every plate element has four edges, but the important feature of elements is what they are attached to along their two longitudinal edges (i.e., parallel to the direction of the compressive force) (Figure 5.22). Along a longitudinal edge, an element can: 1) be attached to another element, 2) have a free edge, or 3) have a stiffener (Figure 5.23). When a plate element is continuously attached to another plate element along its edge (as, say, a channel's flange

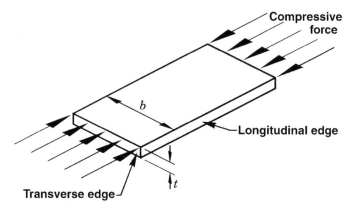

Figure 5.22 Edges of a plate element.

joined along one edge to the web), the Aluminum *Specification* calls that edge "supported." The AISC and AISI specifications (39, 40) differ from aluminum terminology by calling such an edge "stiffened." The Aluminum *Specification* instead uses the term *stiffener* to mean additional material attached to an element that serves to increase the element's capacity to carry compressive loads. Such stiffeners may occur along an otherwise free longitudinal edge (an *edge stiffener*, Figure 5.24a) or between the element edges (an *intermediate stiffener*, Figure 5.24b).

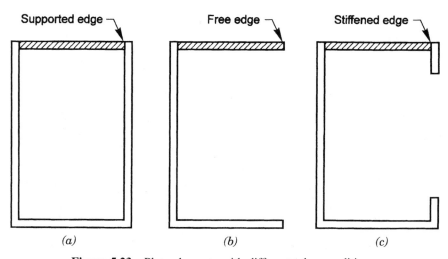

Figure 5.23 Plate elements with different edge conditions.

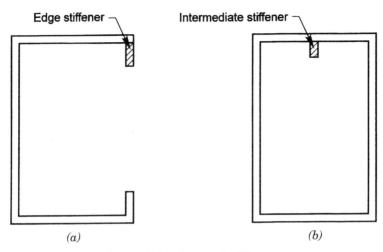

Figure 5.24 Types of stiffeners.

Many configurations of element edge conditions and stiffeners are imaginable (and are used, especially in aluminum, where the extrusion process makes a great variety of shapes readily achievable). For example, an element might have one edge supported and the other edge free, or it might have one edge supported, one edge with a stiffener, and an intermediate stiffener in the middle of the element. As might be expected, the elements we have design rules for are only part of this universe, but a significant part. Provisions are given in the Aluminum *Specification* for the elements shown in Figure 5.25.

Local Buckling for Poets   As we did for overall buckling in Section 5.2.1, let's start with an overview of local buckling behavior. Once again, the behavior is classified in three regimes of slenderness, but we'll see that slenderness has a completely different meaning when applied to an element than it did for overall buckling.

First, imagine a stocky flat plate in axial compression, with a thickness (*t*) not much less than its width (*b*), and supported along one longitudinal edge (Figure 5.26). The axial load may be increased until the stress in the plate reaches the compressive yield strength of the material. No buckling occurs because the plate is so stocky. This regime is illustrated by the leftmost portion of the curve in Figure 5.29, where the element's strength is the yield strength of the material.

Next, imagine increasing the width of this plate while its thickness remains the same (Figure 5.27). Again, an axial load is applied, but this time as the load increases, the plate buckles just before the average stress reaches yield. The stress increases at the buckles due to their out-of-plane eccentricity, so

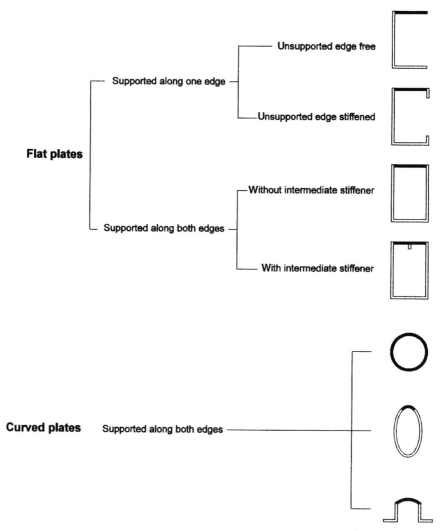

Figure 5.25 Element edge and stiffener conditions for which the Aluminum *Specification* has design provisions.

localized yielding occurs, causing the buckles to remain when the load is removed. Thus, this mode of failure is termed *inelastic buckling* of the element. While the overall buckling we described in Section 5.2.1 was dependent on the length of the column, this local buckling of an element depends on the ratio of its width to its thickness (b/t) and is independent of its length. The center portion of Figure 5.29 shows the regime in which the onset of inelastic buckling limits the strength of an element.

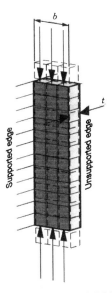

Figure 5.26   Stocky plate element exhibiting compressive yield.

Finally, let's increase the plate width further still and once more load the plate in compression (Figure 5.28). Again, waves or buckles appear, but this time the free edge is so flimsy that it buckles well below yield. Near the supported edge, however, the plate is still sufficiently stiff to support the load. Because the stress is in the *elastic* range, the buckles disappear when the load is removed.

Unlike the ruler loaded as a column that we discussed in Section 5.2.1, a plate with at least one edge supported is capable of supporting additional load after the onset of elastic buckling. This capacity is called *postbuckling strength* and is illustrated by the rightmost portion in Figure 5.29. When buckling begins, compressive stresses are redistributed across the width of the plate, with lower stress where buckling has occurred, and higher stresses along the supported longitudinal edge, which is held straight and in place under the load by the edge support. At maximum load, the average stress on the plate cross section is called the *crippling stress*.

Postbuckling Strength   When designing members with slender elements in compression, you must first reprogram your outlook toward compression behavior. With compact sections, you assume that an axial-compression load results in a uniform compressive stress in the cross section. The stress then, is:

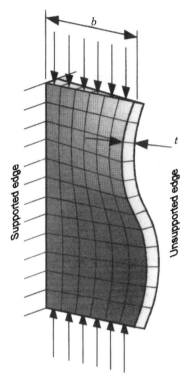

Figure 5.27   Plate element of intermediate slenderness exhibiting inelastic local buckling.

$$f = \frac{P}{A} \tag{5.15}$$

where:

  f = stress
  P = load
  A = area of the cross section.

In actuality, this is a generalization that is no longer valid if any of the cross-sectional elements have buckled. Consider the design of indeterminate frames, for example, where we recognize that regions of greater stiffness attract greater load. These stiffness principles are equally valid in evaluating the compressive force on an individual member; it's just that with a compact section, sufficiently little difference between the stiffness of the individual elements exists to bother with this differentiation. Slender elements, on the other hand, may vary dramatically in their relative stiffness. Stiffness may

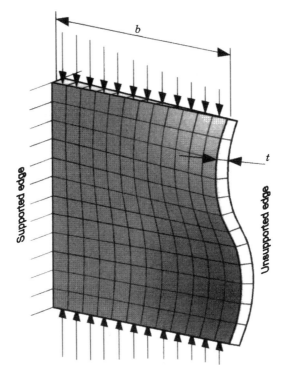

Figure 5.28 Very slender plate element exhibiting elastic local buckling.

also vary significantly across the width of an individual element. This is what we observed in Figure 5.28, where the element is much stiffer near the supported edge than at the unsupported edge.

In Figures 5.26, 5.27, and 5.28, we looked at the relative slenderness and corresponding stiffness of an individual element supported along one edge. Now let's assemble the cross section of an I-beam by combining several elements, as shown in Figure 5.30a. The web is a flat plate element supported along both edges by the flanges, and each flange consists of two flat plate elements supported along one edge by the web. This shape is drawn with exaggerated proportions so that it is visually evident that some portions have greater stiffness than others. When a very small load is applied, it is supported by the entire cross section. As the load increases, the flanges begin to buckle along their unsupported edges at a stress level well within the elastic range. This onset of local elastic buckling, as we saw previously when discussing an individual element, does not limit the load-carrying capacity of the section. It does, however, introduce a change in its behavior.

Once buckled, the wavy edges of the flanges are rendered ineffective, and load is redistributed to the stiffer areas nearer the flange-web junction. The

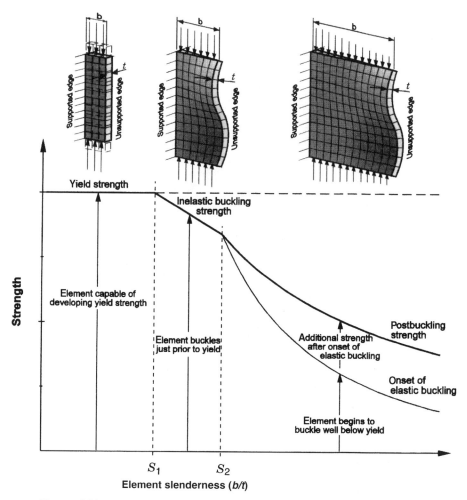

Figure 5.29 The three regimes of local buckling for elements of columns.

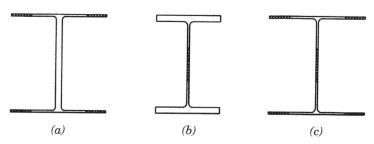

Figure 5.30 Examples of shapes having very slender elements in the postbuckled condition.

area of the cross section that is actually carrying load is, thus, reduced. As the load increases, the buckles extend further across the flanges, and the force continues to be redistributed to the shrinking effective area. Eventually, the combination of the reduced effective area supporting an increased load results in failure of the cross section.

Note that the web might experience elastic buckling rather than the flanges as suggested by Figure 5.30b, or all of the elements as shown in Figure 5.30c. It is not a matter of flange behavior versus web behavior, but rather an application of the principle that the regions of greatest stiffness attract more of the load. This redistribution of force has two significant results:

1) It allows the member to support additional load beyond the onset of elastic buckling
2) It results in a reduced effective section in this postbuckled regime.

While taking advantage of the additional load capacity, we must also remember to account for the reduced area. We will see that the reduced effective section in the postbuckled condition may require you to recalculate capacities that were based on the full cross-sectional properties, including deflection and overall buckling strength of the column.

What's This Have to Do With the Aluminum Specification? The three variations on plate slenderness illustrated in Figures 5.26, 5.27, and 5.28 correspond to the three equations for allowable stress given in Aluminum *Specification* Section 3.4.8, which addresses flat elements supported along one edge:

3.4.8(a), for stocky plates, which yield across their full cross section at failure

3.4.8(b), for plates of intermediate slenderness, which undergo inelastic buckling when loaded to failure

3.4.8(c), for slender plates, which buckle elastically and are then capable of supporting additional load until they reach failure at their crippling stress.

These equations are highlighted in Aluminum *Specification* Table 3.4-3, reproduced here as Figure 5.31. They are also shown highlighted in Figure 5.32 evaluated for the 6061-T6 alloy.

The equations used in the Aluminum *Specification* for plate element strengths are of the same form as those for columns, but the column slenderness ratio L/r is replaced by the corresponding slenderness ratio, b/t, (also called the *aspect ratio*) of the plate element. The slenderness ratio for curved elements is expressed as R/t, where R is the radius to the midthickness of the curved element, and t is its thickness. A factor placed in front of the b/t

GENERAL FORMULAS FOR DETERMINING ALLOWABLE STRESSES FROM SECTION 3.4

Type of Stress	Type of Member or Component	Sub-Sec.	Allowable Stress
TENSION, axial	Any tension member	1	F_{ty}/n_y or $F_{tu}/(k_t n_u)$
TENSION IN BEAMS, extreme fiber, net section	Rectangular tubes, structural shapes bent around strong axis	2	F_{ty}/n_y or $F_{tu}/(k_t n_u)$
	Round or oval tubes	3	$1.17 F_{ty}/n_y$ or $1.24 F_{tu}/(k_t n_u)$
	Shapes bent about weak axis, bars, plates	4	$1.30 F_{ty}/n_y$ or $1.42 F_{tu}/(k_t n_u)$
BEARING	On rivets and bolts	5	$2F_{tu}/n_u$
	On flat surfaces and pins and on bolts in slotted holes	6	$2F_{tu}/(1.5 n_u)$

*For tubes with circumferential welds, equations of Sections 3.4.10, 3.4.12, and 3.4.16.1 apply for $R/t \leq 20$.

Type of Stress	Type of Member or Component	Sub-Sec.	Allowable Stress Slenderness ≤ S_1	Slenderness Limit S_1	Allowable Stress Slenderness Between S_1 and S_2	Slenderness Limit S_2	Allowable Stress Slenderness ≥ S_2
COMPRESSION IN COLUMNS, axial, gross section	All columns	7	$\dfrac{F_{cy}}{n_y}$	$\dfrac{kL}{r} = \dfrac{B_c - \dfrac{n_u F_{cy}}{n_y}}{D_c}$	$\dfrac{1}{n_u}\left(B_c - D_c \dfrac{kL}{r}\right)$	$\dfrac{kL}{r} = C_c$	$\dfrac{\pi^2 E}{n_u\left(\dfrac{kL}{r}\right)^2}$
COMPRESSION IN COMPONENTS OF COLUMNS, axial, gross section	Flat plates supported along one edge - columns buckling about a symmetry axis	8	$\dfrac{F_{cy}}{n_y}$	$\dfrac{b}{t} = \dfrac{B_p - \dfrac{n_u F_{cy}}{n_y}}{5.1 D_p}$	$\dfrac{1}{n_u}\left(B_p - 5.1 D_p \dfrac{b}{t}\right)$	$\dfrac{b}{t} = \dfrac{k_1 B_p}{5.1 D_p}$	$\dfrac{k_2 \sqrt{B_p E}}{n_u(5.1 b/t)}$
	Flat plates supported along one edge - columns not buckling about a symmetry axis	8.1	$\dfrac{F_{cy}}{n_y}$	$\dfrac{b}{t} = \dfrac{B_p - \dfrac{n_u F_{cy}}{n_y}}{5.1 D_p}$	$\dfrac{1}{n_u}\left(B_p - 5.1 D_p \dfrac{b}{t}\right)$	$\dfrac{b}{t} = \dfrac{C_p}{5.1}$	$\dfrac{\pi^2 E}{n_u(5.1 b/t)^2}$
COMPRESSION IN	Flat plates with both edges supported	9	$\dfrac{F_{cy}}{n_y}$	$\dfrac{b}{t} = \dfrac{B_p - \dfrac{n_u F_{cy}}{n_y}}{1.6 D_p}$	$\dfrac{1}{n_u}\left(B_p - 1.6 D_p \dfrac{b}{t}\right)$	$\dfrac{b}{t} = \dfrac{k_1 B_p}{1.6 D_p}$	$\dfrac{k_2 \sqrt{B_p E}}{n_u(1.6 b/t)}$
COMPONENTS OF COLUMNS gross section	Flat plates with one edge supported and other edge with stiffener	9.1	See Section 3.4.9.1				
	Flat plates with both edges supported and with an intermediate stiffener	9.2	See Section 3.4.9.2				
	Curved plates supported on both edges, walls of round or oval tubes	10*	$\dfrac{F_{cy}}{n_y}$	$\dfrac{R_b}{t} = \left(\dfrac{B_t - \dfrac{n_u F_{cy}}{n_y}}{D_t}\right)^2$	$\dfrac{1}{n_u}\left(B_t - D_t \sqrt{\dfrac{R_b}{t}}\right)$	$\dfrac{R_b}{t} = C_t$	$\dfrac{\pi^2 E}{16 n_u\left(\dfrac{R_b}{t}\right)\left(1 + \dfrac{\sqrt{R_b t}}{35}\right)^2}$

Figure 5.31 General requirements of the Aluminum *Specification* for local buckling of column elements.

Allowable Stresses for BUILDING and Similar Type Structures

6061-T6, -T651, -T6510, -T6511

Extrusions up thru 1 in., Sheet & Plate, Pipe, Standard Structural Shapes, Drawn Tube, Rolled Rod and Bar, 6351-T5 Extrusions

WHITE BARS apply to nonwelded members and to welded members at locations farther than 1.0 in. from a weld.

SHADED BARS apply within 1.0 in. of a weld.

Equations that straddle the shaded and unshaded areas apply to both.

*For tubes with circumferential welds, equations of Sections 3.4.10, 3.4.12, and 3.4.16.1 apply for $R_b/t \leq 20$.

Type of Stress	Type of Member or Component	Sec. 3.4.	Allowable Stress	
TENSION, axial	Any tension member	1	19	11†
TENSION IN BEAMS, extreme fiber, net section	Rectangular tubes, structural shapes bent around strong axis	2	19	11†
	Round or oval tubes	3	24	13.5†
	Shapes bent about weak axis, bars, plates	4	28	16†
BEARING	On rivets and bolts	5	39	25†
	On flat surfaces and pins and on bolts in slotted holes	6	26	16†

Type of Stress	Type of Member or Component	Sec. 3.4.	Allowable Stress Slenderness ≤ S_1	Slenderness Limit S_1	Allowable Stress Slenderness Between S_1 and S_2	Slenderness Limit S_2	Allowable Stress Slenderness ≥ S_2
COMPRESSION IN COLUMNS, axial, gross section	All columns	7	—	$kL/r = 0$	$20.2 - 0.126\,(kL/r)$	$kL/r = 66$	$51{,}000/(kL/r)^2$
			12†	—	12†	$kL/r = 65$†	$51{,}000/(kL/r)^2$†
COMPRESSION IN COMPONENTS OF COLUMNS, gross section	Flat plates supported along one edge - columns buckling about a symmetry axis	8	21	$b/t = 2.7$	$23.1 - 0.79\,(b/t)$	$b/t = 10$	$154/(b/t)$
			12†	—	12†	$b/t = 13$†	$154/(b/t)$†
	Flat plates supported along one edge - columns not buckling about a symmetry axis	8.1	21	$b/t = 2.7$	$23.1 - 0.79\,(b/t)$	$b/t = 12$	$1970/(b/t)^2$
			12†	—	12†	$b/t = 13$†	$1970/(b/t)^2$†
	Flat plates with both edges supported	9	21	$b/t = 8.4$	$23.1 - 0.25\,(b/t)$	$b/t = 33$	$490/(b/t)$
			12†	—	12†	$b/t = 41$†	$490/(b/t)$†
	Flat plates with one edge supported and other edge with stiffener	9.1	See Section 3.4.9.1				
	Flat plates with both edges supported and with an intermediate stiffener	9.2	See Section 3.4.9.2				
	Curved plates supported on both edges, walls of round or oval tubes	10*	21	$R_b/t = 2.2$	$22.2 - 0.80\sqrt{R_b/t}$	$R_b/t = 141$	$\dfrac{3200}{(R_b/t)(1 + \sqrt{R_b/t}/35)^2}$
			12†	$R_b/t = 9.0$	$13.5 - 0.30\sqrt{R_b/t}$	$R_b/t = 290$	

January 2000

Figure 5.32 Requirements of the Aluminum *Specification* for local buckling of column elements, for the specific case of alloy 6061-T6.

155

term for flat plate elements reflects the edge support conditions and the compressive stress distribution (uniform stress or varying stress across the width). For example, for uniform compressive stress, for a plate supported along both edges the equivalent slenderness ratio is 1.6 b/t, while for a plate supported along only one edge the equivalent slenderness ratio is 5.1 b/t.

The expression for the strength of stocky plates used in the Aluminum *Specification* is the same as for stocky columns (Equation 5.10), since yielding is not a function of slenderness:

$$\text{yield strength} = F_{cy} \tag{5.16}$$

where:

F_{cy} = compressive yield strength

The expression for inelastic buckling of flat plate components of columns is:

$$\text{inelastic buckling strength} = B_p - D_p(k\ b/t) \tag{5.17}$$

where:

B_p = stress at which the flat plate inelastic buckling strength line graphed versus slenderness (b/t) intersects the y-axis
D_p = slope of the flat plate inelastic buckling strength line
k = constant dependent on the support at the edges of the plate.

The expression for elastic buckling in the most slender regime is:

$$\text{elastic buckling strength} = \frac{\pi^2\ E}{(kb/t)^2} \tag{5.18}$$

where:

E = modulus of elasticity

When postbuckling strength is recognized, which is done for some (but not all) types of elements, the ultimate strength of the element is:

$$\text{postbuckling strength} = \frac{k\ \sqrt{B_p E}}{b/t} \tag{5.19}$$

where:

k = a constant
E = modulus of elasticity
B_p = buckling constant for a plate element
b/t = width-to-thickness ratio of the element.

The equations for the slenderness regimes are plotted for alloy 6061-T6 in Figure 5.33. Note that for b/t ratios slightly greater than S_2, the postbuckling curve actually imposes a lower limit on strength than would be predicted by local buckling (Equation 5.18). This is just an anomaly of the empirical nature of the equations and has no cosmic significance.

When Postbuckling Strength Is Recognized Table 5.3 lists the Aluminum *Specification* sections that address various configurations of plate edge support and stiffeners for elements that are components of columns. This table also identifies which *Specification* sections recognize postbuckling strength and which do not.

In earlier editions of the Aluminum *Specification*, postbuckling strength was not recognized for flat plates supported along one edge under axial compression (for example, the flange of an I-beam). In the 6th edition, postbuckling strength for such elements was recognized, but only when they occur in columns whose overall member buckling axis is an axis of symmetry. (Determining the overall member buckling axis was discussed in Section 5.2.1 above.) The reason for this is illustrated by Figure 5.34a, which shows a column whose buckling axis is not an axis of symmetry. If postbuckling strength were recognized, the tips of the flanges could buckle locally and would no longer be an effective part of the section, shifting the neutral axis for the remaining section resisting the load. The point of application of the

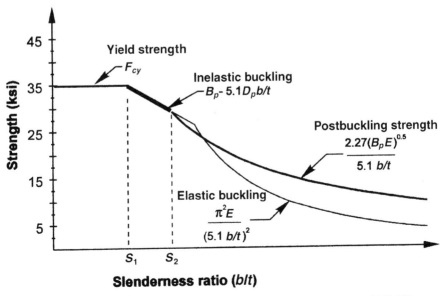

Figure 5.33 The local buckling equations for columns of alloy 6061-T6.

TABLE 5.3 Aluminum *Specification* Sections for Local Buckling of Components of Columns

Specification Number	Type of Element	Effective Slenderness	Postbuckling Strength Recognized?
3.4.8	Flat plate—one edge supported (in a section with buckling axis = axis of symmetry)	$5.1b/t$	Yes
3.4.8.1	Flat plate—one edge supported (in a section with buckling axis ≠ axis of symmetry)	$5.1b/t$	No
3.4.9.1	Flat plate—one edge supported, one with stiffener		Yes
3.4.9	Flat plate—both edges supported	$1.6b/t$	Yes
3.4.9.2	Flat plate—both edges supported and with intermediate stiffener		No
3.4.10	Curved plate—both edges supported		No

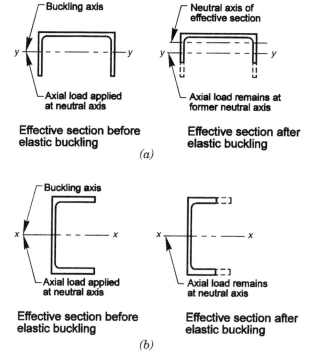

Figure 5.34 Comparison of buckling axis to the axis of symmetry for a channel.

load, however, might not shift—say, if the end connection were bolted—and, thus, an eccentricity and corresponding bending moment, unaccounted for in the design, would be introduced.

This situation is avoided in sections that would buckle about an axis of symmetry because the neutral axis would not shift since equal reductions in the effective section would take place on either side of the neutral axis (Figure 5.34b). Postbuckling strength may, therefore, be used in designing channels when overall member buckling would occur about the x-axis of the section in Figure 5.34b, but not when the y-axis is the buckling axis. Cruciforms governed by torsional buckling buckle about the longitudinal axis of the member, which is not an axis of symmetry for the cross section, and so cannot realize postbuckling strength, even though the section is symmetric about other axes.

Factors of Safety *Factors of safety* for elements of columns, as with overall column behavior, depend on the slenderness regime. When the slenderness ratio (b/t for flat elements and R/t for curved elements) is less than or equal to S_1, the factor of safety on yield strength is used ($n_y = 1.65$). The factor of safety on ultimate strength ($n_u = 1.95$) is used to determine the allowable stress of elements with a slenderness ratio greater than S_1 and thereby subject to buckling.

Appearance Limitations Someone (probably with a degree in architecture) may find elastic buckling of an element unacceptable for aesthetic reasons, even though the element is structurally adequate and can safely carry more load while buckled. In cases where the appearance of buckles must be avoided, Aluminum *Specification,* Section 4.7.1, Local Buckling Stresses, can be used to calculate the buckling stress. Table 4.7.1-1 of the *Specification* lists only the elastic buckling strength (F_{cr}) of those elements that have postbuckling strength recognized in the *Specification*. For the other *Specification* sections, where use of postbuckling strength is not allowed, the appearance of buckling is inherently avoided. When postbuckling strength is otherwise allowed but is to be avoided only for reasons of appearance, the factor of safety against local buckling to be used is given in *Specification* Table 3.4-1 as $n_a = 1.2$. This does not relieve you from also providing a factor of safety of 1.95 against postbuckling strength, however. The allowable stress is, thus, the lesser of:

$$F_c = \frac{F_{cr}}{n_a} = \frac{F_{cr}}{1.2} \quad \text{and} \tag{5.20}$$

$$F_c = \frac{F_u}{n_u} = \frac{F_u}{1.95} \tag{5.21}$$

where:

F_c = allowable compressive stress for an element for which the appearance of buckling is to be avoided

F_{cr} = local elastic buckling stress (given in Aluminum *Specification* Table 4.7.1-1)

F_u = crippling, or ultimate, strength of an element for which postbuckling strength is recognized (from Equation 5.19).

Plate Widths and Thicknesses   Plate width and thickness definitions would seem to be fairly straightforward, but let's not jump to conclusions. The AISC *Specification* (39) defines plate widths in Section B.5.1, Classification of Steel Sections, and the definition varies by the shape of the cross section and the element in question. For example, the width b of an outstanding flange of an I-beam is half the full flange width, whereas the width of an I-beam web is the clear distance between flanges (Figure 5.35). The Aluminum *Specification* takes a different approach, defining the width of almost all plate elements as

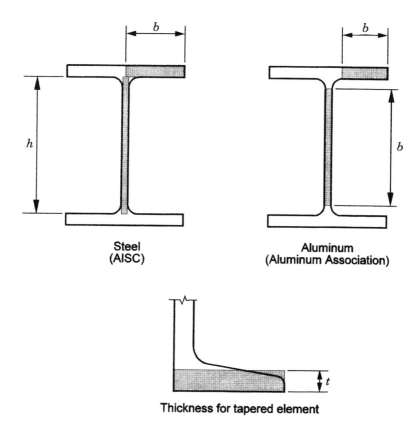

Steel
(AISC)

Aluminum
(Aluminum Association)

Thickness for tapered element

Figure 5.35   Comparison of AISC and Aluminum Association definitions of element width.

the width of its flat portion. This may also be understood as the clear width less any corner radii (Figure 5.35), except that a corner radius deduction is limited to four times the thickness of the element in question. The only plate element in the Aluminum *Specification* for which the width is not the flat width is a web in shear; here, the width is defined as the clear height, that is, the distance between the flanges. The corner radii should not be deducted from this width. This is illustrated in *Specification* Figures 3.4.18-1 and 3.4.19-1.

The provisions of the Aluminum *Specification* are based on elements of constant thickness. However, an average thickness may be used for the thickness of tapered flanges, as shown in Figure 5.35. (See *Specification* Section 3.2, Nomenclature for *t*. Tapered flanges are used in the old American Standard channels and I-beams, as well as other shapes that became standard when shapes were produced by rolling.) Extruding enables designers to vary the thickness across the width of an element in ways other than a constant taper (see Figure 5.36 for examples), but the Aluminum *Specification* does not yet provide a means of calculating the strength of such an element.

The thickness used in design typically does include the thickness of aluminum cladding since the effect of the cladding on strength is accounted for by using different (slightly reduced) minimum mechanical properties for *alclad* alloy products (see Section 3.2). This approach is different from that used in AISI cold-formed steel *Specification* Section A.1.2(f), where thickness used in design excludes galvanized coatings. For both aluminum and cold-formed steel, the thickness of any non-metallic coating, such as paint, is not included in the thickness of an element. And, as for cold-formed steel, when aluminum sheet metal is bent to form a section, the slight reduction in thickness that occurs at bends due to stretching of the material may be ignored for purposes of design.

How to Check a Column for Local Buckling

Example 5.3: What are the allowable stresses for the elements of a 6061-T6 Aluminum Association standard I-beam I 12 x 11.7 used as a column?

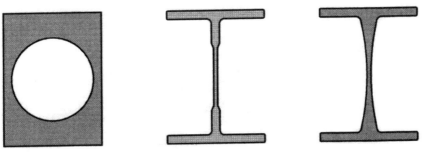

Figure 5.36 Examples of nonuniform element thickness in extrusions.

Dimensions for this shape are given in *Aluminum Design Manual*, Part VI, Table 11, and shown on Figure 5.21 here. You can divide the cross section of this I-beam into 5 elements: two for each of the two flanges, and one being the web. The flange elements are all the same: flat plates supported along one edge (where they join to the web); these are addressed in Aluminum *Specification* Section 3.4.8. The web is a flat plate with both edges supported (where it joins to the flanges) and addressed in Aluminum *Specification* Section 3.4.9. Thus, you have two unique elements to investigate: flange and web.

The flat width of each flange element is:

$$\frac{[(7 \text{ in.}) - (0.29 \text{ in.}) - 2(0.4 \text{ in.})]}{2} = 2.955 \text{ in.}$$

The flange-to-web fillet radius is 0.4 in. $< 4t = 4(0.47 \text{ in.}) = 1.88$ in., so the entire radius may be deducted from the width. The width-to-thickness, or slenderness, ratio is:

$$\frac{b}{t} = \frac{2.955 \text{ in.}}{0.47 \text{ in.}} = 6.29$$

Referencing the *Aluminum Design Manual*, Part VII, Design Aids Table 2-21 (Figure 5.32), which gives allowable stresses for 6061-T6 aluminum, we find, for *Specification* Section 3.4.8, that this slenderness ratio is between the slenderness limits S_1 (2.7) and S_2 (10). (This means the flange strength is based on inelastic buckling.) The allowable stress for the flanges can be determined from the equation found in Figure 5.32:

$$23.1 - 0.79(b/t) = 23.1 - 0.79(6.29) = 18.1 \text{ ksi.}$$

The flat width of the web is:

$$(12 \text{ in.}) - 2[(0.4 \text{ in.}) + (0.47 \text{ in.})] = 10.26 \text{ in.}$$

The flange-to-web fillet radius is 0.4 in. $< 4t = 4(0.29 \text{ in.}) = 1.16$ in., so the entire radius may be deducted from the width. The width-to-thickness, or slenderness, ratio is:

$$\frac{b}{t} = \frac{10.26 \text{ in.}}{0.29 \text{ in.}} = 35.38$$

Referring again to Figure 5.32, we find that, for *Specification* Section 3.4.9, $35.38 > 33 = S_2$, so the allowable stress for the web is:

$$\frac{490}{b/t} = 13.8 \text{ ksi}$$

From Table 5.3, we see that postbuckling strength is recognized in Section 3.4.9, and since the web's slenderness ratio is greater than S_2, the web strength is a postbuckling strength.

The local buckling allowable stress of the cross section could be conservatively taken as the lesser of the allowable stresses for the flange (18.1) and web (13.8), or 13.8 ksi. However, a less conservative approach for determining the allowable stress of the elements of the column, called the *weighted average allowable stress method,* is permitted by the Aluminum *Specification.*

Weighted Average Allowable Compressive Stress   Aluminum *Specification* Section 4.7.2, Weighted Average Allowable Compressive Stress, permits an alternate calculation of local buckling strength, recognizing the nonuniform distribution of stress in the cross section. This method determines the local buckling allowable stress as the weighted average allowable stress for the individual elements, where the allowable stress for each element is weighted in accordance with the ratio of the area of the element to the total area of the section. Perhaps a simpler way of understanding this is to consider it to be a method by which the strength of each element is calculated, and the strength of the section is taken as the sum of the strength of the elements.

The weighted average allowable stress is determined by the following equation:

$$F_{ca} = \frac{A_1 F_{c1} + A_2 F_{c2} + \cdots + A_n F_{cn}}{A} \tag{5.22}$$

where:

A_i = area of element i
F_{ci} = allowable stress of element i
A = area of the cross section ($= A_1 + A_2 + \cdots + A_n$),
F_{ca} = the weighted average allowable stress.

The weighted average method may be applied to elements of any slenderness and is illustrated as a continuation of Example 5.3.

Example 5.4: What is the weighted average allowable stress for the I-beam of Example 5.3 above?

The allowable stress for the flange was determined in Example 5.3 to be 18.1 ksi, and the allowable stress for the web was 13.8 ksi. The area of the flanges is:

$$2(7 \text{ in.})(0.47 \text{ in.}) = 6.58 \text{ in}^2$$

The area of the web is:

$$(0.29 \text{ in.})[12 - 2(.47)] \text{ in.} = 3.21 \text{ in}^2$$

The weighted average allowable stress is:

$$F_{ca} = \frac{(18.1)(6.58) + (13.8)(3.21)}{6.58 + 3.21} = 16.7 \text{ ksi}$$

This value is 21% higher than the 13.8 ksi stress obtained by using the minimum of the flange and web allowable stresses.

As we mentioned previously and the above examples illustrate, the local buckling strengths are only a function of the dimensions of the member cross section, not its length or end supports. Therefore, you can determine local buckling strengths for cross sections and tabulate them for convenience. We've done this for you in Appendix D for Aluminum Association Standard I-beams and Aluminum Association Standard channels.

Effective Width   Another approach for determining the postbuckling strength of sections is the *effective width method.* Aluminum *Specification* Section 4.7.6 employs only the effective width method to compute the deflections of sections with very slender elements. We'll see that this usually is not a concern with extrusions.

The effective width method is required in the AISI's cold-formed steel *Specification* to determine both stresses and deflections for light-gauge steel design, whereas the Aluminum *Specification* uses the much simpler weighted average approach to calculate allowable stress. The steel approach requires that you know the stress in an element to calculate its effective width. To know the stress, however, you must already know the effective width of each element in order to calculate the section's properties. Consequently, the procedure is iterative.

While the weighted average method is simpler and gives results for strength similar to those of the effective width method, it does not afford a way to calculate deflections in the postbuckled regime. The effective width method must, therefore, be used to calculate deflections of members with elements in the postbuckled condition, but the Aluminum *Specification* avoids the iterative process because the stress at design loads is already given by the weighted average method. Aluminum *Specification* Section 4.7.6 obtains the effective width of an element by solving in terms of the calculated stress:

$$b_e = b \sqrt{F_{cr}/f_c} \tag{5.23}$$

where:

b_e = effective width of an element
b = width of the element
F_{cr} = elastic (local) buckling stress for the element (calculated per Aluminum *Specification* Section 4.7.1, also given here in Table 5.4
f_c = compressive stress on the element.

By using this equation to determine the dimensions of the reduced effective area, you can calculate the section properties in order to determine deflections in the postbuckled condition.

As we discussed earlier, postbuckling strength and the associated reduced effective section are considered only when the slenderness ratio of an element is greater than S_2. Deflections of sections with stockier elements will be accurately predicted by the weighted average method. When the slenderness ratio of any element in a section is above S_2, however, some of this element may be buckled and rendered ineffective before the allowable load is reached. This reduction in effective area will result in greater deflections than would be predicted using the full section properties.

Aluminum extrusions usually do not contain these very slender elements because the extrusion process is generally unsuited for producing them. For example, the greatest width-to-thickness ratio for a web of any Aluminum Association Standard channel or I-beam is about 35, just slightly greater than 33, the S_2 lower limit for elastic local buckling of 6061-T6 elements. Furthermore, the S_2 limits for the flanges are not exceeded in these standard channels and I-beams when extruded in 6061-T6 and 6063-T6 alloys. Therefore, limits on width-to-thickness ratios for extruded elements are not usually a concern.

On the other hand, sections made by bending or roll-forming sheet may readily be fabricated with extremely slender elements (Figure 5.37). Consequently, when aluminum formed-sheet sections are used with elements having

TABLE 5.4 Elastic Local Buckling Stresses for Elements

Aluminum *Specification* Section	Elastic Local Buckling Stress (F_{cr})
3.4.8, 3.4.15	$F_{cr} = \dfrac{\pi^2 E}{(5.1 b/t)^2}$
3.4.9, 3.4.16	$F_{cr} = \dfrac{\pi^2 E}{(1.6 b/t)^2}$
3.4.9.1, 3.4.16.2	$F_{cr} = \dfrac{(n_y F_c)^2}{F_{cy}}$
3.4.18	$F_{cr} = \dfrac{\pi^2 E}{(0.67 h/t)^2}$
3.4.19	$F_{cr} = \dfrac{\pi^2 E}{(0.29 h/t)^2}$

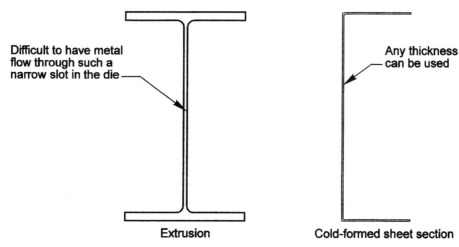

Difficult to have metal
flow through such a
narrow slot in the die

Any thickness
can be used

Extrusion   Cold-formed sheet section

Figure 5.37   Practical limits on element slenderness of extruded and cold-formed shapes.

width-to-thickness ratios exceeding the S_2 slenderness limit, you should do calculations (based on the effective width method) to assure that deflections are not excessive. An example of this is included in Section 10.1, Cold-Formed Aluminum Construction.

The Aluminum *Specification* does not impose an upper limit on the slenderness ratio for elements. AISI cold-formed steel *Specification* Section B1, however, limits the width-to-thickness ratio for cold-formed steel sections to a maximum b/t of 200.

Interaction Between Local Buckling and Overall Buckling   A subtle but potentially dangerous effect that must be guarded against is an interaction between local buckling and overall member buckling. This is another situation

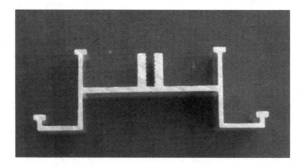

Figure 5.38   Extruded shape with stiffeners.

that arises from the reduced effective area in the postbuckled condition. The elastic buckling of some portions of the cross section results in greater stress in the remaining area, thereby making the column more susceptible to overall buckling.

Aluminum *Specification* Section 4.7.4, Effect of Local Buckling on Column Strength, accounts for the interaction between local buckling and overall buckling. You're probably wondering: when do I need to worry about this? First, the interaction is potentially a problem only for shapes containing elements for which postbuckling strength is recognized. (You can tell if it is by consulting Table 5.4.) Second, the element with postbuckling strength recognized must be more slender than the S_2 slenderness limit, since it's only then that the postbuckling strength kicks in. These requirements help narrow the field of consideration a bit. The final requirement for the interaction to govern is that the elastic buckling stress of an element must be less than the overall column buckling stress.

Example 5.5: Check the effect of local buckling on the strength of the column considered in Examples 5.2, 5.3, and 5.4.

Let's start by asking whether it's possible that the interaction could limit the overall column strength. We see from Table 5.4 that the web does have postbuckling strength recognized and a slenderness ratio of 35.38, which is greater than $S_2 = 33$. Having determined that this shape both utilizes postbuckling strength and is in the postbuckling slenderness regime, we need to check whether the onset of buckling occurs before the column reaches its overall allowable buckling stress (F_c). We calculated F_c in Example 5.2 as 17.7 ksi. The criteria that must be satisfied in order for the interaction to be checked is given by Aluminum *Specification* Equation 4.7.4-2:

$$\frac{F_{cr}}{n_u} < F_c \tag{5.24}$$

where:

n_u = factor of safety on ultimate strength = 1.95
F_{cr} = local buckling stress of the web, (from Table 5.4).

$$F_{cr} = \frac{\pi^2 E}{(1.6b/t)^2} = \frac{\pi^2 (10,100)}{[1.6(35.38)]^2} = 31.1 \text{ ksi}$$

Since (31.1 ksi)/1.95 = 15.9 ksi < 17.7 ksi, we need to calculate F_{rc}, the allowable stress for a column with buckled elements, using Aluminum *Specification* equation 4.7.4-1:

$$F_{rc} = \frac{(F_{ec})^{1/3} (F_{cr})^{2/3}}{n_u}$$

where:
$$F_{ec} = \frac{\pi^2 E}{(kL/r)^2} = \frac{\pi^2 (10,100)}{20^2} = 249 \text{ ksi}$$

So:

$$F_{rc} = \frac{(249)^{1/3} (31.1)^{2/3}}{1.95} = 31.9 \text{ ksi}$$

Since F_{rc} = 31.9 ksi > 17.7 ksi = F_c, the interaction does not govern and the overall column allowable compressive stress is 17.7 ksi. We can now compare the overall allowable stress (17.7 ksi) to the allowable stress of the elements (by the weighted average method used in Example 5.4, 16.7 ksi) and use the lesser of these (16.7 ksi) as the allowable stress for the column.

Stiffeners In Table 5.3, we listed Aluminum *Specification* Sections 3.4.9.1 and 3.4.9.2 that deal with stiffeners. You can just plug and chug on these provisions with no further thought if you prefer. If you're curious about how they work, the next two paragraphs are for you.

Specification Section 3.4.9.1 addresses flat plates with one edge supported and the other with a stiffener (Figure 5.23c). The allowable compressive stress F_c is the lesser of:

$$F_c = \frac{F_{cy}}{n_y} \quad \text{and} \quad F_c = F_{UT} + (F_{ST} - F_{UT})\rho_{ST}$$

where:

F_{cy} = compressive yield strength
F_{UT} = allowable stress for the element calculated according to *Specification* Section 3.4.8 as if no stiffener were present
F_{ST} = allowable stress for the element calculated according to *Specification* Section 3.4.9 as if the element were supported along both longitudinal edges
ρ_{ST} = ratio dependent on the radius of gyration of the stiffener and the slenderness ratio of the element stiffened, determined as:

$$\rho_{ST} = 1 \qquad\qquad\qquad \text{for } b/t \leq S/3$$

$$\rho_{ST} = \frac{r_S}{9t\left(\dfrac{b/t}{S} - \dfrac{1}{3}\right)} \leq 1.0 \qquad \text{for } S/3 < b/t \leq S$$

$$\rho_{ST} = \frac{r_S}{1.5t\left(\dfrac{b/t}{S} + 3\right)} \leq 1.0 \qquad \text{for } S < b/t < 2S$$

where:

r_s = radius of gyration of the stiffener about the midthickness of the element being stiffened
b = width of the element being stiffened
t = thickness of the element being stiffened

$S = 1.28 \sqrt{\dfrac{E}{F_{cy}}}$, a slenderness limit

The approach is to calculate the radius of gyration of the stiffener about the midthickness of the flat plate being stiffened and use it to calculate a ratio (ρ_{ST}) that ranges from 0 to 1.0. Next, calculate the strength of the flat plate as the strength of a flat plate without an edge stiffener (using *Specification* Section 3.4.8, like Figure 5.23b) plus the ratio ρ_{ST} times the difference between the strength of an unstiffened plate and a plate supported on both edges (using *Specification* Section 3.4.9, like Figure 5.23a). The allowable stress for the element being stiffened also cannot exceed the allowable stress for the stiffener itself, which is calculated in accordance with *Specification* Section 3.4.8 since it's an element supported on one edge, with the other edge free.

For the alloys and tempers in Aluminum *Specification*, S ranges from 17 for the strongest alloy (2014-T6, with F_{cy} = 59 ksi [405 MPa]) to 43 for the weakest alloy (Alclad 3003-H12, with F_{cy} = 9 ksi [62 MPa]). S is 21.7 for 6061-T6, for which F_{cy} = 35 ksi [240 MPa].

When is a plate a stiffener and when is it an element? In other words, at what point is a stiffener so big that it becomes an element? Figure 5.39 illustrates the problem. While it seems clear that the plates attached to the tips of the flanges in Figure 5.39a are stiffeners, it seems equally clear that the plates at the tips of the flanges in Figure 5.39c are webs in their own right, qualifying the flange as supported on both edges. In Figure 5.39b, this isn't quite so clear. Are those things at the ends of the flanges stiffeners or elements?

The answer is: If the stiffener length (D_s) exceeds 80% of the width of the element being stiffened (b), you can't apply *Specification* Section 3.4.9.1 because its limit of applicability is $D_s/b \leq 0.8$. In such a case, you could

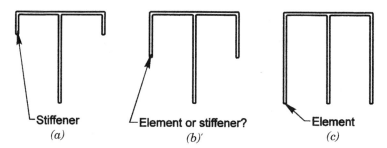

Stiffener
(a)

Element or stiffener?
(b)

Element
(c)

Figure 5.39   When is a plate an element, and when is it a stiffener?

conservatively treat the flange as supported on only one edge and disregard any stiffening effect of the long stiffener, but the *Specification* doesn't offer a way to check the stiffener because you can't even be sure that it's supported on one edge, so it might buckle. This might not have any concrete effect on the strength of the section, but it wouldn't inspire much confidence in on-lookers.

Specification Section 3.4.9.2 addresses flat plates with both edges supported and an intermediate stiffener (Figure 5.24b). This situation is handled by calculating an equivalent slenderness ratio for the stiffened flat plate (λ_s) based on the area and moment of inertia of the stiffener:

Figure 5.40   A built-up aluminum beam. (Courtesy of Conservatek Industries, Inc.)

$$\lambda_s = 4.62(b/t)\sqrt{\frac{1 + A_s/bt}{1 + \sqrt{1 + 10.67\,I_o/bt^3}}}$$

where:

A_s = area of the stiffener

I_o = moment of inertia of a section composed of the stiffener and one-half the width of the adjacent plate sub-elements, taken about its neutral axis

b = *one-half* of the width of the element being stiffened

t = thickness of the element being stiffened.

Notice that the area of the stiffener is defined differently in the determination of I_o than it is in calculating A_s and that b is half the width of the element being stiffened, the distance from the stiffener to a supported edge. Also, as the area of the stiffener approaches zero, I_o approaches $bt^3/12$, which results in $\lambda_s = 3.0b/t$. This is reassuring since the slenderness of an element with a width $2b$ and supported on both edges is $2(1.6b/t) = 3.2b/t$.

A Final Observation If only one thing stands out from this section, it should be that while columns cannot support additional load once they experience overall buckling, they can have additional strength after elements of their cross section have undergone local buckling. The reason is that elements are supported along at least one edge by virtue of their being a part of a section, and this support remains even when a portion of the element buckles. Taking advantage of the postbuckling strength of elements requires designers to expend considerable additional effort, but it can yield the reward of higher design strengths.

5.3 MEMBERS IN BENDING

Just as the term *column* was used as shorthand for any member in axial compression, *beam* is the term used as an abbreviation for a member in bending. Used in this sense, a beam need not necessarily be a horizontal member between columns in a building frame; it can be any member subjected to loads acting transverse to its longitudinal axis.

Bending is associated with three different types of stress in a beam: tension on one side of the neutral axis, compression on the other, and shear. Since we've already discussed axial tension and compression in previous sections, we've already established some understanding of how aluminum beams are designed. So armed, let's tackle each of these stresses as they act on beams.

5.3.1 Bending Yielding and Fracture

The limit states for bending tension are the same as those for axial tension: *fracture* and *yielding*. In other words, we don't want the beam to break or sag excessively by yielding, so we need to determine the bending moments that will cause these to occur.

For aluminum as for steel, we assume that plane sections through the beam remain plane during bending. The *strain* in the cross section, thus, varies linearly with distance from the neutral axis no matter what moment is applied. Within the *elastic* range, the bending stress at any point in the cross section is, then, also a linear function of the distance of that point from the neutral axis. The elastic bending stress is:

$$f_b = \frac{Mx}{I}$$

where:

f_b = bending stress at a point in the section
M = bending moment
x = distance from the point to the neutral axis
I = moment of inertia of the net section about the bending axis.

The maximum tensile stress occurs at the most distant point in the section (extreme fiber) on the tension side of the neutral axis of the beam and is calculated by the expression:

$$f_b = \frac{Mc_t}{I}$$

where:

c_t = distance from the point in the section that is farthest from the neutral axis on the tension side of the axis to the neutral axis.

This neat, linear arrangement of stress is upset when the extreme fiber stress exceeds the yield strength. Since the strain remains a linear function of distance from the neutral axis, the stress distribution plotted versus depth has the same shape as the stress-strain curve. The moment that can be resisted is equal to the force resultant of the stress distribution times the distance between the force resultants (Figure 5.41). The shape of the stress-strain curve, thus, helps determine the bending moment that can be resisted once the extreme fiber stress exceeds the yield stress.

When the extreme fiber stress reaches the yield stress, yielding across the entire cross section is prevented by the material closer to the neutral axis, which is at a lower strain and, consequently, at a lower stress. Similarly, when the extreme fiber stress reaches the tensile ultimate stress, fracture across the whole section is prevented by the material closer to the neutral axis and, therefore, at a lower stress. So, the actual yield and fracture limit state strengths are also a function of the distribution of the material of the cross section about the neutral axis.

Concentrating the material of a cross section away from the neutral axis increases the section's moment of inertia, thereby decreasing the extreme fiber

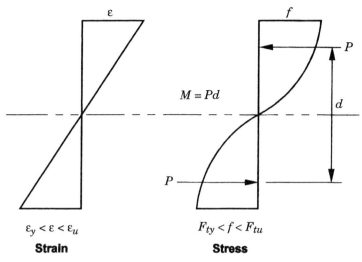

$$M = Pd$$

$\varepsilon_y < \varepsilon < \varepsilon_u$

Strain

$F_{ty} < f < F_{tu}$

Stress

Figure 5.41 Stress and strain diagrams for a section stressed beyond yield at the extreme fiber.

stress. Thus, if an I-beam and a solid rectangular bar of the same area and depth are subjected to identical bending moments, the I-beam will have a lower extreme fiber stress. It will require a greater bending moment, then, to cause the extreme fiber of the I-beam to yield than for the bar. Upon reaching yield of the extreme fiber, however, more of the I-beam's cross section will be near the yield stress because its area is concentrated near the extreme fiber. The rectangular bar, having more of its area distributed near the neutral axis, has more reserve strength beyond yield of the extreme fiber than the I-beam does. Although concentrating the area of a cross section away from the neutral axis increases the moment capacity at the extreme fiber yield stress, it also decreases the reserve moment capacity beyond extreme fiber yield.

The *shape factor* accounts for the effect of the shape of the cross section, as well as the shape of its stress-strain curve. There's a shape factor for yielding and another for fracture. The shape factor for yielding is the ratio of the moment at which the section is fully yielded to the moment at which initial yielding occurs. Similarly, the shape factor for fracture is the ratio of the ultimate moment the section can support to the moment at which the extreme fiber stress first reaches ultimate strength.

The Aluminum *Specification* takes advantage of shape factors by recognizing different strengths for different beam shapes. For example, the allowable tensile stress for an I-beam bent about its strong axis is F_{ty}/n_y, while the allowable tensile stress for a rectangular bar is $1.3F_{ty}/n_y$. If you're accustomed to the Steel *Specification*, you've been using similar factors, although it's a little less obvious there. For compact steel sections, the allowable tensile stress for I-beams bent about the strong axis is $0.66\,F_y$, while for bars, it's $0.75\,F_y$.

The factor of safety and the shape factor have all been rolled into one number for steel, but the idea is the same. A more direct comparison of the aluminum and steel specifications can be drawn from Table 5.5. Aluminum's stress-strain curve is slightly different from that of steel, so we don't expect the effect of shape factors used for the two materials to be identical, but it is similar.

The shape factor for an I-beam depends on the dimensions of the section, but it is greater than 1.0 for practical I-shapes. For hot-rolled steel, the dimensions of I-beam shapes are limited to the standard shapes. This enables steel designers to calculate and use a shape factor for steel I-beams, which accounts for the difference in allowable stress for steel in axial tension ($0.6F_y$) and bending tension ($0.66F_y$). In aluminum, however, anyone who can afford several hundred dollars for a die can employ I-beam shapes other than the standard shapes appearing in the *Aluminum Design Manual*. Consequently, the Aluminum *Specification* does not use a shape factor greater than 1.0 for I-shapes.

Because the ultimate strength of aluminum alloys can also govern bending capacity, additional checks are required for aluminum beams against ultimate strength. A summary of the aluminum requirements for both yielding and fracture is given in Table 5.6. These expressions are obtained from Aluminum *Specification* Table 3.4-3, which is reproduced and highlighted here as Figure 5.42. Values for alloy 6061-T6 from *Aluminum Design Manual*, Part VII, Table 2-21, are highlighted in Figure 5.43. For beams that don't *buckle* (such as a compact shape bent about its weak axis), yielding is the only compressive limit state. In the Aluminum *Specification*, compressive yielding is treated in the same way as tensile yielding, except the compressive yield strength (F_{cy}), rather than the tensile yield strength (F_{ty}), is used in the calculation of bending strength.

Weighted Average Tensile Stress in Beams Aluminum *Specification* Section 4.7.3 allows the weighted average allowable stress method (described earlier in Section 5.2.2 for compressive stress) to be applied to the tensile flange of

TABLE 5.5 Comparison of Aluminum and Steel Shape Factors

Shape	Steel Allowable Stress	Aluminum Allowable Stress on Yielding
I-beams and channels bent about strong axis	$1.1 (0.6F_y) = 0.66F_y$	$0.61F_y = 1.0 (F_y/1.65)$
I-beams bent about weak axis, rectangular plates, solid bars	$1.25 (0.6F_y) = 0.75F_y$	$0.79F_y = 1.3 (F_y/1.65)$
Rectangular tubes	$1.1 (0.6F_y) = 0.66F_y$	$0.61F_y = 1.0 (F_y/1.65)$
Circular tubes	$1.1 (0.6F_y) = 0.66F_y$	$0.71F_y = 1.17 (F_y/1.65)$

TABLE 5.6 Allowable Stresses for Aluminum Beams

Shape	Compressive Yielding[1]	Tensile Yielding	Tensile Fracture[2]
(Aluminum *Specification* Section Number Shown Below)			
Structural shapes bent about strong axis	F_{cy}/n_y 3.4.11	F_{ty}/n_y 3.4.2	$F_{tu}/k_t n_u$ 3.4.2
Shapes bent about weak axis, rectangular plates, solid bars	$1.30\ F_{cy}/n_y$ 3.4.13	$1.30\ F_{ty}/n_y$ 3.4.4	$1.42\ F_{tu}/k_t n_u$ 3.4.4
Rectangular tubes	F_{cy}/n_y 3.4.14	F_{ty}/n_y 3.4.2	$F_{tu}/k_t n_u$ 3.4.2
Round or oval tubes	$1.17\ F_{cy}/n_y$ 3.4.12	$1.17\ F_{ty}/n_y$ 3.4.3	$1.24\ F_{tu}/k_t n_u$ 3.4.3

Notes

[1] For compression in beams, buckling must also be checked.

[2] k_t is the coefficient for tension members that factors for notch sensitivity. See *Aluminum Specification* Table 3.4-2.

beams. For example, the weighted average tensile bending stress for trapezoidal formed-sheet beams can be calculated (Figure 5.44). For this purpose, the flange is defined as the actual flange and that portion of the web farther than two-thirds the extreme fiber distance (c_t) from the neutral axis. At midspan between roof purlins, a cross section of the beam is as shown in Figure 5.45. The allowable stress for the actual flange (segment bc on Figure 5.45) may be calculated from *Specification* Section 3.4.2 (for which the shape factor is 1.0), and the allowable stress for the web portion of the flange (segments ab and cd on Figure 5.45) may be calculated from *Specification* Section 3.4.4 (for which the shape factor is 1.3 on yield and 1.42 on ultimate strength).

When the applied load tends to deflect the beam's tensile flange toward the neutral axis (also called *flange curling,* as shown in Figure 5.46), the ability of the beam to carry moment is diminished. To address this, the Aluminum *Specification* limits the allowable stress for the tensile flange to the magnitude of the allowable compressive stress for the same flange if it were in compression. This will govern only if the tensile flange is wider than the compression flange.

5.3.2 Bending Buckling

If the compressive stresses due to bending cannot cause buckling of the beam, they can be dealt with in the same way as tensile bending stresses, as we noted above. Once buckling rears its ugly head, however, it's a whole new ball game. But how do we know when buckling can occur? Just as for axial compression, there are two areas of concern in bending compression.

GENERAL FORMULAS FOR DETERMINING ALLOWABLE STRESSES FROM SECTION 3.4

Type of Stress	Type of Member or Component	Sub-Sec.	Allowable Stress
TENSION, axial	Any tension member	1	F_{ty}/n_y or $F_{tu}/(k_t n_u)$
TENSION IN BEAMS, extreme fiber, net section	Rectangular tubes, structural shapes bent around strong axis	2	F_{ty}/n_y or $F_{tu}/(k_t n_u)$
	Round or oval tubes	3	$1.17F_{ty}/n_y$ or $1.24F_{tu}/(k_t n_u)$
	Shapes bent about weak axis, bars, plates	4	$1.30F_{ty}/n_y$ or $1.42F_{tu}/(k_t n_u)$
BEARING	On rivets and bolts	5	$2F_{tu}/n_u$
	On flat surfaces and pins and on bolts in slotted holes	6	$2F_{tu}/(1.5n_u)$

*For tubes with circumferential welds, equations of Sections 3.4.10, 3.4.12, and 3.4.16.1 apply for $R/t \leq 20$.

Type of Stress	Type of Member or Component	Sub-Sec.	Allowable Stress Slenderness ≤ S_1	Slenderness Limit S_1	Allowable Stress Between S_1 and S_2	Slenderness Limit S_2	Allowable Stress Slenderness ≥ S_2
COMPRESSION IN COLUMNS, axial, gross section	All columns	7	$\dfrac{F_{cy}}{n_y}$	$\dfrac{kL}{r} = \dfrac{B_c - \frac{n_u F_{cy}}{n_y}}{D_c}$	$\dfrac{1}{n_u}\left(B_c - D_c\dfrac{kL}{r}\right)$	$\dfrac{kL}{r} = C_c$	$\dfrac{\pi^2 E}{n_u\left(\frac{kL}{r}\right)^2}$
COMPRESSION IN COMPONENTS OF COLUMNS gross section	Flat plates supported along one edge - columns buckling about a symmetry axis	8	$\dfrac{F_{cy}}{n_y}$	$\dfrac{b}{t} = \dfrac{B_p - \frac{n_u F_{cy}}{n_y}}{5.1D_p}$	$\dfrac{1}{n_u}\left(B_p - 5.1D_p\dfrac{b}{t}\right)$	$\dfrac{b}{t} = \dfrac{k_1 B_p}{5.1D_p}$	$\dfrac{k_2\sqrt{B_p E}}{n_u(5.1b/t)}$
	Flat plates supported along one edge - columns not buckling about a symmetry axis	8.1	$\dfrac{F_{cy}}{n_y}$	$\dfrac{b}{t} = \dfrac{B_p - \frac{n_u F_{cy}}{n_y}}{5.1D_p}$	$\dfrac{1}{n_u}\left(B_p - 5.1D_p\dfrac{b}{t}\right)$	$\dfrac{b}{t} = \dfrac{C_p}{5.1}$	$\dfrac{\pi^2 E}{n_u(5.1b/t)^2}$
	Flat plates with both edges supported	9	$\dfrac{F_{cy}}{n_y}$	$\dfrac{b}{t} = \dfrac{B_p - \frac{n_u F_{cy}}{n_y}}{1.6D_p}$	$\dfrac{1}{n_u}\left(B_p - 1.6D_p\dfrac{b}{t}\right)$	$\dfrac{b}{t} = \dfrac{k_1 B_p}{1.6D_p}$	$\dfrac{k_2\sqrt{B_p E}}{n_u(1.6b/t)}$
	Flat plates with one edge supported and other edge with stiffener	9.1	See Section 3.4.9.1				
	Flat plates with both edges supported and with an intermediate stiffener	9.2	See Section 3.4.9.2				
	Curved plates supported on both edges, walls of round or oval tubes	10*	$\dfrac{F_{cy}}{n_y}$	$\dfrac{R_b}{t} = \left(\dfrac{B_t - \frac{n_u F_{cy}}{n_y}}{D_t}\right)^2$	$\dfrac{1}{n_u}\left(B_t - D_t\sqrt{\dfrac{R_b}{t}}\right)$	$\dfrac{R_b}{t} = C_t$	$\dfrac{\pi^2 E}{16n_u\left(\frac{R_b}{t}\right)\left(1 + \frac{\sqrt{R_b/t}}{35}\right)^2}$

Figure 5.42 General requirements of the Aluminum *Specification* for yielding and fracture of beams.

Type of Stress	Type of Member or Component	Sub-Sec.	Allowable Stress Slenderness ≤ S₁	Slenderness Limit S₁	Allowable Stress Slenderness Between S₁ and S₂	Slenderness Limit S₂	Allowable Stress Slenderness ≥ S₂
COMPRESSION IN BEAMS, extreme fiber gross section	Single web beams bent about strong axis	11	$\dfrac{F_{cy}}{n_y}$	$\dfrac{L_b}{r_y} = \dfrac{1.2(B_c - F_{cy})}{D_c}$	$\dfrac{1}{n_y}\left(B_c - \dfrac{D_c}{1.2}\dfrac{L_b}{r_y}\right)$	$\dfrac{L_b}{r_y} = 1.2\,C_c$	$\dfrac{\pi^2 E}{n_y\left(\dfrac{L_b}{1.2 r_y}\right)^2}$
	Round or oval tubes	12*	$\dfrac{1.17 F_{cy}}{n_y}$	$\dfrac{R_b}{t} = \left(\dfrac{B_{tb} - 1.17 F_{cy}}{D_{tb}}\right)^2$	$\dfrac{1}{n_y}\left(B_{tb} - D_{tb}\sqrt{\dfrac{R_b}{t}}\right)$	$\dfrac{R_b}{t} = \left[\dfrac{n_{tb} B_{tb}}{n_y}\middle/\left(\dfrac{n_{tb} D_{tb}}{n_y}\right)\right]^2$	Same as Section 3.4.10
	Solid rectangular and round section beams	13	$\dfrac{1.3 F_{cy}}{n_y}$	$\dfrac{d}{t}\sqrt{\dfrac{L_b}{d}} = \dfrac{B_{tb} - 1.3 F_{cy}}{2.3 D_{tb}}$	$\dfrac{1}{n_y}\left(B_{tb} - 2.3 D_{tb}\sqrt{\dfrac{d}{t}\sqrt{\dfrac{L_b}{d}}}\right)$	$\dfrac{d}{t}\sqrt{\dfrac{L_b}{d}} = \dfrac{C_{br}}{2.3}$	$\dfrac{\pi^2 E}{5.29 n_y\left(\dfrac{d}{t}\right)^2\dfrac{L_b}{d}}$
	Rectangular tubes and box sections	14	$\dfrac{F_{cy}}{n_y}$	$\dfrac{L_b S_c}{0.5\sqrt{I_y J}} = \left(\dfrac{B_c - F_{cy}}{1.6 D_c}\right)^2$	$\dfrac{1}{n_y}\left(B_c - 1.6 D_c\sqrt{\dfrac{L_b S_c}{0.5\sqrt{I_y J}}}\right)$	$\dfrac{L_b S_c}{0.5\sqrt{I_y J}} = \left(\dfrac{C_c}{1.6}\right)^2$	$\dfrac{\pi^2 E}{2.56 n_y}\dfrac{L_b S_c}{0.5\sqrt{I_y J}}$
COMPRESSION IN COMPONENTS OF BEAMS, (component under uniform compression), gross section	Flat plates supported on one edge	15	$\dfrac{F_{cy}}{n_y}$	$\dfrac{b}{t} = \dfrac{B_p - F_{cy}}{5.1 D_p}$	$\dfrac{1}{n_y}\left(B_p - 5.1 D_p\dfrac{b}{t}\right)$	$\dfrac{b}{t} = \dfrac{k_1 B_p}{5.1 D_p}$	$\dfrac{k_2\sqrt{B_p E}}{n_y(5.1 b/t)}$
	Flat plates with both edges supported	16	$\dfrac{F_{cy}}{n_y}$	$\dfrac{b}{t} = \dfrac{B_p - F_{cy}}{1.6 D_p}$	$\dfrac{1}{n_y}\left(B_p - 1.6 D_p\dfrac{b}{t}\right)$	$\dfrac{b}{t} = \dfrac{k_1 B_p}{1.6 D_p}$	$\dfrac{k_2\sqrt{B_p E}}{n_y(1.6 b/t)}$
	Curved plates supported on both edges	16.1*	$\dfrac{1.17 F_{cy}}{n_y}$	$\dfrac{R_b}{t} = \left(\dfrac{B_t - 1.17 F_{cy}}{D_t}\right)^2$	$\dfrac{1}{n_y}\left(B_t - D_t\sqrt{\dfrac{R_b}{t}}\right)$	$\dfrac{R_b}{t} = C_t$	$\dfrac{\pi^2 E}{16 n_y\left(\dfrac{R_b}{t}\right)\left(1 + \dfrac{\sqrt{R_b/t}}{35}\right)^2}$
	Flat plates with one edge supported and the other edge with stiffener	16.2		See Section 3.4.16.2			
	Flat plates with both edges supported and with an intermediate stiffener	16.3		See Section 3.4.16.3			
COMPRESSION IN COMPONENTS OF BEAMS, (component under bending in own plane) gross section	Flat plates with compression edge free, tension edge supported	17	$\dfrac{1.3 F_{cy}}{n_y}$	$\dfrac{b}{t} = \dfrac{B_{br} - 1.3 F_{cy}}{3.5 D_{br}}$	$\dfrac{1}{n_y}\left(B_{br} - 3.5 D_{br}\dfrac{b}{t}\right)$	$\dfrac{b}{t} = \dfrac{C_{br}}{3.5}$	$\dfrac{\pi^2 E}{n_y(3.5 b/t)^2}$
	Flat plate with both edges supported	18	$\dfrac{1.3 F_{cy}}{n_y}$	$\dfrac{h}{t} = \dfrac{B_{br} - 1.3 F_{cy}}{0.67 D_{br}}$	$\dfrac{1}{n_y}\left(B_{br} - 0.67 D_{br}\dfrac{h}{t}\right)$	$\dfrac{h}{t} = \dfrac{k_1 B_{br}}{0.67 D_{br}}$	$\dfrac{k_2\sqrt{B_{br} E}}{n_y(0.67 h/t)}$
	Flat plate with horizontal stiffener, both edges supported	19	$\dfrac{1.3 F_{cy}}{n_y}$	$\dfrac{h}{t} = \dfrac{B_{br} - 1.3 F_{cy}}{0.29 D_{br}}$	$\dfrac{1}{n_y}\left(B_{br} - 0.29 D_{br}\dfrac{h}{t}\right)$	$\dfrac{h}{t} = \dfrac{k_1 B_{br}}{0.29 D_{br}}$	$\dfrac{k_2\sqrt{B_{br} E}}{n_y(0.29 h/t)}$
SHEAR IN WEBS, gross section	Unstiffened flat webs	20	$\dfrac{F_{sy}}{n_y}$	$\dfrac{h}{t} = \dfrac{B_s - F_{sy}}{1.25 D_s}$	$\dfrac{1}{n_y}\left(B_s - 1.25 D_s\dfrac{h}{t}\right)$	$\dfrac{h}{t} = \dfrac{C_s}{1.25}$	$\dfrac{\pi^2 E}{n_y(1.25 h/t)^2}$
	Stiffened flat webs $a_e = a_1/\sqrt{1 + 0.7(a_1/a_2)^2}$	21	$\dfrac{F_{sy}}{n_y}$	$\dfrac{a_e}{t} = \dfrac{B_s - (n_a F_{sy}/n_y)}{1.25 D_s}$	$\dfrac{1}{n_a}\left(B_s - 1.25 D_s\dfrac{a_e}{t}\right)$	$\dfrac{a_e}{t} = \dfrac{C_s}{1.25}$	$\dfrac{\pi^2 E}{n_a(1.25 a_e/t)^2}$

Allowable Stresses for BUILDING and Similar Type Structures

6061-T6, -T651, -T6510, -T6511

Extrusions up thru 1 in., Sheet & Plate, Pipe, Standard Structural Shapes, Drawn Tube, Rolled Rod and Bar, 6351-T5 Extrusions

WHITE BARS apply to nonwelded members and to welded members at locations farther than 1.0 in. from a weld.

SHADED BARS apply within 1.0 in. of a weld.

Equations that straddle the shaded and unshaded areas apply to both.

*For tubes with circumferential welds, equations of Sections 3.4.10, 3.4.12, and 3.4.16.1 apply for $R_b/t \le 20$.

Type of Stress	Type of Member or Component	Sec. 3.4.	Allowable Stress (white)	Allowable Stress (shaded)
TENSION, axial	Any tension member	1	19	11†
TENSION IN BEAMS, extreme fiber, net section	Rectangular tubes, structural shapes bent around strong axis	2	19	11†
	Round or oval tubes	3	24	13.5†
	Shapes bent about weak axis, bars, plates	4	28	16†
BEARING	On rivets and bolts	5	39	23†
	On flat surfaces and pins and on bolts in slotted holes	6	26	16†

Type of Stress	Type of Member or Component	Sec. 3.4.	Allowable Stress Slenderness ≤ S_1	Slenderness Limit S_1	Allowable Stress Slenderness Between S_1 and S_2	Slenderness Limit S_2	Allowable Stress Slenderness ≥ S_2
COMPRESSION IN COLUMNS, axial, gross section	All columns	7	—	$kL/r = 0$	$20.2 - 0.126\,(kL/r)$	$kL/r = 66$	$51{,}000/(kL/r)^2$
			12†	—	12†	$kL/r = 65$†	$51{,}000/(kL/r)^2$†
COMPRESSION IN COMPONENTS OF COLUMNS gross section	Flat plates supported along one edge - columns buckling about a symmetry axis	8	21	$b/t = 2.7$	$23.1 - 0.79\,(b/t)$	$b/t = 10$	$154/(b/t)$
			12†	—	12†	$b/t = 13$†	$154/(b/t)$†
	Flat plates supported along one edge - columns not buckling about a symmetry axis	8.1	21	$b/t = 2.7$	$23.1 - 0.79\,(b/t)$	$b/t = 12$	$1970(b/t)^2$
			12†	—	12†	$b/t = 13$†	$1970(b/t)^2$†
COMPRESSION IN COMPONENTS OF COLUMNS gross section	Flat plates with both edges supported	9	21	$b/t = 8.4$	$23.1 - 0.25\,(b/t)$	$b/t = 33$	$490(b/t)$
			12†	—	12†	$b/t = 41$†	$490(b/t)$†
	Flat plates with one edge supported and other edge with stiffener	9.1	See Section 3.4.9.1				
	Flat plates with both edges supported and with an intermediate stiffener	9.2	See Section 3.4.9.2				
	Curved plates supported on both edges, walls of round or oval tubes	10*	21	$R_b/t = 2.2$	$22.2 - 0.80\sqrt{R_b/t}$	$R_b/t = 141$	$\dfrac{3200}{(R_b/t)(1 + \sqrt{R_b/t}/35)^2}$
			12†	$R_b/t = 9.0$	$13.5 - 0.50\sqrt{R_b/t}$	$R_b/t = 290$	$490(b/t)$†

Type of Stress	Type of Member or Component	Sec. 3.4.	Allowable Stress Slenderness ≤ S₁	Slenderness Limit S₁	Allowable Stress Slenderness Between S₁ and S₂	Slenderness Limit S₂	Allowable Stress Slenderness ≥ S₂
COMPRESSION IN BEAMS, extreme fiber gross section	Single web beams bent about strong axis	11	21 / 10.5	$L_b/r_y = 23$ / —	$23.9 - 0.124 L_b/r_y$ / 10.5	$L_b/r_y = 79$ / $L_b/r_y = 91$	$\dfrac{87,000}{(L_b/r_y)^2}$
	Round or oval tubes	12*	25 / 12	$R_b/t = 28$ / $R_b/t = 53$	$39.3 - 2.70\sqrt{R_b/t}$ / $18.6 - 1.00\sqrt{R_b/t}$	$R_b/t = 81$ / $R_b/t = 134$	Same as Section 3.4.10
	Solid rectangular and round section beams	13	28 / 13.5	$d/t\sqrt{L_b/d} = 13$ / —	$40.5 - 0.93 d/t\sqrt{L_b/d}$ / 13.5	$d/t\sqrt{L_b/d} = 29$ / $d/t\sqrt{L_b/d} = 29$	$\dfrac{11,400}{(d/t)^2(L_b/d)}$
	Rectangular tubes and box sections	14	21 / 10.5	$L_b S_c/.5\sqrt{I_y J} = 146$ / —	$23.9 - 0.24\sqrt{L_b S_c/.5\sqrt{I_y J}}$ / 10.5	$L_b S_c/.5\sqrt{I_y J} = 1700$ / $L_b S_c/.5\sqrt{I_y J} = 2290$	$\dfrac{24,000}{(L_b S_c/.5\sqrt{I_y J})}$
	Flat plates supported on one edge	15	21 / 10.5	$b/t = 6.8$ / —	$27.3 - 0.93 b/t$ / 10.5	$b/t = 10$ / $b/t = 17$	$183/(b/t)$ / **183/(b/t)**
COMPRESSION IN COMPONENTS OF BEAMS, (component under uniform compression), gross section	Flat plates with both edges supported	16	21 / 13.5	$b/t = 22$ / —	$27.3 - 0.29 b/t$ / 10.5	$b/t = 33$ / $b/t = 55$	$580/(b/t)$ / **580(b/t)**
	Curved plates supported on both edges	16.1*	25 / 12	$R_b/t = 1.6$ / $R_b/t = 1.3$	$26.2 - 0.94\sqrt{R_b/t}$ / $12.4 - 0.35\sqrt{R_b/t}$	$R_b/t = 141$ / $R_b/t = 340$	$\dfrac{3800}{(R_b/t)(1+\sqrt{R_b/t}/35)^2}$
	Flat plates with one edge supported and the other edge with stiffener	16.2	See Section 3.4.16.2				
	Flat plates with both edges supported and with an intermediate stiffener	16.3	See Section 3.4.16.3				
COMPRESSION IN COMPONENTS OF BEAMS, (component under bending in own plane) gross section	Flat plates with compression edge free, tension edge supported	17	28 / 13.5	$b/t = 8.9$ / —	$40.5 - 1.41 b/t$ / 13.5	$b/t = 19$ / $b/t = 19$	$4,900/(b/t)^2$ / **4,000/(b/t)²**
	Flat plate with both edges supported	18	28 / 13.5	$h/t = 46$ / —	$40.5 - 0.27 h/t$ / 13.5	$h/t = 75$ / $h/t = 113$	$1520/(h/t)$ / **1520(h/t)**
	Flat plate with horizontal stiffener, both edges supported	19	28 / 13.5	$h/t = 107$ / —	$40.5 - 0.117 h/t$ / 13.5	$h/t = 173$ / $h/t = 260$	$3500/(h/t)$ / **3500(h/t)**
SHEAR IN WEBS, gross section	Unstiffened flat webs	20	12 / 6	$h/t = 36$ / —	$15.6 - 0.099 h/t$ / 6	$h/t = 65$ / $h/t = 81$	$39,000/(h/t)^2$ / **39,000/(h/t)²**
	Stiffened flat webs $a_e = a_1/\sqrt{1 - 0.7(a_1/a_2)^2}$	21	12 / 6	— / —	12 / 6	$a_1/t = 66$ / $a_1/t = 94$	$53,000/(a_e/t)^2$ / **53,000/(a_e/t)²**

Figure 5.43 Requirements of the Aluminum *Specification* for yielding and fracture of beams, for the specific case of alloy 6061-T6.

179

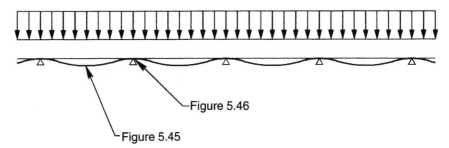

Figure 5.44 Continuous load applied over multiple spans to a trapezoidal formed-sheet beam.

The first is *overall buckling* of the beam as a member. This is called *lateral* or *lateral torsional buckling.* In this failure mode, the material on the compression side of the neutral axis buckles laterally. Meanwhile, the tensile force on the tension side of the bending axis tends to hold the tension flange straight. The net effect is a combination of lateral displacement and twisting (torsion). An example is an I-beam bent about its strong axis (Figure 5.47).

The second type of buckling that can affect a beam is *local buckling,* which occurs when an element of the beam is slender enough to buckle under the compressive stress imposed on it by bending. (In the Aluminum *Specification,* elements of beams are called *components of beams.*) An example is local buckling of the flange of an I-beam due to the nearly uniform compressive stress acting on it (Figure 5.48). These buckles are considered to be local because they occur over a length roughly equal to the width of the element. As you might expect, the strength of elements of beams that are under es-

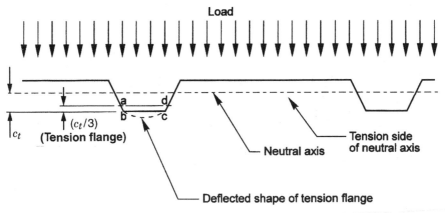

Figure 5.45 Cross section of a trapezoidal formed-sheet beam at midspan, where the bottom flange is in tension.

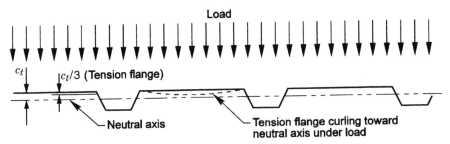

Figure 5.46 Cross section of a trapezoidal formed-sheet beam at an interior support, where the loaded (top) flange is in tension.

sentially uniform compression, like the flange of an I-beam bent about its major axis, is very similar to the strength of elements of columns.

Just as we did for axial compression, we must check the possibility of both types of buckling—overall and local buckling—for a given member. The possibility of local buckling isn't a quirk unique to aluminum. It's just that the Aluminum *Specification* addresses the full spectrum of element slenderness, unlike the Steel *Specification*, which places limits on the ratios of width to thickness of elements and then relegates exceptions to an appendix (B). All cross sections, regardless of how slender their component elements, may be designed with the Aluminum *Specification*. Now we'll show how.

Lateral Buckling Lateral buckling can occur only if the beam in question has a cross section with a larger moment of inertia about one principal axis than the other, and the beam is bent about the axis of the greater moment of

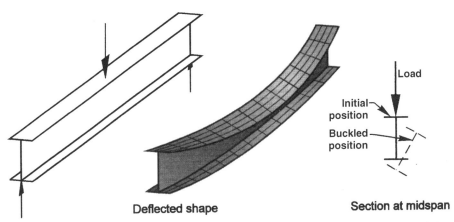

Figure 5.47 Lateral buckling of an I-beam. Note that lateral displacement of the compression flange results in twisting of the section.

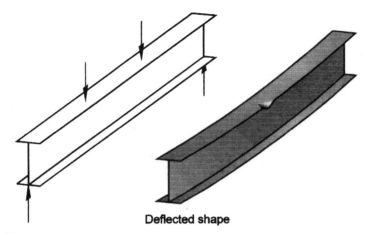

Figure 5.48   Local buckling of the compression flange of an I-beam.

inertia. For example, a wooden 2 × 4 spanning between two points of support and laying on the 4-inch face won't buckle no matter how much load you apply (Figure 5.49). There is no position it can assume with respect to bending that would be more stable than its initial orientation. Stand it up on the 2-inch face, however, and you can load it to the buckling point fairly readily. Similarly, round tubes cannot buckle laterally under bending moment because their section modulus is the same about a vertical or a horizontal axis. (For this reason, the Aluminum *Specification* bending strengths for round tube

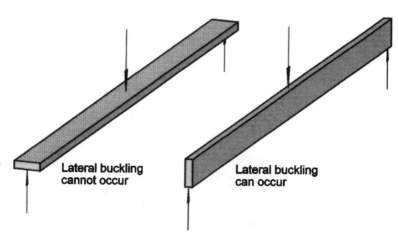

Figure 5.49   A 2 × 4 subject to bending about each of its principal axes.

beams are independent of the length of the beam; see *Specification* Section 3.4.12.)

Once you deal with the cases where overall buckling need not be considered, you have to consider situations where it can occur. After all, you're more likely to place a beam so that it is bent about its strong axis since the beam has a greater moment of inertia to resist loads this way. Lateral buckling strengths can be more conveniently expressed if the type of beam shape (I-beam, rectangular plate, or box section) is given. The Aluminum *Specification,* therefore, provides separate expressions for each of these cross-sectional shapes and summarizes them in *Specification* Table 3.4-3. This table is reproduced here as Figure 5.50 with the lateral buckling portions highlighted. The form of the equations for 6061-T6 is highlighted in Figure 5.51.

As mentioned above, the provisions for round tubes (*Specification* Section 3.4.12) are not a function of the length of the beam because these shapes do not buckle laterally. The other provisions (*Specification* Sections 3.4.11, 13, and 14) *are* a function of beam length since these shapes can buckle laterally. The beam length to be used in calculating beam buckling strength is denoted L_b in the Aluminum *Specification,* just as for hot-rolled steel. L_b is the length of the beam between points of lateral support of the compression flange or restraint against twisting of the cross section (not the length of the compression flange between points of inflection).

You can stop at this point, use these provisions just as they appear in Aluminum *Specification* Sections 3.4.11, 13, and 14, and skip ahead to local buckling of beam elements. However, doing so will be conservative, especially when:

1) The bending moment near the middle of the unbraced segment is less than the maximum moment in the segment
2) L_b/r_y is greater than about 50 for single web beams
3) The transverse load acts in a direction away from the shear center of the beam. An example of this is shown in Figure 5.52. Here the load tends to counteract the lateral buckling.

If, on the other hand, you're willing to invest some more effort to design more efficiently, read on.

Accounting for Variation in Bending Moment (C_b) A beam with a constant moment over its length (Figure 5.53a) is more likely to buckle than the same beam with equal and opposite bending moments at its ends and a point of inflection at midspan (Figure 5.53b). This effect of variation of bending moment (also called *moment gradient*) along the span is addressed by the bending coefficient C_b. The Aluminum *Specification* enables you to account for this by reducing the unbraced length of the beam (L_b) for Sections 3.4.11, 3.4.13, and 3.4.14, using the bending coefficient C_b.

Figure 5.50 General requirements of the Aluminum *Specification* for overall (lateral) buckling of beams.

Type of Stress	Type of Member or Component	Sec. 3.4.	Allowable Stress Slenderness ≤ S_1	Slenderness Limit S_1	Allowable Stress Slenderness Between S_1 and S_2	Slenderness Limit S_2	Allowable Stress Slenderness ≥ S_2
COMPRESSION IN BEAMS, extreme fiber gross section	Single web beams bent about strong axis	11	21	$L_t/r_y = 23$	$23.9 - 0.124 L_t/r_y$	$L_t/r_y = 79$	$\dfrac{87,000}{(L_b/r_y)^2}$
	Round or oval tubes	12*	25	$R_b/t = 28$	$39.3 - 2.70\sqrt{R_b/t}$	$R_b/t = 81$	Same as Section 3.4.10
	Solid rectangular and round section beams	13	28	$d/t\sqrt{L_b/d} = 13$	$40.5 - 0.93 d/t\sqrt{L_b/d}$	$d/t\sqrt{L_b/d} = 29$	$\dfrac{11,400}{(d/t)^2 (L_b/d)}$
	Rectangular tubes and box sections	14	21	$L_b S_c / .5\sqrt{I_y J} = 146$	$23.9 - 0.24[L_b S_c / .5\sqrt{I_y J}]$	$L_b S_c / .5\sqrt{I_y J} = 1700$	$\dfrac{24,000}{(L_b S_c / .5\sqrt{I_y J})}$
	Flat plates supported on one edge	15	21	$b/t = 6.8$	$27.3 - 0.93\, b/t$	$b/t = 10$	$182/(b/t)$
COMPRESSION IN COMPONENTS OF BEAMS, (component under uniform compression), gross section	Flat plates with both edges supported	16	21	$b/t = 22$	$27.3 - 0.29\, b/t$	$b/t = 33$	$580/(b/t)$
	Curved plates supported on both edges	16.1*	25	$R_b/t = 1.6$	$26.2 - 0.94\sqrt{R_b/t}$	$R_b/t = 141$	$(R_b/t)(1 + \sqrt{R_b/t}/35)^2$
	Flat plates with one edge supported and the other edge with stiffener	16.2		See Section 3.4.16.2				
	Flat plates with both edges supported and with an intermediate stiffener	16.3		See Section 3.4.16.3				
COMPRESSION IN COMPONENTS OF BEAMS, (component under bending in own plane) gross section	Flat plates with compression edge free, tension edge supported	17	28	$b/t = 8.9$	$40.5 - 1.41\, b/t$	$b/t = 19$	$4,900/(b/t)^2$
	Flat plate with both edges supported	18	28	$h/t = 46$	$40.5 - 0.27\, h/t$	$h/t = 75$	$1520/(h/t)$
	Flat plate with horizontal stiffener, both edges supported	19	28	$h/t = 107$	$40.5 - 0.117\, h/t$	$h/t = 173$	$3500/(h/t)$
SHEAR IN WEBS, gross section	Unstiffened flat webs	20	12	$h/t = 36$	$15.6 - 0.099\, h/t$	$h/t = 65$	$39,000/(h/t)^2$
	Stiffened flat webs $a_e = a_1/\sqrt{1 + 0.7(a_1/a_2)^2}$	21	12	—	12	$a_e/t = 66$	$53,000/(a_e/t)^2$

Figure 5.51 Requirements of the Aluminum *Specification* for overall (lateral) buckling of beams, for the specific case of alloy 6061-T6.

VII-67

185

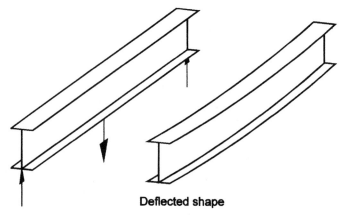

Figure 5.52 Bending load acting in a direction away from the shear center of the beam.

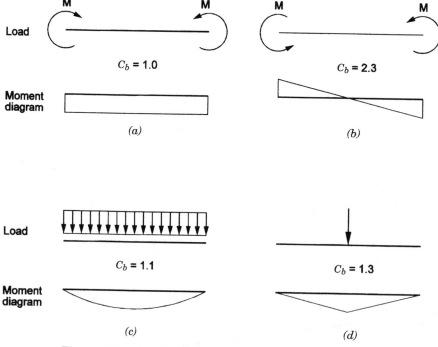

Figure 5.53 Bending coefficient C_b for selected beam cases.

The C_b term is also used by the Steel *Specification* for this purpose. The equation used for C_b in the Aluminum *Specification* for doubly symmetric sections is the same as that used in the 2nd edition of the LRFD Steel *Specification*:

$$C_b = \frac{12.5 \, M_{MAX}}{2.5 \, M_{MAX} + 3 \, M_A + 4 \, M_B + 3 \, M_C} \tag{5.25}$$

where:

M_{MAX} = absolute value of maximum moment in the unbraced beam segment

M_A = absolute value of moment at quarter-point of the unbraced beam segment

M_B = absolute value of moment at midpoint of the unbraced beam segment

M_C = absolute value of moment at three-quarter-point of the unbraced beam segment.

Section 4.9 of the Aluminum *Specification* gives C_b factors for various cases and some others are given here in Figure 5.53. C_b ranges from 1.0 to 2.3.

For single web beams, such as I-beams and channels (*Specification* Section 3.4.11), the unbraced length of the beam (L_b) is divided by $\sqrt{C_b}$. For solid rectangular and round section beams (*Specification* Section 3.4.13) and rectangular tubes (*Specification* Section 3.4.14), the unbraced length of the beam is divided by C_b. To put this in perspective, when C_b is at its greatest value (2.3), you derive the biggest benefit, factoring L_b by 0.66 for single web beams and by 0.43 for the other cases.

One other note: Remember that even though we use the moments at the quarter points along the span to calculate C_b, M_{MAX} is the bending moment used to calculate the compressive bending stress to be compared to the allowable bending stress.

Although the LRFD Steel *Specification* uses the identical equation for C_b as the Aluminum *Specification*, the 9th edition of the Steel ASD *Specification* uses an equation that gives just slightly different results:

$$C_b = 1.75 + 1.05 \, (M_1/M_2) + 0.3 \, (M_1/M_2)^2 \leq 2.3$$

where:

M_1 = smaller of the end moments

M_2 = larger of the end moments.

M_1/M_2 is positive when M_1 and M_2 have the same sign (*reverse curvature bending*) and negative when M_1 and M_2 have opposite signs (*single curvature bending*).

A More Precise Lateral Buckling Expression   Aluminum *Specification* Section 3.4.11 approximates the resistance to lateral torsional elastic buckling of single web beams as:

$$\text{elastic buckling strength} = \frac{\pi^2 E}{\left(\dfrac{L_b}{1.2 r_y}\right)^2} \qquad (5.26)$$

In this equation, $1.2 r_y$ (r_y being the minor axis radius of gyration of the beam) replaces a more complicated (and more accurate) expression for the lateral torsional resistance of the beam. The strength is more accurately calculated using an effective minor axis radius of gyration (r_{ye}) that includes the resistance of the beam cross section to torsion and to warping. Resistance to warping contributes significantly to the lateral buckling strength of open sections, such as I-beams. The Aluminum *Specification* enables you to include the warping and torsion resistance effects on lateral buckling strength of single web beams and beams containing tubular shapes by using *Specification* Section 4.9.1, and more accurately still using Section 4.9.3. A comparison of the results from *Specification* Sections 3.4.11, 4.9.1, and 4.9.3 is made in *Aluminum Design Manual*, Part VIII, Example 28, for a singly symmetric single web beam. The results for this example are summarized in Table 5.7. The effect of moment gradient (C_b) is not included in Table 5.7 in order to isolate the effect of using the more precise provisions (Sections 4.9.1 and 4.9.3) for lateral buckling strength.

The effect of changes in the r_y value used is magnified because it is squared in the elastic buckling strength expression (Equation 5.26).

The provisions of *Specification* Section 4.9.1 apply to doubly symmetric sections and sections symmetric about the bending axis. For a section of such beams at either: 1) brace or support points, or 2) between brace or support points if the span is subjected only to either end moments or transverse loads applied at the neutral axis:

$$r_{ye} = \frac{\sqrt{C_b}}{1.7} \sqrt{\frac{I_y d}{S_c}} \sqrt{1 + 0.152 \frac{J}{I_y} \left(\frac{k_y L_b}{d}\right)^2}$$

To check beam spans between brace or support points for beams subjected

TABLE 5.7   Comparison of Methods for Determining Lateral Buckling Strength in Example 28 of the *Aluminum Design Manual*

Section Used	r_y (in.)	Buckling Strength (ksi)
3.4.11	0.50	3.4
4.9.1	0.89	10.9
4.9.3	0.95	12.5

to transverse loads applied on the top or bottom flange (where the load is free to move laterally with the beam if the beam buckles):

$$r_{ye} = \frac{\sqrt{C_b}}{1.7} \sqrt{\frac{I_y d}{S_c} \left(\pm 0.5 + \sqrt{1.25 + 0.152 \frac{J}{I_y} \left(\frac{k_y L_b}{d} \right)^2} \right)} \qquad (5.27)$$

where:

r_{ye} = effective radius of gyration to be used in *Specification* Section 3.4.11 in place of r_y

C_b = bending moment coefficient (discussed above)

I_y = beam's minor axis moment of inertia

S_c = compression side section modulus of the beam for major axis bending

J = torsion constant of the cross section (see Section 5.4)

k_y = effective length coefficient for the compression flange about the minor axis (see below)

L_b = length of the beam between bracing points or between a brace point and the free end of a cantilever beam. Bracing points are the points at which the compression flange is restrained against lateral movement or twisting.

d = depth of the beam.

In Equation 5.27, the minus sign in front of the 0.5 term is used when the transverse load acts toward the shear center of the beam cross section, and the plus sign is used when the load acts away from the shear center. This provision accounts for the effect of the point of load application. For example, a simple span beam with a gravity load hung from the bottom flange (for which +0.5 is used in Equation 5.27) will sustain a higher bending moment than the same beam with the load applied to the top (compression) flange (for which −0.5 is used).

As these equations show, the more accurate equations for lateral buckling in Aluminum *Specification* Section 4.9 include an effective length factor (k_y) to be applied to L_b, the unbraced length. This k_y factor depends on the restraint against rotation about the minor axis at the ends of the unbraced length. (This is the vertical axis for a beam with its strong axis horizontal.) If both ends are restrained against rotation, k_y is 0.5; if they are not restrained, k_y is 1.0. In reality, full restraint is difficult to achieve, so k_y is typically assumed to be 1.

Since *Specification* Section 3.4.11 allows the use of C_b in modifying L_b, and Section 4.9 also includes C_b in the expression for effective r_y, you need to be careful to avoid using C_b twice. If you use r_{ye} from *Specification* Section 4.9 in Section 3.4.11, then you can't also modify L_b by $1/\sqrt{C_b}$. Use it once only—no double dipping. Also, if rotation about the weak axis is considered restrained (so a k_y less than 1 is used), C_b should be taken as 1 because Equation 5.25 overestimates C_b when a k_y value less than 1 is used.

Local Buckling of Beam Elements   The discussion in Section 5.2.2 on the local buckling of elements of columns applies here. If you haven't read that yet, you should now. Definitions of elements, their widths, and edge supports are the same whether an element is in a beam or a column.

As we mentioned previously, the strength of beam elements is very similar to the strength of column elements. The only difference arises because the stress distribution on a beam element may not be uniform compression, but rather a variation in compression over the width of the element. Consider, for example, the web of an I-beam bent about its major axis. The stress distribution is shown in Figure 5.54. There is both tension and compression in the element, but it is the compression that can cause local buckling.

The Aluminum *Specification,* therefore, divides the local buckling provisions for beam elements into two groups: beam elements under uniform compression, and beam elements under in-plane bending. These provisions are highlighted in Figure 5.55, which shows Aluminum *Specification* Table 3.4-3. The equations are evaluated for alloy 6061-T6 in the highlighted portion of Figure 5.56, from *Aluminum Design Manual*, Part VII, Table 2-21. The *Specification* Sections that deal with beam elements are summarized in Tables 5.8 and 5.9.

If a horizontal stiffener is used for an element designed in accordance with *Specification* Section 3.4.19, (for example, on the web of an I-beam), you can find the requirements for its design in Aluminum *Specification* Section 4.5.

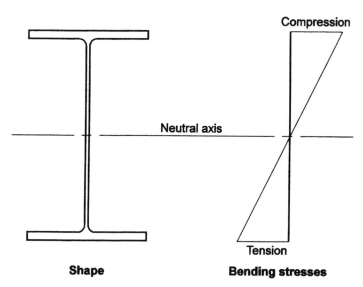

Shape   **Bending stresses**

Figure 5.54   Stress distribution in the web of an I-beam bent about its strong axis.

Type of Stress	Type of Member or Component	Sub-Sec.	Allowable Stress Slenderness ≤ S₁	Slenderness Limit S₁	Allowable Stress Slenderness between S₁ and S₂	Slenderness Limit S₂	Allowable Stress Slenderness ≥ S₂
COMPRESSION IN BEAMS, extreme fiber gross section	Single web beams bent about strong axis	11		$\dfrac{L_b}{r_y} = 1.2\,C_c$	$\dfrac{1}{n_y}\left(B_c - \dfrac{D_c L_b}{1.2\,r_y}\right)$	$\dfrac{L_b}{r_y} = 1.2\,C_c$	$\dfrac{\pi^2 E}{n_y\left(\dfrac{L_b}{1.2\,r_y}\right)^2}$
	Round or oval tubes	12*	$\dfrac{1.17 F_{cy}}{n_y}$	$\dfrac{R_b}{t} = \left(\dfrac{B_{tb} - 1.17 F_{cy}}{D_{tb}}\right)^2$	$\dfrac{1}{n_y}\left(B_{tb} - D_{tb}\sqrt{\dfrac{R_b}{t}}\right)$	$\dfrac{R_b}{t} = \left[\dfrac{\frac{n_s B_b - B_t}{n_s}}{\left(\frac{2 D_{tb} - D_t}{n_s}\right)}\right]^{2b}$	Same as Section 3.4.10
	Solid rectangular and round section beams	13	$\dfrac{1.3 F_{cy}}{n_y}$	$\dfrac{d}{t}\sqrt{\dfrac{L_b}{d}} = \dfrac{B_{br} - 1.3 F_{cy}}{2.3 D_{br}}$	$\dfrac{1}{n_y}\left(B_{br} - 2.3 D_{br}\,\dfrac{d}{t}\sqrt{\dfrac{L_b}{d}}\right)$	$\dfrac{d}{t}\sqrt{\dfrac{L_b}{d}} = \dfrac{C_{br}}{2.3}$	$\dfrac{\pi^2 E}{5.29\,n_y\left(\dfrac{d}{t}\right)^2\dfrac{L_b}{d}}$
	Rectangular tubes and box sections	14	$\dfrac{F_{cy}}{n_y}$	$\dfrac{L_b S_c}{0.5\sqrt{I_y J}} = \dfrac{(B_c - F_{cy})^2}{1.6 D_c}$	$\dfrac{1}{n_y}\left(B_c - 1.6 D_c\sqrt{\dfrac{L_b S_c}{0.5\sqrt{I_y J}}}\right)$	$\dfrac{L_b S_c}{0.5\sqrt{I_y J}} = \left(\dfrac{C_c}{1.6}\right)^2$	$\dfrac{\pi^2 E}{2.56\,n_y\left(\dfrac{L_b S_c}{0.5\sqrt{I_y J}}\right)}$
COMPRESSION IN COMPONENTS OF BEAMS, (component under uniform compression), gross section	Flat plates supported on one edge	15	$\dfrac{F_{cy}}{n_y}$	$\dfrac{b}{t} = \dfrac{B_p - F_{cy}}{5.1 D_p}$	$\dfrac{1}{n_y}\left(B_p - 5.1 D_p\,\dfrac{b}{t}\right)$	$\dfrac{b}{t} = \dfrac{k_1 B_p}{5.1 D_p}$	$\dfrac{k_2\sqrt{B_p E}}{n_y(5.1\,b/t)}$
	Flat plates with both edges supported	16	$\dfrac{F_{cy}}{n_y}$	$\dfrac{b}{t} = \dfrac{B_p - F_{cy}}{1.6 D_p}$	$\dfrac{1}{n_y}\left(B_p - 1.6 D_p\,\dfrac{b}{t}\right)$	$\dfrac{b}{t} = \dfrac{k_1 B_p}{1.6 D_p}$	$\dfrac{k_2\sqrt{B_p E}}{n_y(1.6\,b/t)}$
	Curved plates supported on both edges	16.1*	$\dfrac{1.17 F_{cy}}{n_y}$	$\dfrac{R_b}{t} = \left(\dfrac{B_t - 1.17 F_{cy}}{D_t}\right)^2$	$\dfrac{1}{n_y}\left(B_t - D_t\sqrt{\dfrac{R_b}{t}}\right)$	$\dfrac{R_b}{t} = C_t$	$\dfrac{\pi^2 E}{16 n_y\left(\dfrac{R_b}{t}\right)\left(1 + \dfrac{\sqrt{R_b/t}}{35}\right)^2}$
	Flat plates with one edge supported and the other edge with stiffener	16.2	See Section 3.4.16.2				
	Flat plates with both edges supported and with an intermediate stiffener	16.3	See Section 3.4.16.3				
COMPRESSION IN COMPONENTS OF BEAMS, (component under bending in own plane) gross section	Flat plates with compression edge free, tension edge supported	17	$\dfrac{1.3 F_{cy}}{n_y}$	$\dfrac{b}{t} = \dfrac{B_{br} - 1.3 F_{cy}}{3.5 D_{br}}$	$\dfrac{1}{n_y}\left(B_{br} - 3.5 D_{br}\,\dfrac{b}{t}\right)$	$\dfrac{b}{t} = \dfrac{C_{br}}{3.5}$	$\dfrac{\pi^2 E}{n_y(3.5\,b/t)^2}$
	Flat plate with both edges supported	18	$\dfrac{1.3 F_{cy}}{n_y}$	$\dfrac{h}{t} = \dfrac{B_{br} - 1.3 F_{cy}}{0.67 D_{br}}$	$\dfrac{1}{n_y}\left(B_{br} - 0.67 D_{br}\,\dfrac{h}{t}\right)$	$\dfrac{h}{t} = \dfrac{k_1 B_{br}}{0.67 D_{br}}$	$\dfrac{k_2\sqrt{B_{br} E}}{n_y(0.67\,h/t)}$
	Flat plate with horizontal stiffener, both edges supported	19	$\dfrac{1.3 F_{cy}}{n_y}$	$\dfrac{h}{t} = \dfrac{B_{br} - 1.3 F_{cy}}{0.29 D_{br}}$	$\dfrac{1}{n_y}\left(B_{br} - 0.29 D_{br}\,\dfrac{h}{t}\right)$	$\dfrac{h}{t} = \dfrac{k_1 B_{br}}{0.29 D_{br}}$	$\dfrac{k_2\sqrt{B_{br} E}}{n_y(0.29\,h/t)}$
SHEAR IN WEBS, gross section	Unstiffened flat webs	20	$\dfrac{F_{sy}}{n_y}$	$\dfrac{h}{t} = \dfrac{B_s - F_{sy}}{1.25 D_s}$	$\dfrac{1}{n_y}\left(B_s - 1.25 D_s\,\dfrac{h}{t}\right)$	$\dfrac{h}{t} = \dfrac{C_s}{1.25}$	$\dfrac{\pi^2 E}{n_y(1.25\,h/t)^2}$
	Stiffened flat webs $a_e = a_1/\sqrt{1 + 0.7(a_1/a_2)^2}$	21	$\dfrac{F_{sy}}{n_y}$	$\dfrac{a_e}{t} = \dfrac{B_s - (a_e F_{sy}/n_y)}{1.25 D_s}$	$\dfrac{1}{n_a}\left(B_s - 1.25 D_s\,\dfrac{a_e}{t}\right)$	$\dfrac{a_e}{t} = \dfrac{C_s}{1.25}$	$\dfrac{\pi^2 E}{n_a(1.25\,a_e/t)^2}$

Figure 5.55 General requirements of the Aluminum *Specification* for local buckling of beam elements.

I-A-25

191

Type of Stress	Type of Member or Component	Sec. 3.4.	Allowable Stress Slenderness ≤ S₁	Slenderness Limit S₁	Allowable Stress Slenderness Between S₁ and S₂	Slenderness Limit S₂	Allowable Stress Slenderness ≥ S₂
COMPRESSION IN BEAMS, extreme fiber gross section	Single web beams bent about strong axis	11	21	$L_b/r_y = 23$	$23.9 - 0.124 L_b/r_y$	$L_b/r_y = 79$	$\dfrac{87,000}{(L_b/r_y)^2}$
			12†		13†		
	Round or oval tubes	12*	25	$R_b/t = 28$	$39.3 - 2.70\sqrt{R_b/t}$	$R_b/t = 81$	Same as Section 3.4.10
			14†	$R_b/t = 51$	$23.5 - 1.39\sqrt{R_b/t}$	$R_b/t = 13†$	
	Solid rectangular and round section beams	13	28	$d/t\sqrt{L_b/d} = 13$	$40.5 - 0.93 d/t\sqrt{L_b/d}$	$d/t\sqrt{L_b/d} = 29$	$\dfrac{11,400}{(d/t)^2(L_b/d)}$
			16†	$d/t\sqrt{L_b/d} = 26$	$40.5 - 0.93 d/t\sqrt{L_b/d}$	$d/t\sqrt{L_b/d} = 29$	
	Rectangular tubes and box sections	14	21	$L_b S_c / \sqrt{I_y J} = 146$	$23.9 - 0.24\sqrt{L_b S_c / \sqrt{I_y J}}$	$L_b S_c / \sqrt{I_y J} = 1700$	$\dfrac{24,000}{L_b S_c / \sqrt{I_y J}}$
			12†				
COMPRESSION IN COMPONENTS OF BEAMS, (component under uniform compression), gross section	Flat plates supported on one edge	15	21	$b/t = 6.8$	$27.3 - 0.93 b/t$	$b/t = 10$	$182/(b/t)$
			12†				
	Flat plates with both edges supported	16	21	$b/t = 22$	$27.3 - 0.29 b/t$	$b/t = 33$	$580/(b/t)$
			12†				
	Curved plates supported on both edges	16.1*	25	$R_b/t = 1.6$	$26.2 - 0.94\sqrt{R_b/t}$	$R_b/t = 141$	$\dfrac{3800}{(R_b/t)(1 + \sqrt{R_b/t/35})^2}$
			14†	$R_b/t = 2.5$	$147 - 0.044\sqrt{R_b/t}$	$R_b/t = 490$	
	Flat plates with one edge supported and the other edge with stiffener	16.2			See Section 3.4.16.2		
	Flat plates with both edges supported and with an intermediate stiffener	16.3			See Section 3.4.16.3		
COMPRESSION IN COMPONENTS OF BEAMS, (component under bending in own plane) gross section	Flat plates with compression edge free, tension edge supported	17	28	$b/t = 8.9$	$40.5 - 1.41 b/t$	$b/t = 19$	$4,900/(b/t)^2$
			16†	$b/t = 12$	$40.5 - 1.41 b/t$		
	Flat plate with both edges supported	18	28	$h/t = 46$	$40.5 - 0.27 h/t$	$h/t = 75$	$1520/(h/t)$
			16†				
	Flat plate with horizontal stiffener, both edges supported	19	28	$h/t = 107$	$40.5 - 0.117 h/t$	$h/t = 173$	$3500/(h/t)$
			16†				
SHEAR IN WEBS, gross section	Unstiffened flat webs	20	12	$h/t = 36$	$15.6 - 0.099 h/t$	$h/t = 65$	$39,000/(h/t)^2$
			12†				
	Stiffened flat webs $a_e = a_1 / \sqrt{1 + 0.7(a_1/a_2)^2}$	21	12	—	12	$a_1/t = 66$	$53,000/(a_1/t)^2$
			12†				

Figure 5.56 Requirements of the Aluminum *Specification* for local buckling of beam elements, for the specific case of alloy 6061-T6.

TABLE 5.8   Aluminum *Specification* **for Local Buckling of Components of Beams, Component Under Uniform Compression**

Section	Type of Element	Effective Slenderness	Postbuckling Strength Recognized?
3.4.15	Flat plate—one edge supported	$5.1b/t$	Yes
3.4.16.2	Flat plate—one edge supported, one with stiffener		Yes
3.4.16	Flat plate—both edges supported	$1.6b/t$	Yes
3.4.16.3	Flat plate—both edges supported and with intermediate stiffener		No
3.4.16.1[1]	Curved plate—both edges supported		No

[1] *Specification* Section 3.4.16.1 applies to curved components of beams other than tubes, which are covered by Section 3.4.12. An example of an element covered by 3.4.16.1 is the curved portion of a corrugated sheet beam.

This ensures that the stiffener is adequately sized and placed so the provisions of *Specification* Section 3.4.19 are valid for the element being stiffened.

You may be wondering if deciding if an element is under uniform compression or in-plane bending isn't clear cut. In reality, no element in a beam is under purely uniform compression because the compressive stress varies as the distance to the neutral axis varies. However, in the case of the flange of a channel bent about its strong axis, the compression is nearly uniform. The compression varies only slightly across the thickness of the flange, and not at all across the flange width, so we consider the compression uniform. But what about a stiffener on the edge of this flange, as shown in Figure 5.57?

TABLE 5.9   Aluminum *Specification* **for Local Buckling of Components of Beams, Component Under In-Plane Bending**

Section	Type of element	Effective Slenderness	Postbuckling Strength Recognized?
3.4.17	Flat plate—tension edge supported, compression edge free	$3.5b/t$	No
3.4.18	Flat plate—both edges supported	$0.67h/t$	Yes
3.4.19 (4.5)[1]	Flat plate with longitudinal stiffener—both edges supported	$0.29h/t$	Yes

[1] *Specification* Section 4.5 includes requirements for longitudinal (horizontal) stiffeners necessary for Section 3.4.19 to be used.

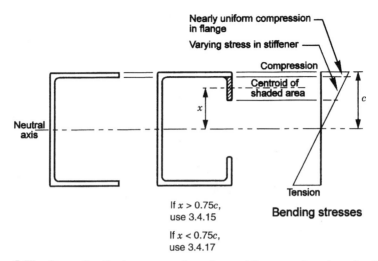

Figure 5.57   Stress distribution assumptions for a stiffener on the edge of a flange.

The rule for such an element cited in the *Aluminum Design Manual*, Illustrative Example 23, is: If the distance from the centroid of the element in question to the beam's neutral axis is greater than 75% of the distance (c) from the neutral axis to the extreme fiber, treat the element as if it were under uniform compression (use *Specification* Section 3.4.15). Otherwise, treat it as if it were under in-plane bending (use *Specification* Section 3.4.17). If the free edge of the element is in tension (Figure 5.58), you may consider both edges to be supported (use *Specification* Section 3.4.18).

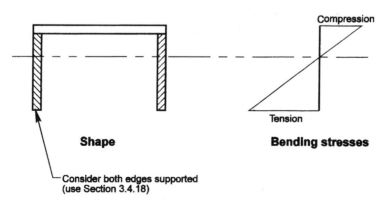

Figure 5.58   An element subject to in-plane bending, with its free edge in tension.

As we mentioned previously for columns, the local buckling strengths are only a function of the dimensions of the member cross section, not its length or end supports. Therefore, you can determine local buckling strengths for elements of beams (like elements of columns) for cross sections and tabulate them for convenience. We've done this for you in Appendix D for Aluminum Association Standard I-beams and Aluminum Association Standard Channels.

Weighted Average Allowable Compressive Stress for Beams As for columns, the lowest local buckling strength of any element of the beam section can be conservatively used as the local buckling strength for the section. The Aluminum *Specification* permits you, however, to calculate the local buckling allowable stress as the weighted average allowable stress of the elements of the section. Through this method, you weight the allowable stress for each element in accordance with the ratio of its area to the total area of the cross section. Using the weighted average allowable stress method produces higher allowable stresses than simply using the lowest strength of the elements.

For purposes of calculating the weighted average allowable stress for a beam, a *flange* is defined as that portion of the cross section that lies more than two-thirds the distance from the neutral axis to the extreme fiber. (This definition of flange does not apply to the weighted average allowable stress for columns, for which the flange is only the actual flange and does not include any part of the web.) Weighted average allowable stresses for standard Aluminum Association shapes used as beams are given in Appendix D.

Interaction Between Local Buckling and Lateral Buckling Just as in column design, where we considered the potential for local buckling to reduce the overall buckling strength, we must consider this interaction for beams. (If you haven't read this part of Section 5.2.2, you should now). As for columns, this potential is a consequence of using the postbuckling strength of beam elements, such as a flange. The method of accounting for this interaction is given in *Specification* Section 4.7.5 and is similar to the treatment for columns, but the *Specification* only requires it be considered for single web beams. The interaction needs to be checked only when:

$$\frac{F_{cr}}{n_y} < F_c$$

where:

n_y = factor of safety on yield strength = 1.65
F_{cr} = local buckling stress (from Table 5.4)
F_c = beam's lateral buckling stress as determined from Section 3.4.11 or 4.9

When this condition is satisfied, the allowable stress may not exceed:

$$F_{rb} = \frac{(F_{eb})^{1/3} (F_{cr})^{2/3}}{n_y}$$

where:

F_{eb} = elastic lateral buckling stress of the beam, which is:

$$F_{eb} = \frac{\pi^2 E}{\left(\dfrac{L_b}{1.2r_y}\right)^2}$$

Plastic Design   *Plastic design* is a design method that utilizes members with the ability to maintain a full plastic moment through large rotations, so that a hinge mechanism can develop, thereby allowing moment redistribution in the structure. Because the members must be able to undergo large rotations without failing by local buckling, they must be *compact* (that is, composed of stocky elements [flanges and webs]). Also, on the tensile side of the neutral axis, the members must be able to sustain large rotations without fracturing, so the material must be ductile.

AISC specifications (38, 39) have compactness requirements to ensure this, but the cold-formed steel (40) and aluminum specifications do not. Since cold-formed steel members are typically composed of slender elements made of sheet, they are generally incapable of developing plastic hinges without local buckling. Members covered by the Aluminum *Specification* may be composed of stocky elements (for example, extrusions having webs and flanges that are thick in relation to their width), or slender elements, which is typical of cold-formed aluminum sheet. Therefore, certain types of aluminum members could develop plastic hinges, but the criteria for their proportions have not been established. Clark (109, Section 10-12) suggests slenderness ratio limits for elements of I, channel, tube, and other shapes for plastic bending that would preclude local buckling, but this has not been incorporated into the Aluminum *Specification*. Eurocode 9 (102) provides criteria for classifying sections capable of forming a plastic hinge in its Annex G.

Web Crippling   Web crippling is a local buckling phenomenon that occurs at a point of concentrated transverse load on a beam (Figure 5.59). The difference between the local buckling of a flat plate element of a beam, such as the flange, and local buckling of a web due to web crippling is that web crippling results from transverse loading along just a short portion of the edge of the element. The Aluminum *Specification* gives provisions for web crippling in Section 4.7.7. The equations given in *Specification* Section 4.7.7 are general and may be applied, for example, to sections formed from bent sheet or to extruded I-beams or channels. A common example is an I-beam web at a beam support. For this case (an extrusion with a single web parallel to the applied transverse load and a fillet at the web-flange juncture [e.g., Figure 5.21]), the equation boils down to:

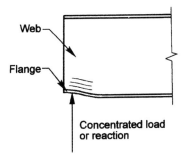

Figure 5.59 Crippling in the web of a beam.

$$P_c = \frac{kt^2\,(0.46F_{cy} + 0.02\,\sqrt{EF_{cy}})(N + C)}{n_y C_3}$$

where:

P_c = allowable reaction or concentrated load
k = 1 for interior reactions or concentrated loads
   = 1.2 for end reactions
t = web thickness
F_{cy} = minimum compressive yield strength of web
E = compressive modulus of elasticity
N = length of bearing at reaction or concentrated load
C = 5.4 in. [140 mm] for interior reactions and concentrated loads
   = 1.3 in. [33 mm] for end reactions
n_y = 1.65 (factor of safety for yielding)
C_3 = 0.4 in. [10 mm]

This equation is derived from the Aluminum *Specification* expressions by setting θ, the angle between the web plane and the bearing surface plane, equal to 90°, and R_i, the inside bend radius at the web flange juncture, equal to 0, as they are for this case. (See *Aluminum Design Manual*, Illustrative Example 4.) For tube members with wall thickness t, the allowable reaction should be calculated for each web separately, rather than using $2t$.

An alternative to making the web heavy enough to resist web crippling is to stiffen it. The required moment of inertia of the stiffener (I_b) is given in Aluminum *Specification* Section 4.6.2:

$$I_b = I_s + \frac{P_{bs}h^2 n_u}{\pi^2 E}$$

where:

I_s = moment of inertia required to resist shear buckling (see *Specification* Section 4.6.1)

P_{bs} = concentrated load on the stiffener

h = clear height of the web between flanges

n_u = factor of safety = 1.95

E = compressive modulus of elasticity

This equation requires that the elastic buckling strength of the stiffener treated as a column of height h is adequate to resist the concentrated load. It does not account for any bracing by attaching the stiffener to the web, and it assumes the stiffener buckles elastically. The yield limit state should also be checked for the stiffeners, and bearing stress between the stiffener and the flange should not exceed that provided by *Specification* Section 3.4.6 (bearing on flat surfaces).

Combined Web Crippling and Bending   When the concentrated load tending to cause web crippling occurs at the same place as a bending moment acting on the beam, the web is subjected to a combined stress state, as shown in Figure 5.60. This condition may be more severe than when either the concentrated load or bending moment act alone. The interaction equation for web crippling and bending is:

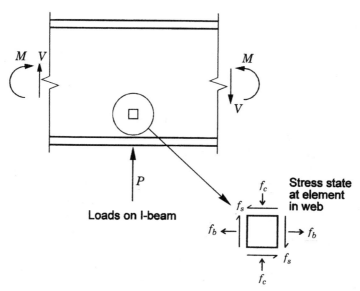

Figure 5.60   Combined stress state for web crippling and bending in the web of a beam.

$$\left(\frac{M}{M_a}\right)^{1.5} + \left(\frac{P}{P_c}\right)^{1.5} \le 1.0$$

where:

M = bending moment applied to the member at the point of application of interior reaction or concentrated load

M_a = allowable bending moment for the member if bending moment alone were applied to the member

P = interior reaction or concentrated load applied to each web

P_c = allowable interior reaction or allowable concentrated load per web calculated according to *Specification* Section 4.7.7.

5.3.3 Bending Shear

Shear also presents the potential for buckling because, as Mohr's circle demonstrates, a plane at 45° to a plane of pure shear is in direct compression (Figure 5.61). So, when an element is loaded to capacity in shear, buckling waves tend to be produced at an angle to horizontal.

The longitudinal and transverse shear stress f_s can be calculated as:

$$f_s = \frac{VQ}{It}$$

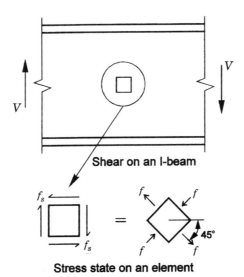

Figure 5.61   Stress state due to shear in the web of a beam.

where:

V = shear force
Q = first moment of the cross-sectional area farther from the neutral axis than the point at which shear stress is calculated
I = moment of inertia of the entire cross-sectional area about the neutral axis
t = thickness of the section at the point at which shear stress is calculated.

In an I-beam, shear stresses are small in the flanges, larger in the web, and a maximum at the neutral axis. The maximum shear stress is often approximated as:

$$f_s = \frac{V}{A_w}$$

where:

V = shear force
A_w = area of the web = ht_w
h = clear height of the web with no deductions for web-flange fillets (Figure 5.35)
t_w = thickness of the web

Flat Plate Elements in Shear   The Aluminum *Specification* includes two provisions for determining allowable shear stress: one for flat plate unstiffened webs (*Specification* Section 3.4.20), and another for flat plate stiffened webs (*Specification* Section 3.4.21). These are highlighted in Figure 5.62, reproduced from Aluminum *Specification* Table 3.4-3. Figure 5.63 is reprinted from *Aluminum Design Manual,* Part VII, Table 2-21, showing the form of these equations for alloy 6061-T6. The slenderness ratio of a flat plate web is h/t, where h is the clear height of the web between the flanges with no deduction for fillets (see Figure 5.35), and t is the thickness of the web.

   In order for the web to be considered stiffened so that the provisions of *Specification* Section 3.4.21 can be used, the stiffeners must meet the requirements of Section 4.6.1, Stiffeners for Web Shear. (These stiffeners aren't vertical unless the beam is horizontal. They're more accurately called *transverse stiffeners* in the Steel *Specification*, Section F4.)

   Neither *Specification* Sections 3.4.20 nor 3.4.21 has postbuckling strength directly included. However, the buckling strength of stiffened webs (3.4.21) indirectly recognizes postbuckling strength by use of a lower factor of safety (1.2 versus 1.65) on the elastic buckling strength. True postbuckling behavior of webs is also called *tension field action*, where the beam is considered to act like a truss (Figure 5.64). The buckled portions of the web are disregarded,

Type of Stress	Type of Member or Component	Sub-Sec.	Allowable Stress Slenderness ≤ S₁	Slenderness Limit S₁	Allowable Stress Slenderness Between S₁ and S₂	Slenderness Limit S₂	Allowable Stress Slenderness ≥ S₂
COMPRESSION IN BEAMS, extreme fiber gross section	Single web beams bent about strong axis	11		$\dfrac{L_b}{r_y} = \dfrac{1.2(B_c - F_{cy})}{D_c}$	$\dfrac{1}{n_y}\left(B_c - \dfrac{D_c L_b}{1.2 r_y}\right)$	$\dfrac{L_b}{r_y} = 1.2 C_c$	$\dfrac{\pi^2 E}{n_y\left(\dfrac{L_b}{1.2 r_y}\right)^2}$
	Round or oval tubes	12*	$\dfrac{1.17 F_{cy}}{n_y}$	$\dfrac{R_b}{t} = \left(\dfrac{B_{tb} - 1.17 F_{cy}}{D_{tb}}\right)^2$	$\dfrac{1}{n_y}\left(B_{tb} - D_{tb}\sqrt{\dfrac{R_b}{t}}\right)$	$\dfrac{R_b}{t} = \left(\dfrac{n_a B_{tb} - B_t}{n_a D_{tb} - D_t}\right)^2$	Same as Section 3.4.10
	Solid rectangular and round section beams	13	$\dfrac{1.3 F_{cy}}{n_y}$	$\dfrac{d}{t}\sqrt{\dfrac{L_b}{d}} = \dfrac{B_{br} - 1.3 F_{cy}}{2.3 D_{br}}$	$\dfrac{1}{n_y}\left(B_{br} - 2.3 D_{br}\dfrac{d}{t}\sqrt{\dfrac{L_b}{d}}\right)$	$\dfrac{d}{t}\sqrt{\dfrac{L_b}{d}} = \dfrac{C_{br}}{2.3}$	$\dfrac{\pi^2 E}{5.29 n_y\left(\dfrac{d}{t}\right)^2\dfrac{L_b}{d}}$
	Rectangular tubes and box sections	14	$\dfrac{F_{cy}}{n_y}$	$\dfrac{L_b S_c}{0.5\sqrt{I_y J}} = \left(\dfrac{B_c - F_{cy}}{1.6 D_c}\right)^2$	$\dfrac{1}{n_y}\left(B_c - 1.6 D_c\sqrt{\dfrac{L_b S_c}{0.5\sqrt{I_y J}}}\right)$	$\dfrac{L_b S_c}{0.5\sqrt{I_y J}} = \left(\dfrac{C_c}{1.6}\right)^2$	$\dfrac{\pi^2 E}{2.56 n_y\dfrac{L_b S_c}{0.5\sqrt{I_y J}}}$
COMPRESSION IN COMPONENTS OF BEAMS, (component under uniform compression), gross section	Flat plates supported on one edge	15	$\dfrac{F_{cy}}{n_y}$	$\dfrac{b}{t} = \dfrac{B_p - F_{cy}}{5.1 D_p}$	$\dfrac{1}{n_y}\left(B_p - 5.1 D_p\dfrac{b}{t}\right)$	$\dfrac{b}{t} = \dfrac{k_1 B_p}{5.1 D_p}$	$\dfrac{k_2\sqrt{B_p E}}{n_y(5.1 b/t)}$
	Flat plates with both edges supported	16	$\dfrac{F_{cy}}{n_y}$	$\dfrac{b}{t} = \dfrac{B_p - F_{cy}}{1.6 D_p}$	$\dfrac{1}{n_y}\left(B_p - 1.6 D_p\dfrac{b}{t}\right)$	$\dfrac{b}{t} = \dfrac{k_1 B_p}{1.6 D_p}$	$\dfrac{k_2\sqrt{B_p E}}{n_y(1.6 b/t)}$
	Curved plates supported on both edges	16.1*	$\dfrac{1.17 F_{cy}}{n_y}$	$\dfrac{R_b}{t} = \left(\dfrac{B_t - 1.17 F_{cy}}{D_t}\right)^2$	$\dfrac{1}{n_y}\left(B_t - D_t\sqrt{\dfrac{R_b}{t}}\right)$	$\dfrac{R_b}{t} = C_t$	$\dfrac{\pi^2 E}{16 n_y\left(\dfrac{R_b}{t}\right)\left[1 + \left(\dfrac{\sqrt{R_b/t}}{35}\right)^2\right]}$
	Flat plates with one edge supported and the other edge with stiffener	16.2			See Section 3.4.16.2		
	Flat plates with both edges supported and with an intermediate stiffener	16.3			See Section 3.4.16.3		
COMPRESSION IN COMPONENTS OF BEAMS, (component under bending in own plane) gross section	Flat plates with compression edge free, tension edge supported	17	$\dfrac{1.3 F_{cy}}{n_y}$	$\dfrac{b}{t} = \dfrac{B_{br} - 1.3 F_{cy}}{3.5 D_{br}}$	$\dfrac{1}{n_y}\left(B_{br} - 3.5 D_{br}\dfrac{b}{t}\right)$	$\dfrac{b}{t} = \dfrac{C_{br}}{3.5}$	$\dfrac{\pi^2 E}{n_y(3.5 b/t)^2}$
	Flat plate with both edges supported	18	$\dfrac{1.3 F_{cy}}{n_y}$	$\dfrac{h}{t} = \dfrac{B_{br} - 1.3 F_{cy}}{0.67 D_{br}}$	$\dfrac{1}{n_y}\left(B_{br} - 0.67 D_{br}\dfrac{h}{t}\right)$	$\dfrac{h}{t} = \dfrac{k_1 B_{br}}{0.67 D_{br}}$	$\dfrac{k_2\sqrt{B_{br} E}}{n_y(0.67 h/t)}$
	Flat plate with horizontal stiffener, both edges supported	19	$\dfrac{1.3 F_{cy}}{n_y}$	$\dfrac{h}{t} = \dfrac{B_{br} - 1.3 F_{cy}}{0.29 D_{br}}$	$\dfrac{1}{n_y}\left(B_{br} - 0.29 D_{br}\dfrac{h}{t}\right)$	$\dfrac{h}{t} = \dfrac{k_1 B_{br}}{0.29 D_{br}}$	$\dfrac{k_2\sqrt{B_{br} E}}{n_y(0.29 h/t)}$
SHEAR IN WEBS, gross section	Unstiffened flat webs	20	$\dfrac{F_{sy}}{n_y}$	$\dfrac{h}{t} = \dfrac{B_s - F_{sy}}{1.25 D_s}$	$\dfrac{1}{n_y}\left(B_s - 1.25 D_s\dfrac{h}{t}\right)$	$\dfrac{h}{t} = \dfrac{C_s}{1.25}$	$\dfrac{\pi^2 E}{n_y(1.25 h/t)^2}$
	Stiffened flat webs $a_e = a_1/\sqrt{1 + 0.7(a_1/a_2)^2}$	21	$\dfrac{F_{sy}}{n_y}$	$\dfrac{a_e}{t} = \dfrac{B_s - (n_a F_{sy}/n_y)}{1.25 D_s}$	$\dfrac{1}{n_a}\left(B_s - 1.25 D_s\dfrac{a_e}{t}\right)$	$\dfrac{a_e}{t} = \dfrac{C_s}{1.25}$	$\dfrac{\pi^2 E}{n_a(1.25 a_e/t)^2}$

Figure 5.62 General requirements of the Aluminum *Specification* for shear in beams.

201

Type of Stress	Type of Member or Component	Sec. 3.4.	Allowable Stress Slenderness ≤ S₁	Slenderness Limit S₁	Allowable Stress Slenderness Between S₁ and S₂	Slenderness Limit S₂	Allowable Stress Slenderness ≥ S₂
COMPRESSION IN BEAMS, extreme fiber gross section	Single web beams bent about strong axis	11	21	$L_b/r_y = 23$	$23.9 - 0.124 L_b/r_y$	$L_b/r_y = 79$	$\dfrac{87,000}{(L_b/r_y)^2}$
	Round or oval tubes	12*	25 (14†)	$R_b/t = 28$	$39.3 - 2.70\sqrt{R_b/t}$	$R_b/t = 81$	Same as Section 3.4.10
	Solid rectangular and round section beams	13	28	$d/t\sqrt{L_b/d} = 13$	$40.5 - 0.93\, d/t\sqrt{L_b/d}$	$d/t\sqrt{L_b/d} = 29$	$\dfrac{11,400}{(d/t)^2(L_b/d)}$
	Rectangular tubes and box sections	14	21	$L_b S_c / S\sqrt{I_y J} = 146$	$23.9 - 0.24\sqrt{L_b S_c / S\sqrt{I_y J}}$	$L_b S_c / S\sqrt{I_y J} = 1700$	$\dfrac{24,000}{(L_b S_c / S\sqrt{I_y J})}$
COMPRESSION IN COMPONENTS OF BEAMS, (component under uniform compression), gross section	Flat plates supported on one edge	15	21	$b/t = 6.8$	$27.3 - 0.93\, b/t$	$b/t = 10$	$182/(b/t)$
	Flat plates with both edges supported	16	21	$b/t = 22$	$27.3 - 0.29\, b/t$	$b/t = 33$	$580/(b/t)$
	Curved plates supported on both edges	16.1*	25 (14†)	$R_b/t = 1.6$	$26.2 - 0.94\sqrt{R_b/t}$	$R_b/t = 141$	$\dfrac{3800}{(R_b/t)(1 + \sqrt{R_b/t}/35)^2}$
	Flat plates with one edge supported and the other edge with stiffener	16.2			See Section 3.4.16.2		
	Flat plates with both edges supported and with an intermediate stiffener	16.3			See Section 3.4.16.3		
COMPRESSION IN COMPONENTS OF BEAMS, (component under bending in own plane) gross section	Flat plates with compression edge free, tension edge supported	17	28	$b/t = 8.9$	$40.5 - 1.41\, b/t$	$b/t = 19$	$4,900/(b/t)^2$
	Flat plate with both edges supported	18	28	$h/t = 46$	$40.5 - 0.27\, h/t$	$h/t = 75$	$1520/(h/t)$
	Flat plate with horizontal stiffener, both edges supported	19	28	$h/t = 107$	$40.5 - 0.117\, h/t$	$h/t = 173$	$3500/(h/t)$
SHEAR IN WEBS, gross section	Unstiffened flat webs	20	12	$h/t = 36$	$15.6 - 0.099\, h/t$	$h/t = 65$	$39,000/(h/t)^2$
	Stiffened flat webs $a_e = a_1/\sqrt{1 + 0.7(a_1/a_2)^2}$	21	12	—	12	$a_e/t = 66$	$53,000/(a_e/t)^2$

Figure 5.63 Requirements of the Aluminum *Specification* for shear in beams, for the specific case of alloy 6061-T6.

202

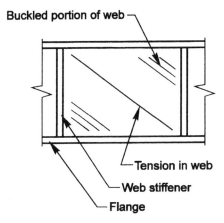

Figure 5.64   Tension field action in the elastically buckled web of a beam.

and the flanges act like top and bottom chords of a truss. The transverse stiffeners are treated as vertical members of a horizontal truss, and the bands of the web that are in tension are treated as the diagonal members of the truss. While the Steel *Specification* has provisions for tension field action, the Aluminum *Specification* does not.

If you recall that tensile stresses used for design are based on the lesser of the tensile yield and ultimate strengths, each with an appropriate factor of safety, you may be wondering why the shear ultimate strength (F_{su}) is not used to calculate shear capacity. For most of the alloys listed in the Aluminum *Specification*, the shear ultimate strength is greater than 1.95/1.65 (the ratio of ultimate to yield factor of safety) times the shear yield strength, so the ultimate strength won't govern. Some alloys, such as 3003-H16, would have slightly lower allowable shear stresses if the factor of safety n_u (1.95) applied to ultimate strength were checked, as well as n_y applied to shear yield. For 3003-H16, F_{su}/n_u = (14 ksi)/1.95 = 7.2 ksi, while F_{sy}/n_y = (12 ksi)/1.65 = 7.3 ksi. This difference is negligible.

Curved Elements in Shear   What about curved elements in shear? The Aluminum *Specification* does not include provisions for curved elements in Table 3.4-3 with the other elements. *Specification* Section 4.2 (Torsion and Shear in Tubes), however, lets you adapt Section 3.4.20 (for flat webs in shear) to curved elements in shear if you come up with an equivalent h/t ratio based on the radius, thickness, and length of the tube:

$$\frac{h}{t} = 2.9 \left(\frac{R_b}{t}\right)^{5/8} \left(\frac{L_s}{R_b}\right)^{1/4}$$

where:

R_b = mid-thickness radius of round tube or maximum mid-thickness radius of an oval tube

t = tube thickness

L_s = length of tube between circumferential stiffeners or overall length if no circumferential stiffeners are present.

This equation is a simplification that is up to 40% conservative for thin-walled tubes and 20% unconservative for thick-walled tubes. A more precise approach is given in "Design of Aluminum Tubular Members" by Clark and Rolf (97).

Table 5.10 is a summary table of bending shear requirements in the Aluminum *Specification*.

5.4   TORSION

Torsion occurs in a member when an applied transverse load does not act through the cross section's *shear center* or when a torque is applied to the member's longitudinal axis. When the cross section is circular, such as a solid round bar or round tube, or when the member is not restrained from warping, the torsion is called *St. Venant torsion,* and the torsional stresses are shear stresses only (illustrated by the free end of the cantilever beam; Figure 5.65c). When the member is restrained from warping, such as the fixed end of the cantilever beam in Figure 5.65b, the resulting torsion at the point of warping restraint is called warping torsion. The warping stresses are shear (different from the St. Venant shear stresses) and longitudinal stresses (Figure 5.65b). These longitudinal stresses will directly add to bending stresses if the member is also subjected to major or minor axis bending moments, as Aluminum *Specification* Section 4.3 reminds us. Along the length of the cantilever beam, both St. Venant and warping torsional stresses are present.

TABLE 5.10   Shear Stress Provisions in the Aluminum *Specification*

Specification Section	Element Description
3.4.20	Flat plate element
3.4.21 (4.6.1)[1]	Flat plate element with transverse stiffener (Requirements for transverse stiffeners)
4.2	Curved element

[1] *Specification* Section 4.6.1 includes requirements for transverse (vertical) stiffeners necessary for Section 3.4.21 to be used.

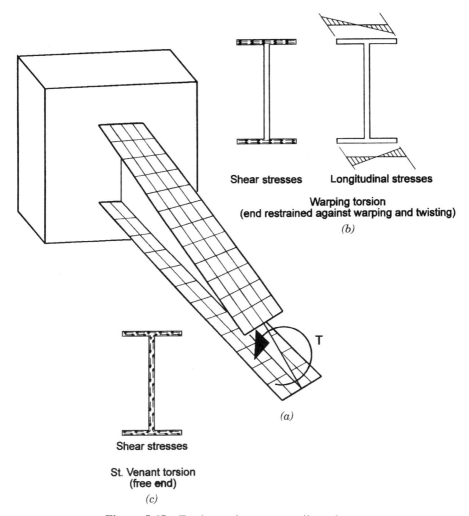

Shear stresses   Longitudinal stresses

Warping torsion
(end restrained against warping and twisting)

(b)

(a)

Shear stresses

St. Venant torsion
(free end)

(c)

Figure 5.65   Torsion acting on a cantilever beam.

Torsional stresses are often difficult to compute and poorly understood. Consequently, the route commonly taken is to avoid them by making sure that transverse loads act through the shear center. When torsion can't be avoided, closed (hollow) shapes are the most efficient selection. We'll look briefly at the two types of torsion stresses to give you a fighting chance of dealing with them. While torsion stresses are usually low, torsional deflections, especially for *open shapes* composed of thin elements, are often significant.

5.4.1 St. Venant Torsion

The rotation St. Venant torsion produces is:

$$\theta = \frac{TL}{GJ} \tag{5.28}$$

where:

θ = angle of rotation (radians)
T = torque
L = length over which twist is permitted
J = torsional constant (called the *polar moment of inertia* for round shapes
$= \pi D^4/32$ where D = the diameter)
G = shear modulus (about 3,800 ksi [26,000 MPa] for aluminum alloys).

The St. Venant shear stress is:

$$f_s = \frac{Tr}{J} \tag{5.29}$$

where:

f_s = shear stress
r = radius of round shape.

For shapes other than solid round bars and round tubes, the torsional constant to be used in Equations 5.28 and 5.29 and the radius to be used in Equation 5.29 are not readily apparent. Fortunately, some reasonable approximations are available. For rectangular tubes (or any closed tube of constant wall thickness):

$$f_s = \frac{T}{2At} \tag{5.30}$$

where:

A = area enclosed by the mean perimeter of the tube
t = tube wall thickness.

For calculating rotation:

$$J = \frac{4A^2t}{S} \tag{5.31}$$

where:

S = mean perimeter of the tube, which is the length of a line at the midthickness of the walls of the tube.

For a rectangular tube with outside dimensions a and b with wall thickness t:

$$J = \frac{2t(a - t)^2 (b - t)^2}{a + b - 2t}$$

For any rectangular solid section or open section composed of a number of rectangular elements, including an I beam, tee, channel, angle, or zee, the stress in each element is:

$$f_s = \frac{Tt}{J} \tag{5.32}$$

where:

t = thickness of the elements

$$J = \sum \left(\frac{bt^3}{3} - 0.21t^4 \right) \tag{5.33}$$

For rectangular elements with $b/t \geq 10$ (in other words, slender), Aluminum *Specification* Section 4.9.1 allows you to disregard the t^4 term in Equation 5.33, which then becomes:

$$J = \sum \frac{bt^3}{3} \tag{5.34}$$

We've provided the torsion constant (J) for some common aluminum shapes with the section properties in Appendix B. These torsion constants include the web-flange fillet in the calculation of J, so they are slightly greater than values given by Equation 5.34.

Torsion constants are independent of the material, so if you have an aluminum shape that is the same as a steel shape for which you can find the torsion constant, you can use the steel torsion constant. Even if your shape is only approximately the same as the steel shape, you can use the steel torsion constant as a check. This also applies to other section properties, such as moments of inertia and warping constants, which we'll discuss under warping torsion below.

What do you do with the torsion stresses once you've calculated them? Aluminum *Specification* Section 4.3 requires that they be added to stresses due to bending, if any, and that the total for each type of stress not exceed the appropriate allowable stress. For example, if a rectangular tube with thin walls is twisted about its longitudinal axis, the walls' may buckle due to St. Venant shear stresses.

Consequently, the walls of the tube must be checked against this dire threat using the provisions of the Aluminum *Specification* Section 3.4.20, Shear in Unstiffened Flat Webs. Should shear stress due to bending also be present, the total shear stress is the sum of the bending and the St. Venant shear stresses. If the thin-walled tube being twisted were round instead of rectangular, you could use Aluminum *Specification* Section 4.2, Torsion and Shear in Tubes to determine an equivalent slenderness ratio (h/t) to be used for the check in Aluminum *Specification* Section 3.4.20.

5.4.2   Warping Torsion

At the free end of the cantilevered I-beam shown in Figure 5.65a, the torsional stresses are the St. Venant stresses discussed above. At the other end of the beam, which is restrained from warping, the stresses are the warping stresses (Figure 5.65b). You can conservatively estimate them by considering the torsional moment to be resisted by a force couple in the beam flanges (95), as shown in Figure 5.66. The shear force in each flange is:

$$V = \frac{T}{d} \tag{5.35}$$

where:

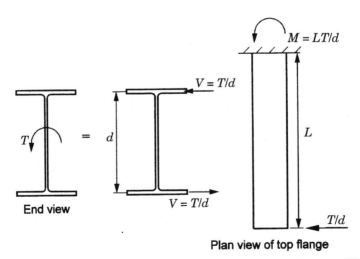

Figure 5.66   Warping stresses considered as a force couple.

V = shear force in each flange
T = torque
d = distance between the midthickness of the flanges.

The shear stress can be calculated for each flange approximately as:

$$f_s = \frac{V}{A_f} \tag{5.36}$$

where:

f_s = shear stress in the flange
A_f = area of a flange.

More precisely, this can be calculated as $f_s = VQ/It$ (see Section 5.3.3), where Q, I, and t are for a single flange bent about the minor axis of the beam.

To conservatively approximate the bending moment acting on a flange, consider the shear force on a flange (V) to act over the length of the cantilever. The moment for the specific case of the cantilever beam in Figure 5.65 is:

$$M = \frac{LT}{d} \tag{5.37}$$

where:

M = bending moment acting on each flange
L = length of the cantilever beam.

The bending stress in each flange can be calculated as M/S, where S is the section modulus for a flange bent about the minor axis of the beam.

A more precise expression for the bending moment (M) in each flange is obtained from:

$$M = \frac{Ta}{d} (\tanh(L/a)) \tag{5.38}$$

where:

$$a = \sqrt{\frac{EC_w}{JG}} = \left(\frac{d}{2}\right)\sqrt{\frac{EI_y}{JG}} \tag{5.39}$$

C_w = warping constant
I_y = minor axis moment of inertia of the section

M approaches TL/d when $L/a \leq 0.30$. In other words, for relatively short cantilevers, the approximate equation ($M = LT/d$) derived above is reasonably accurate. The "tanh" in Equation 5.38 is the hyperbolic tangent function, which your calculator or spreadsheet can probably calculate for you. Warping stresses for other support and load conditions are beyond the scope of this book (in other words, a lot of work), but they can be found elsewhere (95).

5.4.3 A Final Note

If you're asking yourself, "Haven't I seen J and C_w somewhere else?" the answer is: you have. Because torsional buckling of columns involves twisting and warping of the cross section, these same parameters are used to determine the resistance of shapes to buckling as columns (discussed in Section 5.2.1). The torsion (J) and warping (C_w) constants also affect lateral-torsional buckling of beams since this too involves twisting and warping (see Aluminum *Specification* Section 4.9.3). So even if you never need to check torsion, you may need to calculate these properties. As mentioned previously, the values of these properties are tabulated in Appendix B for the Aluminum Association standard I-beams and channels.

5.5 COMBINED STRESSES

The effect of stresses acting in combination may be more severe than the sum of their individual effects. The Aluminum *Specification* provides interaction equations to guard against failure from combined stresses.

Performing one combined stress check does not relieve you from performing others. For example, a beam-column will typically have axial, bending, and shear stresses; the interaction of bending and axial force should be checked in accordance with Section 5.5.1, and the combined effects of bending, axial force, and shear should be checked in accordance with Section 5.5.3.

5.5.1 Combined Axial Compression and Bending

Because the superposition of bending on an axially loaded column can increase the tendency of the column to buckle, Aluminum *Specification* Section 4.1.1 provides interaction equations to quantify this effect. Both of the following equations must be satisfied:

$$\frac{f_a}{F_a} + \frac{C_{mx}f_{bx}}{F_{bx}\left(1 - f_a/F_{ex}\right)} + \frac{C_{my}f_{by}}{F_{by}(1 - f_a/F_{ey})} \leq 1.0 \qquad (5.40)$$

and

$$\frac{f_a}{F_{ao}} + \frac{f_{bx}}{F_{bx}} + \frac{f_{by}}{F_{by}} \leq 1.0 \tag{5.41}$$

When the axial compressive stress is less than or equal to 15% of the allowable axial stress, the *Specification* permits you to neglect the small magnification effect of bending on column buckling and simply use the equation:

$$\frac{f_a}{F_a} + \frac{f_{bx}}{F_{bx}} + \frac{f_{by}}{F_{by}} \leq 1.0 \tag{5.42}$$

In the above equations:

f_a = compressive stress due to axial compression load

f_b = extreme fiber compressive stress due to bending

F_a = allowable compressive stress for the member if it were subjected only to axial compression (see Section 5.2)

F_b = allowable compressive stress for the member if it were subjected only to bending (see Section 5.3.2)

C_m = $0.6 - 0.4(M_1/M_2)$ for members in frames braced against joint translation

or

= 0.85 for members in frames subject to joint translation (sidesway) in the plane of bending, where

M_1/M_2 = ratio of the smaller to larger moments at the ends of that portion of the member unbraced in the plane of bending under consideration. M_1/M_2 is positive when the member is bent in reverse curvature and negative when bent in single curvature.

x = subscript for strong axis bending

y = subscript for weak axis bending

F_{ao} = allowable axial compressive stress of the member as if its overall slenderness ratio were less than S_1 (i.e., the allowable stress for the cross-sectional elements) (see Section 5.2),

F_e = allowable axial compressive stress of the member if its overall slenderness ratio were greater than S_2 (see Section 5.2), which is = $\pi^2 E/n_u(kL/r)^2$ where:

L = unbraced length as a column in the plane of bending (Figure 5.67)

k = effective length factor in the plane of bending

r = radius of gyration about the bending axis (Figure 5.67).

Designers familiar with the AISC and AISI steel specifications (39, 40) will recognize these checks as the same as those for steel. The C_m factor can vary

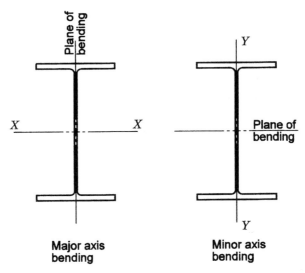

Figure 5.67 Cross section showing the plane of bending, for determining Euler column slenderness kL/r used in F_e.

Figure 5.68 The framing for a custom, curved cladding system will undoubtedly be subject to combined shear, compression, and bending. (Courtesy of Temcor)

between 0.2 (when $M_1/M_2 = 1$, the case when the member is bent in *reverse curvature* with the end moments equal) and 1.0 (when $M_1/M_2 = -1$, the case when the member is bent in *single curvature* with the end moments equal). From this, it's evident that the C_m factor reduces the bending stress terms in the interaction equation as much as 80% depending on the bending curvature. Meanwhile, the $(1 - f_a/F_e)$ term magnifies the bending stress terms in the equation as the compressive stress approaches the elastic buckling stress.

5.5.2 Combined Tension and Bending

As stipulated in Section 4.1.2 of the Aluminum *Specification*, members subjected to axial tension and bending must be proportioned so that:

$$\frac{f_a}{F_t} + \frac{f_{bx}}{F_{bx}} + \frac{f_{by}}{F_{by}} \le 1.0 \qquad (5.43)$$

where:

f_a = tensile stress due to axial tensile load
f_b = extreme fiber tensile stress due to bending
F_t = allowable tensile stress for the member if it were subjected only to axial tension (see Section 5.1)
F_b = allowable tensile stress for the member if it were subjected only to bending (see Section 5.3.1)
x = subscript for strong axis bending
y = subscript for weak axis bending

The axial tension force induces a secondary moment (equal to the axial force times the bending deflection) that opposes the effect of the primary bending moment on the beam. In other words, while the moment acts to bend the member into a curved shape, the tension force acts to straighten it. Consequently, there is no amplification effect as for combined axial compression and bending.

This check is also the same as that used in the AISC and AISI steel specifications (39, 40).

5.5.3 Combined Shear, Compression, and Bending

Unlike the steel specifications, the Aluminum *Specification* (Section 4.4) provides interaction equations for combined shear, compression, and bending.

For webs of shapes composed of rectangular elements, for example, I-beams, channels, and plate girders:

$$\frac{f_a}{F_a} + \left(\frac{f_b}{F_b}\right)^2 + \left(\frac{f_s}{F_s}\right)^2 \leq 1.0 \tag{5.44}$$

For shapes composed of curved elements, such as round tubes:

$$\frac{f_a}{F_a} + \frac{f_b}{F_b} + \left(\frac{f_s}{F_s}\right)^2 \leq 1.0 \tag{5.45}$$

where:

f_a = compressive stress due to axial compression load
F_a = allowable compressive stress for the member if it were subjected only to axial compression (see Section 5.2)
f_b = extreme fiber compressive stress due to bending
F_b = allowable compressive stress for the member if it were subjected only to bending (see Section 5.3.2)
f_s = maximum shear stress
F_s = allowable shear stress for the member if it were subjected only to shear (see Section 5.3.3).

The combined shear, compression, and bending check will only govern the design of members subject to high shear stresses, such as deep beams. When no axial compression occurs, Equation 5.44 is the same as the combined bending and shear interaction equation in the cold-formed steel specification (40). That equation should be used for checking combined bending and shear even when no axial compression is present. Combined web crippling and shear (see Section 5.3.2) may also need to be checked in this case.

Remember that the interaction equations of Section 5.5.1 (combined axial compression and bending) must also be satisfied, whether or not shear is present.

5.5.4 Biaxial and Triaxial Stresses

The stresses and stress combinations addressed by the Aluminum *Specification* and discussed above are generally for longitudinal, uniaxial stress states in prismatic members such as columns. This is partly because the *Specification* predates finite element analysis (FEA) and doesn't directly address issues that arise from such analyses. FEA, on the other hand, can provide triaxial stresses by reporting, in addition to longitudinal stresses, transverse and through-thickness stresses. Many FEA programs calculate a von Mises stress (explained below) from the triaxial stresses at a given element. If you're wondering what to do in situations such as this, read on.

Yielding occurs in ductile materials, such as aluminum, when:

$$(f_1 - f_2)^2 + (f_2 - f_3)^2 + (f_3 - f_1)^2 \geq 2\,F_y^{\,2}$$

where:

f_1, f_2, f_3 = principal stresses (the normal stress on each of three orthogonal surfaces such that the shear stresses on the surfaces are zero)

F_y = yield stress

This equation is called the *von Mises criterion* or *distortion energy criterion.* It predicts that yielding occurs when the distortion energy equals the distortion energy in an axially loaded member at yield. The above equation is for the general triaxial stress state. If stresses are biaxial, $f_3 = 0$, and the equation above predicts yielding when:

$$(f_1 - f_2)^2 + f_2^{\,2} + f_1^{\,2} \geq 2\,F_{ty}^{\,2}$$

The von Mises stress is defined from the von Mises criterion as:

$$\text{Triaxial von Mises stress} = \sqrt{\frac{(f_1 - f_2)^2 + (f_2 - f_3)^2 + (f_3 - f_1)^2}{2}}$$

so that it may be compared directly to the yield stress to determine if yielding occurs. In the biaxial stress state, the von Mises stress becomes:

$$\text{Biaxial von Mises stress} = \sqrt{f_1^{\,2} - f_1 f_2 + f_2^{\,2}}$$

The von Mises criterion isn't just an exotic concept without relevance to the Aluminum *Specification:* it's used in the *Specification* to determine the shear yield strength of aluminum alloys, since there is no established test method to measure shear yield strength. In the case of pure shear, the shear stresses in a biaxial stress element are τ and $-\tau$. Mohr's circle can be used to show that the principal stresses f_1 and f_2 are, then, also τ and $-\tau$, so the von Mises stress is:

$$\sqrt{\tau^2 - \tau(-\tau) + \tau^2} = \tau\,\sqrt{3}$$

When the von Mises stress equals F_{ty}, yielding occurs, so shear yield τ_y is:

$$\tau_y = \frac{F_{ty}}{\sqrt{3}}$$

This is the relationship used in the *Specification's* minimum mechanical property tables as discussed in Section 4.2 of this book.

The von Mises criterion only predicts yielding, and doesn't address fracture or instability. Furthermore, Roark (140) notes that yielding "may occur locally in a member and do no real damage if the volume of material affected is so small or so located as to have only negligible influence on the form and strength of the member as a whole." So just because the von Mises stress has been exceeded at a given point in the structural component doesn't mean the part has reached the limit of its structural usefulness. Also, just because the von Mises stress hasn't exceeded the yield strength doesn't mean the part hasn't buckled. But yielding causing significant distortion of the part is a limit state, so knowing when yielding occurs is useful. This can be summarized as:

Stresses that cause distortion that prevents a member from performing its intended function shall be limited by:

$$\sqrt{\frac{(f_1 - f_2)^2 + (f_2 - f_3)^2 + (f_3 - f_1)^2}{2}} \leq \frac{F_y}{n_y}$$

where:

f_1, f_2, f_3 = principal stresses
F_y = yield strength
n_y = safety factor on yield

6 Orientation to the Aluminum *Specification*

Aluminum has a much briefer commercial history than steel, with the first significant aluminum production in the U.S. in the 1890s and the first aluminum structural applications around 1930. As you might expect, the history of aluminum structural specifications is also shorter than that of steel. The first AISC steel manual was published in 1923; its aluminum counterpart, in 1967. This chapter provides the background of U.S. aluminum structural design specifications.

6.1 BACKGROUND

The earliest design rules for aluminum structures were developed at Alcoa around 1930 and were used to design the aluminum deck and floor beams installed on the Smithfield Street Bridge in Pittsburgh in 1933. In May 1952, the American Society of Civil Engineers (ASCE), which celebrated its centennial that year, published an article titled, "Specifications for Structures of a Moderate Strength Aluminum Alloy of High Resistance to Corrosion." This specification, and similar ones that followed in 1956 and 1962 (130), were the work of the Task Committee on Lightweight Alloys, Committee on Metals, of the Structural Division of the ASCE. At the same time, each of the major producers of aluminum, including Alcoa (1), followed by Reynolds and Kaiser (both of whom began production in the 1940s), developed structural design handbooks for their products.

The *Specifications for Aluminum Structures* was first published in 1967 by the Aluminum Association, the nationally recognized trade organization for the aluminum industry, based on the ASCE work. The Aluminum *Specifications* supplanted the design specifications of the ASCE and the various producers and became part of a five-volume construction-manual series that included:

- Section 1: *Specifications for Aluminum Structures*
- Section 1A: *Commentary on Specifications for Aluminum Structures*
- Section 2: *Illustrative Examples of Design*

217

- Section 3: *Engineering Data for Aluminum Structures*
- Section 5: *Specifications for Aluminum Sheet Metal Work in Building Construction*

Section 3, *Engineering Data for Aluminum Structures*, contained tables of section properties for common aluminum shapes, such as channels, I-beams, angles, and tees, and other useful information for structural design in aluminum. (Whatever was intended for Section 4 has been lost to the ages.) Taken together, the various sections of the construction manual series were an aluminum equivalent of the AISC *Manual of Steel Construction* (39).

Unlike the AISC steel specification, the Aluminum *Specifications* contained two sets of safety factors: one set for buildings, and one for bridges. The Aluminum *Specifications'* safety factors for buildings were roughly consistent with the safety factors used for steel buildings—about $\frac{5}{3}$ for yielding and 2 for ultimate strength. A number of aluminum bridges had been designed and built just before the first publication of the Aluminum *Specifications*, so provisions were made for design of such structures with higher safety factors than those for buildings. This, too, was consistent with the philosophy used for steel bridge design.

The Aluminum *Specifications* were revised in 1971, 1976, 1982, and 1986, but most of the provisions remained unchanged. In the fifth edition (1986), fatigue provisions were significantly expanded to define stress categories similar to those used for steel. A substantial rewrite of the *Specifications* was undertaken for the sixth edition (1994). Changes included:

- Postbuckling strength in flat plate elements of sections supported along one edge was recognized.
- Provisions for edge stiffeners were refined, and new provisions for intermediate stiffeners were added.
- Sections on lateral buckling of beams and web crippling were improved.
- Fatigue strengths were revised based on new data, and provisions were added to address variable amplitude loads.
- Provisions were added for columns subject to torsional or torsional-flexural buckling, and the effective length factor k was introduced.
- Design criteria for self-tapping screws in shear and tension were greatly expanded.
- Yield strengths for longitudinally welded members were revised.
- Two types of testing were identified: tests for determining mechanical properties and tests for determining structural performance, and new procedures for the application of their results were given.

Perhaps even more significantly, the first U.S. aluminum load and resistance factor design (LRFD) specification was added. (Canada, which published *Strength Design in Aluminum* in 1983, (96) already had one.) The Aluminum

Association's LRFD *Specification* applies only to building type structures. The American Association of State Highway and Transportation Officials *LRFD Bridge Design Specifications* governs LRFD for aluminum bridge members; the aluminum section in this specification was developed with the cooperation of the Aluminum Association (35). The LRFD Aluminum *Specification* is discussed in Chapter 11.

The Aluminum *Specification* was revised again in 2000, including a cosmetic change in the title from the plural to the singular. The significant changes were:

- 5052-O, 5083-O, 5086-O, 5454-O, 5456-O, 6066-T6, 6070-T6, and 6463-T6 were added to the minimum mechanical property tables; 3006-H391, 5083-H323, and 5083-H343 were dropped; and metric versions of these tables were added.
- Fatigue provisions for mechanically fastened connections were revised.
- Design strengths were increased for yielding compression limit states by about 10% by eliminating the k_c factor.
- Yielding on the net section was eliminated as a limit state for tension members.
- Bearing design strengths were substantially increased.
- Provisions for slip-critical bolted connections were added.
- Block shear design provisions similar to those for hot-rolled steel were added.
- Screw pull-out strengths were revised in light of additional test data.

The Aluminum *Specification* provisions for member design are explained in Chapter 5.

6.2 THE *ALUMINUM DESIGN MANUAL*

The publishing format of the old construction manual series was changed in 1994 to incorporate most of the aluminum construction manual series material and some new information in the *Aluminum Design Manual*, which contains:

Part IA—Specification for Aluminum Structures—Allowable Stress Design

Part IB—Specification for Aluminum Structures—Load and Resistance Factor Design of Buildings and Similar Type Structures See the comments in Section 6.1 and Chapter 11 regarding LRFD.

Part IIA—Commentary on Specification for Aluminum Structures Allowable Stress Design The commentary provides background information and cites references for provisions in the allowable stress design *Specification* (Part IA).

This information (except for discussion of safety factors) also applies to the LRFD *Specification* (Part IB).

Part IIB—Commentary on Specification for Aluminum Structures Load and Resistance Factor Design   This commentary documents the resistance factors in the LRFD *Specification* (Part IB).

Part III—Design Guide   This part gives general information on the design of aluminum structural members that may be useful in any industry, not just building or bridge construction. The intent is to provide designers of aluminum transportation structures, for example, rail cars, with information about the strength of aluminum structural components. Users may then apply whatever factors of safety are appropriate to the application, which may be different from those used for buildings or bridges.

Part IV—Materials   General information on aluminum alloys is given here, including tables of comparative characteristics and typical applications.

Part V—Material Properties   Minimum and typical mechanical properties are given for a number of commonly used aluminum alloys in this part. All of the provisions of the Aluminum *Specification* are based on using *minimum mechanical properties*. For structural design, consider typical properties to be for general information and amusement only! Also, don't despair if the alloy you need isn't listed in this section: some minimum mechanical properties for other aluminum alloys can be found in the Aluminum Association publication *Aluminum Standards and Data*. See also Chapter 4 of this book, "Material Properties for Design," for more information.

Part VI—Section Properties   Section properties, such as area, moments of inertia, and section moduli are given for various aluminum *shapes*. A relatively new feature is the introduction of a designation system for aluminum shapes; this system hasn't caught on yet commercially. While all of the shapes listed here are theoretically possible to produce in aluminum, not all are available. For example, the 12 in. × 20 in. rectangular tube shown is not listed in any extruder's current catalog, and would be difficult and expensive to manufacture. Such shapes were included under the if-you've-got-the-money-we've-got-the-time theory. Consequently, users would be well advised to check the availability of a particular section with suppliers before investing too much time on design.

This part also gives formulas for calculating section properties, as well as some information useful in determining dimensions of parts fabricated by bending.

Part VII—Design Aids This section contains:

- graphs of allowable stresses
- buckling constants for the alloys and tempers included in the Aluminum *Specification*
- allowable stress tables for common aluminum alloys and some allowable load tables
- beam formulas

The allowable stress tables (Tables 2-2 through 2-23) are extremely useful because they can spare you a lot of number crunching. You should, however, heed some warnings when using them. Many of the tables are groupings of several alloys, such as 6061-T6 and 6351-T5, or product forms, such as extrusions and plate. When this is done, the tables use the lowest minimum strengths of any alloy or product form in the grouping to determine allowable stresses. Consequently, the tables give lower allowable stresses for some alloys than would be calculated from the direct application of the specification provisions. For example, Table 2-21 lists the allowable axial tensile stress for 6061-T6 extrusions and plate both as 19 ksi (*Specification* Section 3.4.1). This is based on the strengths for extrusions, which are less than those of plate, and are:

$$F_{ty} = 35 \text{ ksi} \quad \text{and} \quad F_{tu} = 38 \text{ ksi}$$

from which the allowable axial tensile stress is calculated as the lesser of:

$$F = (35 \text{ ksi})/1.65 = 21.2 \text{ ksi} \quad \text{and}$$
$$F = (38 \text{ ksi})/1.95 = 19.5 \text{ ksi}$$

which is rounded to 19 ksi in the table. However, the strength of 6061-T6 sheet and plate 0.01 in. through 4.00 in. thick is:

$$F_{ty} = 35 \text{ ksi} \quad \text{and} \quad F_{tu} = 42 \text{ ksi}$$

from which the allowable axial tensile stress is calculated as the lesser of:

$$F = (35 \text{ ksi})/1.65 = 21.2 \text{ ksi} \quad \text{and}$$
$$F = (42 \text{ ksi})/1.95 = 21.5 \text{ ksi}.$$

The allowable stress for the plate is, therefore, 21.2 ksi, which is about 12% ($= 21.2/19$) higher than the value given in the table. This example shows how the tables are conservative; they may be too conservative for your taste.

If so, you can calculate the allowable stresses for your particular case. See Chapter 5 for help.

You should note one other item in the *Aluminum Design Manual* allowable stress tables for the various alloys in Part VII. Slenderness limits S_1 and S_2 in the tables differ from the values that are calculated for S_1 and S_2 using the expressions in the Aluminum *Specification*. For example, for 1100-H14, in Table 2-2 Section 3.4.11 (beams), S_1 as calculated from the buckling constants in Table 2-1 and its expression in Part IA, Table 3.4-3, is:

$$S_1 = \frac{1.2(B_c - F_{cy})}{D_c} = \frac{1.2(14.5 - 13)}{0.067} = 27$$

This makes the S_1 value of 24 given in Table 2-2 appear to be in error. However, if S_1 is calculated as the intersection of the expressions given in Table 2-2 for allowable stress for slendernesses less than S_1 and allowable stress for slendernesses between S_1 and S_2:

$$8.0 = 8.8 - 0.034(S_1)$$

$$S_1 = \frac{8.8 - 8.0}{0.034} = 24$$

This is the value for S_1 given in Table 2-2, demonstrating that this second method is the one used to calculate slenderness limits in these tables. Don't be alarmed by the difference, however; the difference in allowable stress calculated for a slenderness equal to S_1 between these two methods is only 0.1 ksi:

$$8.8 - 0.034(27) = 7.9 \text{ ksi}$$

$$8.8 - 0.034(24) = 8.0 \text{ ksi}$$

This is another reason, however, not to carry too many significant figures in your calculations (see Section 6.4 below).

Part VIII—Illustrative Examples This part provides 30 examples showing how the provisions of the Aluminum *Specification* are applied.

Part IX—Guidelines for Aluminum Sheet Metal Work in Building Construction The final part of the *Design Manual* is a revised version of the 3rd edition (1980) of the *Specifications for Aluminum Sheet Metal Work in Building Construction*. Changes are slight: the title was changed to a guideline (as opposed to a specification) to reflect that it is non-mandatory, the document has been metricated, and commentary has been separated from the text.

6.3 TYPES OF STRUCTURES ADDRESSED BY THE ALUMINUM *SPECIFICATION*

The allowable stress design version of the Aluminum *Specification* has two sets of safety factors: one for building type structures and one for bridge type structures. These structure types are not defined in the *Specification*. Its commentary notes that building type structures include the type of structures covered by the American Association of State Highway and Transportation Officials (AASHTO) *Standard Specifications for Structural Supports for Highway Signs, Luminaires and Traffic Signals*, but the commentary offers no other definition or examples of building type structures. You might think that buildings would be easy to identify, but that's not always the case. For example, an industrial building that supports a moving crane might actually be more like a bridge structure. Conversely, a pedestrian bridge is usually considered a building structure, at least for design purposes. In light of the *Specification* commentary, a highway sign bridge would be designed as a building structure.

To help you distinguish the types of structures, the difference can be expressed as: bridge structures take a pounding from dynamic loads; building structures don't. While this definition isn't very technical, it paints an appropriate picture. While many loads can be dynamic—even gusting wind—vehicular traffic loads are usually applied at high rates. Even though AASHTO highway bridge specifications prescribe impact factors, higher safety factors are also considered to be required by bridge designers to provide structures rugged enough for such heavy duty. Similarly, a structure subjected to wave action might also be designed as a bridge type structure due to the pounding action of the loading. In the Aluminum *Specification*, the safety factors prescribed for bridge type structures are nominally $\frac{9}{8}$ times the safety factors prescribed for building type structures, or, expressed another way, about 12% higher than for building structures (see Table 6.1). For example, for yield limit states, the bridge safety factor is 1.85, which is approximately equal to $\frac{9}{8}$ (1.65), an approach taken from past steel bridge design.

In a similar manner, resistance factors are different for buildings than bridges. (If you dare, see Chapter 11 for more on LRFD.) Because the strength is multiplied by the resistance factor in LRFD (as opposed to dividing the strength by the safety factor in allowable stress design), bridge resistance

TABLE 6.1 Safety Margins in the Aluminum *Specification*

Type of Structure	Yield Strength	Ultimate Strength
Building Type	1.65	1.95
Bridge Type	1.85	2.20

factors are *lower* than building structure resistance factors. So the resistance factor on axial tension fracture for buildings is 0.85, while the same resistance factor for bridges is 0.75. (See Table 6.2.)

Fatigue design may also be required for bridges, but this is an entirely additional category of design. (See Section 9.2.) Fatigue design is triggered by the number of times the loads are applied (usually taken as more than 100,000), rather than the rate of application of the loads.

Other classes of structures are designed with different sets of safety factors. For example, lifting devices have traditionally been designed with a safety factor of 5, while pressure vessels and piping use 3.5. (This was recently revised from 4, and the industry is still feeling slightly self-conscious about it, so it's referred to as a change in the "design" factor rather than as a reduction in the safety factor.) The Aluminum *Specification* allowable stress design version Section 1.3.3 lets designers use whatever safety factor they deem appropriate.

All of the above assumes that the stresses remain elastic with the exception of some local yielding. When stresses exceed yield and considerable energy absorption and deformation occur under the design loads, you're in a whole new kind of design. These situations arise, for example, when you design automobiles or guardrails for crashworthiness and building structures for earthquakes. Although these examples demonstrate the significance of this type of design to the aluminum industry, the Aluminum *Specification* is silent on this issue. More information can be found in the Aluminum Association's *Automotive Aluminum Crash Energy Management Manual* (12).

6.4 SIGNIFICANT FIGURES AND THE ALUMINUM *SPECIFICATION*

A word on the precision of design calculations is in order here. Strengths of aluminum alloys are stated in *Aluminum Standards and Data* (11) to no more than three significant figures—for example, for 3003-H16 sheet, the minimum tensile ultimate strength listed is 24.0 ksi—and the modulus of elasticity of the various alloys is given to three figures in Aluminum *Specification* Table 3.3-1. Tolerances on fabricating processes also generally limit the proper expression of typical member dimensions, such as the width of an extrusion, to

TABLE 6.2   Resistance Factors in the Aluminum *Specification* and AASHTO *LRFD Bridge Design Specifications*

Type of Structure	Yield Strength	Ultimate Strength
Building Type	0.95	0.85
Bridge Type	0.90	0.75

three significant figures or fewer. For example, the Aluminum Association standard I-beam (designated I 4 × 2.79) extruded in 6061-T6 alloy has a tolerance on its 3 in. wide flange of 0.024 in., and so could be between 2.976 in. and 3.024 in. wide. Thus, the nominal width can be expressed only as 3.0 in., or to two significant figures. Recognizing the effect of dimensional tolerances, the section properties in Part VI of the *Design Manual* show no more than three significant figures. Consequently, the use of more than three significant figures in calculations involving the strength of aluminum members is unwarranted.

PART III
Design Checks for Structural Components

Aluminum space frame with a suspended aluminum deck. (Courtesy of Conservatek Industries, Inc.)

7 Structural Members

Structural design specifications have evolved over the years into rather byzantine formats. Unless you were intimately familiar with them, you may have felt you needed a guide dog to find your way to all of the provisions applicable to a given case. Complication seems to be the inevitable price of the improved precision and wider applicability specification writers and users seek.

On the other hand, it seems that for a given mode of behavior (e.g., compression, tension, bending) and shape, it would be extremely useful to go to one place and find all of the required design checks listed. This would save users from having to sift through a general listing of provisions, most of which don't apply to the case at hand.

In 1976, Galambos presented a paper at the American Society of Civil Engineers (ASCE) Annual Convention titled "History of Steel Beam Design" (105), in which he traced the development of the steel specifications for beams and suggested a reorganization of what he called "a patchwork of many revisions." The 1986 Steel LRFD *Specification* (and 1989 Steel Allowable Stress Design *Specification*) were based on such a reorganized format in an effort to streamline the design process. If you want to check a beam, for example, you go to the section on beams. There, a brief introduction identifies the special cases for which provisions are found elsewhere, and then the design checks applicable to the more typical beam problems are presented. This chapter is structured in a similar format for aluminum design.

The procedures given in this chapter are for *allowable stress design*; for *load and resistance factor design* (LRFD), follow the same approach but substitute *design stress* for *allowable stress* and compare to stresses produced by *factored loads*. (See Chapter 11 for more on LRFD.)

7.1 TENSION MEMBER DESIGN PROCEDURE

This section applies to *prismatic members* without welds subject to axial tension caused by forces acting through the centroidal axis. Special conditions discussed elsewhere include:

Members subject to combined axial tension and bending (see Section 7.4.2)
Threaded members (see Section 8.1)

Welded members (see Section 9.1)

Members subject to fatigue (see Section 9.2).

The steps to determine the capacity of a tension member are:

1. Determine the *net effective area* (see Section 7.1.1).
2. Determine the allowable stress (see Section 7.1.2).
3. Determine the tensile capacity (see Section 7.1.3).

Discussion of the axial tensile stress provisions of the Aluminum *Specification* can be found in Section 5.1.

7.1.1 Net Effective Area

The net area (A_n) is defined in *Specification* Section 5.1.6 as the sum of the net areas of the elements of the section. The net area of the element is the product of its thickness and its least net width. The net width (w_n) for a chain of holes extending across the element is:

$$w_n = w_g + (s^2/4g) - \Sigma d_h$$

where:

w_g = gross width of the element
s = spacing parallel to the direction of load (pitch)
g = spacing perpendicular to the direction of load (gauge)
d_h = hole diameter for drilled or reamed holes; hole diameter + $\frac{1}{32}$ in. [0.8 mm] for punched holes.

The net effective area is the same as the net area except for single and double angles when bolts or rivets transmit the load through one but not both of the legs (*Specification* Section 5.1.7):

1) For single or double angles connected to one side of a gusset, the net effective area (A_e) is the net area of the connected leg plus one-third of the unconnected leg.
2) For double angles placed back-to-back and connected to both sides of a gusset, the net effective area (A_e) is the net area of the connected legs plus two-thirds of the unconnected legs.

7.1.2 Allowable Stress

The allowable axial tensile stress (F_t) is:

$$F_{tn} = \frac{F_{tu}}{k_t n_u} \quad \text{on the net effective area}$$

and

$$F_{tg} = \frac{F_{ty}}{n_y} \quad \text{on the gross area}$$

where:

F_{tu} = tensile ultimate strength (*Specification* Table 3.3-1)
F_{ty} = tensile yield strength (*Specification* Table 3.3-1)
k_t = coefficient for tension members (*Specification* Table 3.4-2)
n_u = factor of safety on ultimate strength = 1.95 for building type structures (*Specification* Table 3.4-1)
n_y = factor of safety on yield strength = 1.65 for building type structures (*Specification* Table 3.4-1).

For 6061-T6 extrusions, the allowable axial tensile stress is:

$$F_{tn} = (38 \text{ ksi})/[(1.0)(1.95)] = 19.5 \text{ ksi on the net effective area}$$

$$[F_{tn} = (260 \text{ MPa})/[(1.0)(1.95)] = 133 \text{ MPa}]$$

and

$$F_{tg} = (35 \text{ ksi})/1.65 = 21.2 \text{ ksi on the gross area}$$

$$[F_{tn} = (240 \text{ MPa})/(1.65) = 145 \text{ MPa}]$$

The *Aluminum Design Manual*, Part VII, Tables 2-2 through 2-23 provide the lesser of F_{tn} and F_{tg} for specific alloys in the row identified as Section 3.4.1.

7.1.3 Tensile Capacity

The allowable tensile capacity (T) of a member is the lesser of

$$T_n = A_e F_{tn}$$
$$T_g = A_g F_{tg}$$

where:

A_e = net effective area of the member cross section
A_g = gross area
F_{tn} = allowable axial tensile stress on the net section
F_{tg} = allowable axial tensile stress on the gross section

7.2 COMPRESSION MEMBER DESIGN PROCEDURE

This section applies to doubly symmetric *prismatic members* without welds, that are subject only to axial compression through their centroidal axis and that are composed of *elements* of constant thickness or taper. Special conditions discussed elsewhere include:

Welded members (see Section 9.1)

Members with combined axial compression and bending (see Section 7.4.1)

Singly symmetric or nonsymmetric shapes (see Section 5.2)

Shapes with stiffened elements (see Section 5.2.2)

The steps in determining allowable stress are:

1. Determine the overall column slenderness ratio (see Section 7.2.1).
2. Determine the slenderness ratio of the cross-sectional elements (see Section 7.2.2).
3. Perform the allowable stress checks for the provisions of the Aluminum *Specification* that are applicable to your shape (see Section 7.2.3).

We describe these steps for typical doubly symmetric structural shapes, and provide a checklist of the Aluminum *Specification* sections applicable to each. We include the general equations at each step, as well as the specific equations for 6061-T6.

To check the capacity of a certain shape or to compare several commonly available shapes, simply substitute the values for your alloy and shape for the variables in this section. Should you have questions on the provisions as you work through them, refer to Chapter 5.

7.2.1   Overall Column Slenderness Ratio

Start by determining the slenderness ratio for overall buckling. Determine the larger of the major and minor axis slenderness ratios (kL/r) for flexural buckling and record it in Table 7.1.

TABLE 7.1   Overall Column Slenderness Ratio Comparison—Flexural Buckling

Aluminum *Specification* Section	Axis	k	L	r	kL/r
Section 3.4.7.1	Major axis (x)				
Section 3.4.7.1	Minor axis (y)				

where:

k = 1.0 for columns fixed against translation at both ends (for other cases, see *Aluminum Design Manual*. Part III, Figure 3.4-1, reprinted here as Figure 5.20).

L = maximum length between braces restraining deflection normal to the particular axis

r = radius of gyration about that particular axis.

Next, compare the larger of the flexural buckling slenderness ratios to the torsional buckling slenderness ratio, which is determined in accordance with Aluminum *Specification* Section 3.4.7.2. For doubly symmetric shapes, this reduces to the following formula:

$$\lambda_\phi = \sqrt{\dfrac{I_x + I_y}{0.038J + \dfrac{C_w}{(k_\phi L)^2}}} \tag{7.1}$$

where:

λ_ϕ = equivalent slenderness ratio for torsional buckling
I_x, I_y = moments of inertia about the two principal axes
J = torsion constant
C_w = warping constant
k_ϕ = effective length coefficient for torsional buckling
L = length of the column not braced against twisting.

The values for I_x, I_y, J, and C_w for common shapes are provided in Appendix B. The value for k_ϕ is 1.0 when the ends of the unbraced length are prevented from twisting. Values of k_ϕ for other cases are given as the K value for torsional buckling in *Aluminum Design Manual,* Part III, Figure 3.4-1, found here in Figure 5.20.

Singly symmetric and nonsymmetric shapes may further be subject to torsional-flexural buckling (see Section 5.2.1). Doubly symmetric shapes, however, are not subject to torsional-flexural buckling.

7.2.2 Slenderness Ratio of Cross-Sectional Elements

The slenderness of an element of a column is the ratio of width (for flat elements) or radius (for curved elements) to thickness. The design procedures that follow depend on the relationship of the element slenderness ratios to the limits S_1 and S_2. If you have chosen a shape listed in Appendix D, then the slenderness ratios and allowable stresses for the elements have already been computed. For other shapes, however, you must calculate their elements' slenderness ratios.

The slenderness ratios for each element of a cross-section may be recorded in a table like Table 7.2; list as many elements of differing dimensions as there are in the shape.

7.2.3 Allowable Column Stress of Typical Shapes

The provisions of the Aluminum *Specification* involved in the design of doubly symmetric shapes are summarized in Table 7.3, followed by guidance in performing the individual calculations.

TABLE 7.2 Element Slenderness Ratio Determination and Comparison to S_1 and S_2

Element	Width (b) or Radius (R)	Thickness (t)	Ratio b/t or R/t	Aluminum *Specification* Equation For S_1	Aluminum *Specification* Equation for S_2
Flat plate supported along one edge (e.g., I-beam flange or cruciform leg)				Equation 3.4.8-4 ($S_1 = 2.7$ for 6061-T6)	Equation 3.4.8-5 ($S_2 = 10$ fcr 6061-T6)
Flat plate supported along both edges (e.g., I-beam web or either face of a rectangular tube)				Equation 3.4.9-4 ($S_1 = 8.4$ for 6061-T6)	Equation 3.4.9-5 ($S_2 = 33$ for 6061-T6)
Curved plate supported along both edges (e.g., wall of round or oval tubes)				Equation 3.4.10-4 ($S_1 = 2.2$ for 6061-T6)	Equation 3.4.10-5 ($S_2 = 141$ for 6061-T6)

where:

b = width, which is measured from the toe of fillets. (If, however, the fillet radius exceeds $4t$, take the fillet radius as $4t$ to determine b; see Figure 5.35.)

R = radius to the midthickness of curved elements

t = thickness of the element, or the average thickness of tapered elements.

234

TABLE 7.3 Doubly Symmetric Column Shapes—Allowable Stresses

Description of Design Check	Aluminum Specification Section	I-beam	Rectangular Tube	Round or Oval Tube	Cruciform
1) Overall allowable column stress	3.4.7	Gross section	Gross section	Gross section	Gross section
2) Effect of local buckling on column strength (if any cross-sectional element has $b/t > S_2$)	4.7.1 4.7.4	Determine effective section	Determine effective section	Not applicable	Determine effective section
3) Allowable stress of elements					
a) Flat plates with one edge supported	3.4.8	Flange	Not applicable	Not applicable	Each leg
b) Flat plates with both edges supported	3.4.9	Web	Each face	Not applicable	Not applicable
c) Curved plates supported on both edges	3.4.10	Not applicable	Not applicable	Wall	Not applicable
4) Optional provision for elements: weighted average allowable stress method	4.7.2	All elements	All elements	Not applicable	All elements

As we mentioned earlier, the allowable stresses for the individual cross-sectional elements of common aluminum shapes are given in Appendix D. If you select your shape from those given in Appendix D, you will have to calculate only the overall allowable column stress, which depends on the unbraced length for each axis. We have also noted which shapes do not have elements with a slenderness ratio greater than S_2. For these shapes, you do not need to check the effect of local buckling on the column strength (step 2, below).

The calculations involved in these steps follow:

1) **Determine the overall allowable column stress (F_c).** The equation for the allowable column stress is dependent on the overall column slenderness ratio, as shown in Table 7.4. We discuss the variables in the general equations for overall column buckling in Section 5.2.1. The *Aluminum Design Manual*, Part VII, Tables 2-2 through 2-23, presents solutions in terms of kL/r for specific alloys in a manner similar to that shown above for 6061-T6. The row for the overall column equations identified as Specification Number 7 in *Aluminum Design Manual*, Tables 2-2 through 2-23, also includes the corresponding values for S_1 and S_2.

2) **Determine the effect of local buckling on column strength** (not applicable to round or oval tubes). When the slenderness of a flat plate element is greater than S_2, the effective area of that element may be reduced due to local buckling (see Section 5.2.2). The provisions in the Aluminum *Specification* for checking this condition are outlined in the following three steps:

2.a) The element local buckling stress (F_{cr}) is determined from Aluminum *Specification* Section 4.7.1, which includes the equations in Table 7.5 for typical doubly symmetric shapes.

TABLE 7.4   Overall Allowable Column Stress (F_c)

	$kL/r < S_1$	$S_1 < kL/r < S_2$	$S_2 < kL/r$
Aluminum *Specification* Equation	3.4.7-1	3.4.7-2	3.4.7-3
General equation	$F_c = \dfrac{F_{cy}}{n_y}$	$F_c = \dfrac{1}{n_u}\left(B_c - D_c\,\dfrac{kL}{r}\right)$	$F_c = \dfrac{\pi^2 E}{n_u\left(\dfrac{kL}{r}\right)^2}$
Expressed for 6061-T6 (ksi)	$F_c = 21$	$F_c = 20.2 - 0.126\,\dfrac{kL}{r}$	$F_c = \dfrac{51{,}000}{(kL/r)^2}$

where:

$S_1 = 0$ for 6061-T6
$S_2 = 66$ for 6061-T6
kL/r = largest of $(kL/r)_x$, $(kL/r)_y$, and λ_{jb}, as determined in Section 7.2.1

TABLE 7.5 Doubly Symmetric Column Shapes—Local Buckling Stresses for Elements of Columns

Type of Element	Aluminum Specification Section	Equation	Typical Element
Flat plate with one edge supported	3.4.8	$F_{cr} = \dfrac{\pi^2 E}{(5.1b/t)^2}$	I-beam flange, or either leg of a cruciform
Flat plate with both edges supported	3.4.9	$F_{cr} = \dfrac{\pi^2 E}{(1.6b/t)^2}$	I-beam web, or either face of a rectangular tube

2.b) If:

$$\frac{F_{cr}}{n_u} < F_c \tag{7.2}$$

where:

F_{cr} = element local buckling stress (from Table 7.5)
n_u = factor of safety on ultimate strength (= 1.95)
F_c = allowable stress for overall buckling (from Table 7.4).

then Aluminum *Specification* Section 4.7.4 must be checked to determine whether local buckling may reduce the overall column strength, using the following equation:

$$F_{rc} = \frac{(F_{ec})^{1/3}(F_{cr})^{2/3}}{n_u} \tag{7.3}$$

where:

$F_{ec} = \pi^2 E/(kL/r)^2$
kL/r = largest of $(kL/r)_x$, $(kL/r)_y$, and λ_ϕ, as determined in Section 7.2.1.

3) **Determine the allowable stress for the elements** from their respective slenderness ratios and edge support conditions, as outlined in Tables 7.6 and 7.7. We discuss the variables in the general equations for allowable stress of elements supported along one edge in Section 5.2.2. The *Aluminum Design Manual,* Part VII, Tables 2-2 through 2-23, presents solutions in terms of b/t for specific alloys in a similar manner to the illustration shown above for 6061-T6, including providing the corresponding values for S_1 and S_2 in the row identified as Section 3.4.8. We discuss the variables in the general equa-

TABLE 7.6   Allowable Stress for Elements Supported Along One Edge (e.g., I-beam Flange or Cruciform Leg)

	$b/t < S_1$	$S_1 < b/t < S_2$	$S_2 < b/t$
Aluminum *Specification* Equation	3.4.8-1	3.4.8-2	3.4.8-3
General equation	$F_c = \dfrac{F_{cy}}{n_y}$	$F_c = \dfrac{1}{n_u}\left(B_p - 5.1 D_p \dfrac{b}{t}\right)$	$F_c = \dfrac{k_2\sqrt{B_p E}}{n_u\left(5.1\dfrac{b}{t}\right)}$
Expressed for 6061-T6 (ksi)	$F_c = 21$	$F_c = 23.1 - 0.79\dfrac{b}{t}$	$F_c = \dfrac{154}{(b/t)}$

where:

S_1 = 2.7 for 6061-T6
S_2 = 10 for 6061-T6
b/t = slenderness ratio of the element, as determined in Section 7.2.2

tions for allowable stress of elements supported along both edges in Section 5.2.2. The *Aluminum Design Manual,* Part VII, Tables 2-2 through 2-23, present solutions in terms of b/t for specific alloys in a similar manner to the illustration shown above for 6061-T6. The row for the allowable stress equations in the *Aluminum Design Manual* tables, identified as Section 3.4.9, also includes the corresponding values for S_1 and S_2.

The allowable stress for the column (F_a) is the lowest of the allowable stresses calculated for the overall column and its individual elements.

TABLE 7.7   Allowable Stress for Elements Supported Along Both Edges (e.g., I-beam Web or Rectangular Tube Face)

	$b/t < S_1$	$S_1 < b/t < S_2$	$S_2 < b/t$
Aluminum *Specification* Equation	3.4.9-1	3.4.9-2	3.4.9-3
General equation	$F_c = \dfrac{F_{cy}}{n_y}$	$F_c = \dfrac{1}{n_u}\left(B_p - 1.6 D_p \dfrac{b}{t}\right)$	$F_c = \dfrac{k_2\sqrt{B_p E}}{n_u\left(1.6\dfrac{b}{t}\right)}$
Expressed for 6061-T6 (ksi)	$F_c = 21$	$F_c = 23.1 - 0.25\dfrac{b}{t}$	$F_c = \dfrac{490}{(b/t)}$

where:

S_1 = 8.4 6061-T6
S_2 = 33 for 6061-T6
h/t = Slenderness ratio of the I beam web or each face of the tube, as determined in Section 7.2.2

4) **Weighted average allowable stress.** When the allowable stress of one element is significantly lower than those of other elements and is controlling the design of the column, it may be beneficial to employ the weighted method of averaging the allowable stresses for the individual cross-sectional elements.

The weighted average allowable stress method was discussed in Section 5.2.2, and summarized in Equation 5.22. For easy reference, it is repeated below as Equation 7.4:

$$F_{ca} = \frac{A_1 F_{c1} + A_2 F_{c2} + \cdots + A_n F_{cn}}{A} \tag{7.4}$$

where:

F_{ca} = weighted average allowable compressive stress
A_i = the area of element i
F_{ci} = local buckling allowable stress of element i
A = area of the section $(A_1 + A_2 + \cdots + A_n)$.

7.2.4 Summary of Allowable Column Stress

Table 7.8 provides a format for comparing the allowable stresses calculated for the overall column and its individual elements. As you work through the procedure of Section 7.2.3, you can enter the value of each allowable stress into the appropriate cell in Table 7.8.

After you record the values for each of the applicable provisions, you can readily select the allowable column stress (F_a) by inspection. If you have calculated the weighted average allowable stress, take it as the element allowable stress. Otherwise, select the lowest allowable stress calculated for an individual element as the element allowable stress. Compare the element allowable stress to the reduced allowable column stress, if applicable, or to the overall allowable column stress. The allowable column stress (F_a) to be used for design is the lower of the individual element allowable stress and the overall allowable column stress.

·

7.3 BENDING MEMBER DESIGN PROCEDURE

This section applies to single web and box beams bent about their major axis, without welds, that are composed of elements of constant thickness or taper. Special conditions discussed elsewhere include:

Members subject to combined axial compression and bending (see Section 7.4.1)

Members subject to combined axial tension and bending (see Section 7.4.2)

Welded members (see Section 9.1)

Members subject to fatigue (see Section 9.2).

TABLE 7.8 Doubly Symmetric Column Shapes—Summary of Allowable Stresses

Description of Design Check	Aluminum Specification Section	I-beam	Rectangular Tube	Round or Oval Tube	Cruciform
1) Overall allowable column stress	3.4.7				
2) Reduced allowable column stress due to effect of local buckling	4.7.1 4.7.4	(If applicable)	(If applicable)	Not applicable	(If applicable)
3) Allowable stress of individual elements					
a) Flat plates with one edge supported.	3.4.8	(Or weighted average)	Not applicable	Not applicable	(Or weighted average)
b) Flat plates with both edges supported.	3.4.9	(Or weighted average)	(Or weighted average)	Not applicable	Not applicable
c) Curved plates supported on both edges.	3.4.10	Not applicable	Not applicable	Not applicable	Not applicable
4) Weighted average allowable stress method (optional)	4.7.2			Not applicable	

The steps to determine the allowable stresses for a beam are:

1. Determine the allowable bending tensile stress (see Section 7.3.1).
2. Determine the allowable bending compressive stress (see Section 7.3.2).
3. Determine the allowable shear stress (see Section 7.3.3).

Discussion of the beam design provisions of the Aluminum *Specification* can be found in Section 5.3.

7.3.1 Bending Tension

The allowable bending tensile stress (F_{bt}) for structural shapes bent about the strong axis and rectangular tubes is:

$$F_{bt} = \min[F_{tu}/(k_t n_u) \text{ and } F_{ty}/n_y] \tag{7.5}$$

For 6061-T6 extrusions, the allowable bending tensile stress is

$$F_{bt} = \min[(38 \text{ ksi})/[(1.0)(1.95)] \text{ and } (35 \text{ ksi})/(1.65)]$$

$$F_{bt} = \min[19.5 \text{ ksi and } 21.2 \text{ ksi}] = 19.5 \text{ ksi on the net area}$$

$$[F_{bt} = \min[(260 \text{ MPa})/[(1.0)(1.95)] \text{ and } (240 \text{ MPa})/(1.65)]$$

$$F_{bt} = \min[133 \text{ MPa and } 145 \text{ MPa}] = 133 \text{ MPa on the net area}]$$

where:

F_{tu} = tensile ultimate strength (*Specification* Table 3.3-1)
F_{ty} = tensile yield strength (*Specification* Table 3.3-1)
k_t = coefficient for tension members (*Specification* Table 3.4-2)
n_u = factor of safety on ultimate strength = 1.95 for building type structures (*Specification* Table 3.4-1)
n_y = factor of safety on yield strength = 1.65 for building type structures (*Specification* Table 3.4-1)

The *Aluminum Design Manual,* Part VII, Tables 2-2 through 2-23, presents allowable bending tensile stresses for specific alloys in the rows identified as Section 3.4.2, 3.4.3, and 3.4.4.

7.3.2 Bending Compression

To determine the allowable bending compression stress:

1) **Determine the lateral buckling allowable stress.** The lateral buckling allowable stress is determined from the shape of the beam section in accord-

ance with Aluminum *Specification* Sections 3.4.11 through 3.4.14. Allowable stress calculations for single web and box sections are illustrated below. For single web beams bent about the strong axis, the lateral buckling allowable stress is shown in Table 7.9. Lateral buckling allowable stress for tubes bent about the major axis may also be determined using the provisions for single web beams bent about their major axis, as in Aluminum *Specification* Section 3.4.11.

2) **Determine the slenderness ratio for each element of the cross section.**

2.a) Element Under Uniform Compression   The slenderness ratio of flat or tapered plate elements supported on one edge and under uniform compression is b/t

where:

b = distance from the unsupported edge of the element to the toe of the fillet or bend at the supported edge. If the fillet or bend inside radius exceeds $4t$, take the radius as $4t$ to calculate b.

t = average thickness of the element.

The slenderness ratio of flat plate elements supported on both edges without an intermediate stiffener and under uniform compression is b/t

where:

TABLE 7.9   Lateral Buckling Allowable Stresses for Single Web Beams Bent About the Strong Axis

	$L_b/r_y < S_1$	$S_1 < L_b/r_y < S_2$	$S_2 < L_b/r_y$
Aluminum Specification Equation Number	3.4.11-1	3.4.11-2	3.4.11-3
General equation	$F_b = \dfrac{F_{cy}}{n_y}$	$F_b = \dfrac{1}{n_y}\left(B_c - \dfrac{D_c L_b}{1.2r_y}\right)$	$F_b = \dfrac{\pi^2 E}{n_y(L_b/1.2r_y)^2}$
Expressed for 6061-T6 (ksi)	$F_b = 21$	$F_b = 23.9 - 0.124L_b/r_y$	$F_b = 87{,}000/(L_b/r_y)^2$

where:

S_1 = 23 for 6061-T6
S_2 = 79 for 6061-T6
L_b = length of the beam between brace points. L_b may be replaced by $L_b/\sqrt{C_b}$ where C_b is determined in accordance with Aluminum *Specification* Section 4.9.4
r_y = minor axis radius of gyration, or the effective minor axis radius of gyration calculated in accordance with Aluminum *Specification* Section 4.9

TABLE 7.10   Lateral Buckling Allowable Stresses for Rectangular Tubes Bent About the Strong Axis

	$S < S_1$	$S_1 < S < S_2$	$S_2 < S$
Aluminum Specification Equation Number	3.4.14-1	3.4.14-2	3.4.14-3
General equation	$F_b = \dfrac{F_{cy}}{n_y}$	$F_b = \dfrac{1}{n_y}(B_c - 1.6D_c\sqrt{S})$	$F_b = \dfrac{\pi^2 E}{2.56 n_y S}$
Expressed for 6061-T6 (ksi)	$F_b = 21$	$F_b = 23.9 - 0.24\sqrt{S}$	$F_b = 24{,}000/S$

where:

$S_1 = 146$ for 6061-T6
$S_2 = 1{,}700$ for 6061-T6
$$S = \frac{2L_b S_c}{\sqrt{I_y J}}$$
L_b = length of the beam between brace points. L_b may be replaced by L_b/C_b where C_b is determined in accordance with Aluminum *Specification* Section 4.9.4
S_c = major axis section modulus on the compression side of the neutral axis
I_y = minor axis moment of inertia
J = torsion constant

    b = distance between the toe of the fillets or bends of the supporting elements. If the fillet or bend inside radius exceeds $4t$, take the radius as $4t$ to calculate b.
    t = thickness of the element.

2.b) Element Under Bending in Its Own Plane   The slenderness ratio of flat elements under in-plane bending and with both edges supported is h/t

where:

    h = clear width of the element
    t = thickness of the element.

3) Determine the allowable stress for each element of the cross section. The allowable stress is determined from its slenderness ratio, as defined in (2) above), and Aluminum *Specification* Sections 3.4.15 through 3.4.19. Allowable stress calculations for several of these sections are illustrated below.

The allowable bending compression stress for flat plate elements supported along one edge and under uniform compression is shown in Table 7.11.

The allowable bending compression stress for flat plate elements supported along both edges and under uniform compression is shown in Table 7.12.

TABLE 7.11 Allowable Compressive Stresses for Elements of Beams (Flat Plate Elements Supported Along One Edge, e.g., Flange of an I-beam Bent About Its Major Axis)

	$b/t < S_1$	$S_1 < b/t < S_2$	$S_2 < b/t$
Aluminum Specification Equation Number	3.4.15-1	3.4.15-2	3.4.15-3
General Equation	$F_b = \dfrac{F_{cy}}{n_y}$	$F_b = \dfrac{1}{n_y}[B_p - 5.1D_p(b/t)]$	$F_b = \dfrac{k_2\sqrt{B_pE}}{n_y(5.1b/t)}$
Expressed for 6061-T6 (ksi)	$F_b = 21$	$F_b = 27.3 - 0.93b/t$	$F_b = 182/(b/t)$

where:

$S_1 = 6.8$ for 6061-T6
$S_2 = 10$ for 6061-T6
$b/t = $ slenderness ratio of the element

TABLE 7.12 Allowable Compressive Stresses for Elements of Beams (Flat Plate Elements Supported Along Both Edges, e.g., Flange of a Rectangular Tube)

	$b/t < S_1$	$S_1 < b/t < S_2$	$S_2 < b/t$
Aluminum Specification Equation Number	3.4.16-1	3.4.16-2	3.4.16-3
General Equation	$F_b = \dfrac{F_{cy}}{n_y}$	$F_b = \dfrac{1}{n_y}[B_p - 1.6D_p(b/t]$	$F_b = \dfrac{k_2\sqrt{B_pE}}{n_y(1.6b/t)}$
Expressed for 6061-T6 (ksi)	$F_b = 21$	$F_b = 27.3 - 0.29b/t$	$F_b = 580/(b/t)$

where:

$S_1 = 22$ for 6061-T6
$S_2 = 33$ for 6061-T6
$b/t = $ slenderness ratio of the element

The allowable bending compression stress for flat plate elements with both edges supported and under bending in its own plane is shown in Table 7.13.

4) **Determine the weighted average allowable stress of the cross section.** To compute the weighted average allowable stress of the cross section, weight the allowable stress for each element in accordance with the ratio of its area to the total area of the cross section:

$$F_{ba} = \frac{A_1 F_{b1} + A_2 F_{b2} + \cdots + A_n F_{bn}}{A} \tag{7.6}$$

where:

A_i = area of an element
F_{bi} = local buckling allowable stress of an element
A = area of the cross section
F_{ba} = weighted average allowable stress.

For purposes of calculating the weighted average allowable stress for a beam, a *flange* is defined as that portion of the cross section that lies more than two-thirds the distance from the neutral axis to the extreme fiber. Weighted average allowable stresses for standard Aluminum Association shapes used as beams are given in Appendix D.

5) **Check the effect of local buckling on lateral buckling.** If the allowable stress for any element of the section was determined from Aluminum

TABLE 7.13 Allowable Bending Stresses for Elements of Beams (Flat Plate Elements with Both Edges Supported and Under Bending in Its Own Plane, e.g., Web of an I-beam or Rectangular Tube, Bent About Its Major Axis)

	$h/t < S_1$	$S_1 < h/t < S_2$	$S_2 < h/t$
Aluminum Specification Equation Number	3.4.18-1	3.4.18-2	3.4.18-3
General equation	$F_b = \dfrac{1.3 F_{cy}}{n_y}$	$F_b = \dfrac{1}{n_y} [B_{br} - 0.67 D_{br}(h/t)]$	$F_b = \dfrac{k_2 \sqrt{B_{br} E}}{n_y (0.67 h/t)}$
Expressed for 6061-T6 (ksi)	$F_b = 28$	$F_b = 40.5 - 0.27 h/t$	$F_b = 1{,}520/(h/t)$

where:

S_1 = 46 for 6061-T6
S_2 = 75 for 6061-T6
h/t = slenderness ratio of the element

Specification Sections 3.4.15, 3.4.16, 3.4.16.2, 3.4.18, or 3.4.19 *and* the slenderness ratio for that element is greater than S_2 local buckling may reduce the allowable lateral buckling stress.

5.a) The element local buckling stress is determined from Aluminum *Specification* Section 4.7.1, excerpted in Table 7.14 for elements of beams.

5.b) If:

$$\frac{F_{cr}}{n_y} < F_c \tag{7.7}$$

where:

F_{cr} = local buckling stress, as computed in (5.a) above
n_y = factor of safety on yield = 1.65
F_c = allowable lateral buckling stress, as computed in (1) above

Next use Aluminum *Specification* Section 4.7.5 to calculate F_{rb}:

$$F_{rb} = \frac{(F_{eb})^{1/3}(F_{cr})^{2/3}}{n_y} \tag{7.8}$$

where:

F_{eb} = lateral buckling strength of the beam, calculated by using Aluminum *Specification* equation 3.4.11-3 with $n_y = 1.0$:

$$F_{eb} = \frac{\pi^2 E}{\left(\dfrac{L_b}{1.2r_y}\right)^2}$$

6) **Determine the allowable bending compression stress.** The allowable

TABLE 7.14   Local Buckling Stresses for Elements of Beams

Aluminum *Specification* Section	Local Buckling Stress (F_{cr})	Type of Element
3.4.15	$\pi^2 E/(5.1b/t)^2$	Supported along one edge, element in uniform compression
3.4.16	$\pi^2 E/(1.6b/t)^2$	Both edges supported, element in uniform compression
3.4.18	$\pi^2 E/(0.67h/t)^2$	Both edges supported, element under bending in its own plane

bending compression stress (F_b) is the lesser of the lateral buckling allowable stress (step 1) reduced if necessary by the effect of local buckling (step 5), and the weighted average allowable stress (F_{ba}) of the section (step 4).

7.3.3 Shear

1) **Determine the slenderness ratio of the element in shear.** The slenderness ratio of an unstiffened flat plate element in shear is h/t,

where:

h = clear height of the web, with no deductions for web-flange fillets
t = thickness of the element

2) **Determine the allowable shear stress.** The allowable shear stress (F_s) for unstiffened flat plate elements and curved plate elements is shown in Table 7.15.

7.4 COMBINED STRESSES DESIGN PROCEDURE

Refer to Section 5.5 for further explanation.

7.4.1 Combined Axial Compression and Bending

For members subjected to both axial compression and bending, both of the following equations must be satisfied:

TABLE 7.15 Allowable Shear Stresses for Elements of Beams

	$h/t < S_1$	$S_1 < h/t < S_2$	$S_2 < h/t$
Aluminum Specification Equation Number	3.4.20-1	3.4.20-2	3.4.20-3
General equation	$F_s = \dfrac{F_{sy}}{n_y}$	$F_s = \dfrac{1}{n_y}[B_s - 1.25D_s(h/t)]$	$F_s = \dfrac{\pi^2 E}{n_y(1.25h/t)^2}$
Expressed for 6061-T6 (ksi)	$F_s = 12$	$F_s = 15.6 - 0.099h/t$	$F_s = 39,000/(h/t)^2$

where:

$S_1 = 36$ for 6061-T6
$S_2 = 65$ for 6061-T6
h/t = slenderness ratio of the element.

$$\frac{f_a}{F_a} + \frac{C_{mx}f_{bx}}{F_{bx}(1 - f_a/F_{ex})} + \frac{C_{my}f_{by}}{F_{by}(1 - f_a/F_{ey})} \leq 1.0 \qquad (7.9)$$

$$\frac{f_a}{F_{ao}} + \frac{f_{bx}}{F_{bx}} + \frac{f_{by}}{F_{by}} \leq 1.0 \qquad (7.10)$$

When the axial compressive stress is less than or equal to 15% of the allowable axial stress ($f_a \leq 0.15 F_a$), only the following equation must be satisfied:

$$\frac{f_a}{F_a} + \frac{f_{bx}}{F_{bx}} + \frac{f_{by}}{F_{by}} \leq 1.0 \qquad (7.11)$$

In the above equations:

f_a = compressive stress due to axial compression load
f_b = extreme fiber compressive stress due to bending
F_a = allowable compressive stress for the member if it were subjected to only axial compression (see Section 7.2)
F_b = allowable compressive stress for the member if it were subjected to only bending (see Section 7.3)
C_m = $0.6 - 0.4(M_1/M_2)$ for members in frames braced against joint translation, or
= 0.85 for members in frames subject to joint translation (sidesway) in the plane of bending
M_1/M_2 = ratio of the smaller to larger moments at the ends of that portion of the member unbraced in the plane of bending under consideration. M_1/M_2 is positive when the member is bent in reverse curvature, and negative when bent in single curvature
x = subscript for strong axis bending
y = subscript for weak axis bending
F_{ao} = allowable axial compressive stress of the member as if its overall slenderness ratio were less than S_1 (i.e., the allowable stress for the cross-sectional elements; (see Section 7.2)
F_e = allowable axial compressive stress of the member if its overall slenderness ratio were greater than S_2 (see Section 7.2)
= $\pi^2 E/n_u(kL/r)^2$
L = unbraced length as a column in the plane of bending (see Figure 5.67)
k = effective length factor in the plane of bending
r = radius of gyration (see Figure 5.67).

7.4.2 Combined Tension and Bending

Members subjected to axial tension and bending must be proportioned so that:

$$\frac{f_a}{F_t} + \frac{f_{bx}}{F_{bx}} + \frac{f_{by}}{F_{by}} \leq 1.0 \qquad (7.12)$$

where:

f_a = tensile stress due to axial tensile load
f_b = extreme fiber tensile stress due to bending
F_t = allowable tensile stress for the member if it were subjected to only axial tension (see Section 7.1)
F_b = allowable tensile stress for the member if it were subjected to only bending (see Section 7.3)
x = subscript for strong axis bending
y = subscript for weak axis bending.

7.4.3 Combined Shear, Compression, and Bending

Members subjected to shear and bending and members subjected to shear, compression, and bending should be proportioned so that:
For webs of shapes composed of rectangular elements:

$$\frac{f_a}{F_a} + \left(\frac{f_b}{F_b}\right)^2 + \left(\frac{f_s}{F_s}\right)^2 \leq 1.0 \qquad (7.13)$$

For shapes composed of curved elements, such as walls of tubes:

$$\frac{f_a}{F_a} + \frac{f_b}{F_b} + \left(\frac{f_s}{F_s}\right)^2 \leq 1.0 \qquad (7.14)$$

where:

f_a = compressive stress due to axial compression load
F_a = allowable compressive stress for the member if it were subjected to only axial compression (see Section 7.2)
f_b = extreme fiber compressive stress due to bending
F_b = allowable compressive stress for the member if it were subjected to only bending (see Section 7.3)
f_s = maximum shear stress
F_s = allowable shear stress for the member if it were subjected to only shear (see Section 7.3).

The interaction equations of Section 7.4.1 must also be satisfied, whether or not shear is present.

8 Connections

Just when you were starting to get the hang of aluminum structural design, having sized a few members, you realize you have yet to figure out how to connect them. What kind of fasteners are used for aluminum, and what are their design strengths? Can't we just avoid the issue and call for all joints to be full penetration welds and be done with it?

Not so fast. After all, airplanes that log 50 years of service don't have welded frames. We'll consider the benefits and disadvantages of mechanical and welded connections in aluminum and offer some guidelines for their uses. Both aluminum bolting and welding differ from their counterparts in steel, which we'll explain.

We *won't* discuss the third alternative, adhesives, even though adhesives are used in a number of aluminum applications. Adhesives are not commonly used for primary structural connections in construction, and standards for their structural design have not yet been developed. Silicone, polysulfide, and other sealants can adhere well to aluminum, however, and should be considered in place of nonstructural seal welds when it is desirable to keep air or moisture out of joints.

Connections are critical to the integrity of structures, but they tend to be the most imperfectly analyzed component. Also, failure at a connection is more catastrophic than some other failure modes, such as local buckling. Consequently, metal design specifications require greater reliability for connections. This translates into higher factors of safety (20% higher in allowable stress design) and lower resistance factors (in load and resistance factor design [LRFD]) for connections than for members. This is summarized in Table 8.1. No safety margin is given for yielding of connections because such yielding isn't considered a limit state.

8.1 MECHANICAL CONNECTIONS

8.1.1 Introduction

Mechanical connections are the most reliable way to join aluminum structural members, so we'll begin with them. The mechanically fastened joints we'll discuss include those that are connected with rivets, bolts, pins, or screws. Rules for these joints are given in Section 5 of the Aluminum *Specification*.

TABLE 8.1   Safety and Resistance Factors for Members vs. Connections

	Yielding		Fracture or Collapse	
	Safety Factor n_y	Resistance Factor ϕ	Safety Factor n_y	Resistance Factor ϕ
Members	1.65	0.95	1.95	0.85
Connections	–	–	2.34	0.65

Mechanical connectors are more frequently used in aluminum structures than they are in steel for several reasons. Probably the most important is that welding (or any heating above about 450°F [230°C], even of short duration) significantly reduces the strength of artificially aged or strain-hardened aluminum alloys. For example, the heat-affected zone in 6061-T6 has a tensile strength about 40% less than portions of the material not affected by welding. Steel, on the other hand, is often stronger at the weld. The strength of the steel base metal is generally unaffected by welding, and the steel filler alloys are often stronger than the base metal. Also, there are fewer skilled, qualified aluminum welders than steel welders. Last, the inspection and repair of welds can be difficult and expensive. For these reasons, efficient designs of aluminum structures employ as few welds as possible.

Figure 8.1   Aluminum frame with bolted connections. (Courtesy of Conservatek Industries, Inc.)

Aluminum connections are also made with mechanical clinches and snap-fits. Mechanical clinches are used to join thin sheet by locally deforming or shearing the material; various commercial methods are available, but the strength depends on the method and is usually determined from tests. Snap-fits, also called *friction fits,* are made with extrusions that are designed to snap together when deformed elastically. Snap-fits cannot reliably transmit shear along the length of the joint because tolerances on extrusion dimensions are too large for a tight fit.

8.1.2   Types of Fasteners

Four kinds of mechanical fasteners are usually considered for aluminum structures: (1) *rivets,* (2) bolts, (3) pins, and (4) screws.

Rivets   The kind of fastener Rosie the Riveter installed in the World War II is not seen often in construction today, but some are still used. Today, these rivets are usually cold-driven (in other words, at the as-received temperature, without preheating). Alloys 6053-T61, 6061-T6, and 7075-T73 are among the more commonly used; mechanical properties and allowable stresses are provided in Section 8.1.4 below. The rivet (Figure 8.2) is inserted into the hole from one side and is long enough so that there is enough material protruding through the back side of the joint to be forged into the driven head. The preformed head is backed up with a tool called a *bucking bar,* while the other end is pounded by a pneumatic hammer to form a head on the back side of the work. When performed in the shop, installation may also be by continuous pressure using squeeze riveters when this tool can reach both sides of the joint.

Both methods require access to both sides of the parts to be connected. The hole diameter should not be larger than 4% greater than the rivet's nominal diameter (7% for hot-driven rivets), which is best achieved by reaming

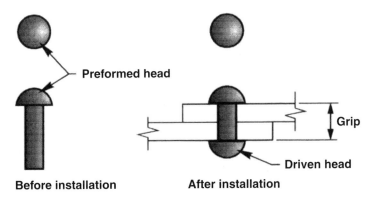

Figure 8.2   Cold-driven rivet.

the assembled parts in place for good hole alignment (110). Because the driven rivet is assumed to fill the hole, the effective area of the rivet in shear is the diameter of the hole, not the nominal diameter of the rivet.

Various head styles are available. The *Aluminum Design Manual*, Part VII, Tables 5-6 through 5-13, give rivet specifications, dimensions, and driving pressures. Don't worry if you never consider using this kind of aluminum rivet; most of us never use steel rivets either. Aluminum riveting is still widely used in certain applications, such as aircraft, rail cars, and tractor trailers. Rivets can resist only shear loads, though, since they are not reliable in tension. They've been largely replaced by bolts.

More common today are *blind rivets,* so called because they can be installed from one side of the work without access to the other side (Figure 8.3). An installation tool pulls the mandrel or pintail, deforming the *rivet body,* which is a sleeve around the mandrel, on the back side of the work to form a head there. These fasteners are sometimes also called *pop rivets* due to the sound made when the mandrel breaks off once the installation is complete. One of the most important parameters in selecting a blind rivet is the total thickness of the material joined, called the *grip,* which dictates the length of the rivet for proper formation of the head on the back side of the material. The most useful blind rivets are those that can accommodate a wide range of grips.

The shear strength of blind rivets is a function of the material in the shear plane. This may be difficult to calculate because rivet body and mandrel material strengths and dimensions may not be accurately known in their installed condition. Consequently, testing is often used to determine these strengths.

The smallest rivet diameter allowed by the Aluminum *Specification* (Sections 5.5.3 and 5.6) for building sheathing and flashing laps is $\frac{3}{16}$ in. [5 mm]. Larger blind fasteners based on the same installation concept are available in

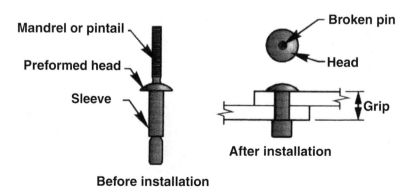

Figure 8.3   Blind rivet.

diameters through $\frac{3}{4}$ in. [20 mm]. These fasteners are especially useful for connections to hollow members.

Blind rivets appropriate for aluminum structures are made of *austenitic stainless steel* or aluminum, and both are commonly available. Usually the rivet body and mandrel are made of different aluminum alloys because they perform different functions. The rivet body must be soft enough to form a head, while the mandrel must be strong enough to deform the body.

Self-piercing rivets, as the name implies, don't require holes in the parts being joined. They're typically used to join overlapping sheets. The rivet is forced through the top sheet and into the bottom sheet without perforating the second sheet. The displaced sheet metal is forced into a cavity in an anvil on the back side of the work, and the tubular end of the rivet is flared by a cone in the center of the anvil cavity to secure the joint. Access to both sides of the joint is required, but shear strength of the joint is comparable to spot-welding. Self-piercing rivets for joining aluminum parts can be made of carbon steel, stainless steel, or aluminum.

Bolts   Aluminum bolts may be hex head or square head, using a hex or square nut, respecively. Dimensions of aluminum bolts, nuts, and threads conform to the same ASME standards as steel parts do. Bolts conform to ASME B18.2.1, except that 1 in. is the largest diameter offered because that's the largest bar available from which they can be made. Threads are in accordance with ASME B1.1. Washers are optional for bolts in holes, but they should be used for bolts in slots. Nuts comply with ASME B18.2.2 (Figure 8.4). The *Aluminum Design Manual,* Part VII, Tables 5-15 through 5-19, give dimensions.

Commonly used aluminum bolt alloys are 2024-T4, 6061-T6, and 7075-T73. (Alloy 6262-T9 is used almost exclusively for nuts.) However, bolts used

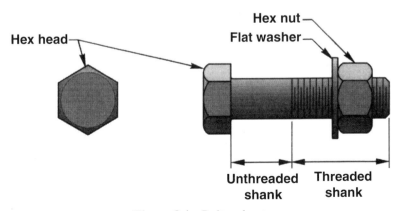

Figure 8.4   Bolt and nut.

in aluminum structures need not be made of aluminum. In fact, other materials, such as 300 series stainless steel and plated and galvanized carbon steel, are specifically allowed by the Aluminum *Specification* Section 5.1.16, Steel Bolts, and frequently used. (See Section 8.1.3 below for more information on material selection.)

The *lockbolt* is another kind of fastener that combines features of rivets and bolts. (See Section 3.3 for more on lockbolts, including installation information.) A lockbolt assembly includes a pin (analogous to a bolt) and a collar (similar to a nut) (Figure 8.5). The collar is swaged onto the pin, tensioning the pin and creating a clamping force between the connected parts. Installation requires access to both sides of the work, but it proceeds quickly and with reliable clamping force, without the need to control torque as conventional threaded fasteners require. Inspection is visual (if it looks right, it is right) and, therefore, readily performed.

Lockbolts are available in stainless or carbon steel or aluminum alloys (2024-T4, 6061-T6, and 7075-T73) and can be manufactured to comply with the MIL-P-23469 specification. The aluminum and stainless lockbolt diameters and grip ranges are shown in Table 8.2. While they require a special installation tool, they are an efficient and economical fastener for aluminum structures.

Pins   Aluminum and stainless steel pins are also very useful to connect aluminum parts. A pin is made from round bar with a hole through each end to receive a *cotter pin.* (Stainless steel cotter pins are often used.) Pins are also available with a head at one end. Because pins do not clamp the connected

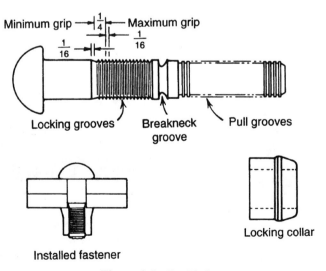

Figure 8.5   Lockbolt.

**TABLE 8.2   Aluminum and Stainless Steel
Lockbolt Diameters and Grip Ranges**

Nominal Diameter (in.)	Grip Range (in.)
$\frac{3}{16}$	$\frac{1}{8}$
$\frac{1}{4}$	$\frac{1}{8}$
$\frac{5}{16}$	$\frac{1}{4}$
$\frac{3}{8}$	$\frac{1}{4}$
$\frac{1}{2}$	$\frac{1}{4}$
$\frac{5}{8}$	$\frac{1}{4}$
$\frac{3}{4}$	$\frac{1}{4}$

parts together, they can be used only where the joint is held together by other means, such as bolts at other locations in the joint or an extrusion, as shown in Figure 8.6. However, aluminum pins are a very economical method of fastening and may be readily fabricated in any bar stock diameter. They need not be limited to applications intended to permit rotation, so more than one pin may be used at the end of a member. The use of extrusions designed to receive pins at connections facilitates a wider use of pins in aluminum construction than in steel.

Screws   Screws are often used in aluminum structures to attach building sheathing, such as corrugated roofing and siding (Figure 8.7). The Aluminum *Specification* does not establish material requirements for screws, but aluminum, such as 7075-T73, or 300 series stainless steel screws are durable. *Tapping,* or *thread-forming, screws* thread their own hole as they are being installed; self-drilling screws drill and tap their own hole (in the proper application). Aluminum and austenitic stainless steel self-drilling screws are available, but don't assume they will drill their own hole in your application. Some may drill through aluminum up to 0.125 in. [3 mm] thick; test them before counting on it. Hardened, plated steel screws more reliably drill their own holes, but they routinely rust if exposed to the elements. Steel screws with a Rockwell hardness of C35 or greater are also prone to hydrogen-assisted stress corrosion cracking when subjected to moisture in dissimilar metal applications; avoid such use with aluminum. Regardless of material, the distance from the end of the screw's drilling point to the beginning of its threads must be greater than the thickness of the material to be drilled.

   The Aluminum *Specification* gives design provisions for screws from 0.164 in. (No. 8) [4.2 mm] to 0.25 in. [6.3 mm] in diameter. However, since the *Specification* also requires that screws for building sheathing and flashing be No. 12 (nominal diameter 0.216 in.) as a minimum, No. 12 is, effectively, the minimum size for use in construction. Screws that attach sheathing to structural framing members are typically No. 14 ($\frac{1}{4}$ in. nominal diameter) $\times$

Figure 8.6 Pinned support for an aluminum dome. (Courtesy of Conservatek Industries, Inc.)

Hex washer
head

Type A sheet
metal screw

Type B self
tapping screw

Figure 8.7 Sheet metal screw.

1 in. long type B 300 series stainless steel tapping screws. Screws used to join roofing or siding sheets to each other at laps are typically No. $14 \times \frac{3}{4}$ in. long type A 300 series stainless steel sheet-metal screws. The type A and type B designations are for types of points and threads most suitable for these applications and both require a predrilled hole. Section 10.1 contains more information on screws for building sheathing.

Washers—either part of the head or separate—are used under the screw head to decrease the likelihood of failure under tensile load by means of the sheet pulling over the head of the screw (see Section 8.1.17 below). Metal washers with neoprene gaskets (called *combination washers*) also seal the joint against leakage and are effective on properly driven screws.

Aluminum nails are also available. Aluminum nails produced to ASTM F1667, *Driven Fasteners: Nails, Spikes, and Staples,* are made of 2024, 5056, 6061, or 6110 alloy and all have a minimum tensile strength of 60 ksi [415 MPa].

Dimensions and other information about many fasteners are available from the Industrial Fastener Institute (IFI) publication *Fastener Standards* (in both English and metric unit versions). A great variety of other types of fasteners, such as *turnbuckles, clevises,* and *eyebolts,* are available in materials suitable for use in aluminum structures, including, in many cases, aluminum and 300 series stainless steel. Availability is a function of quality, quantity, and lead time, so you'll have to consult fastener suppliers.

8.1.3 Fastener Material Selection

In steel design, engineers usually give little consideration to fastener material: steel fasteners are used in steel structures. In aluminum, the choice is less obvious. This is probably because more options exist, as well as because aluminum structures are often provided without protective coatings and in corrosive environments unsuitable for steel alloys. Corrosion of fasteners is a critical concern since fasteners are much smaller than the material they join. Corrosion of fasteners tends to result in a greater decrease in the strength of the structure than the same depth of corrosion in a larger member.

When different materials are used for fasteners and the members they join, *galvanic corrosion* may occur. Different materials, even different alloys of aluminum, have different electrical potential, and in the presence of an electrolyte, such as a wet industrial atmosphere, electric current flows from one material to another. This current tends to corrode the anode and protect the cathode—the principle used in cathodic protection. The greater the difference in electrical potential, the more severe the corrosion; conversely, you can prevent galvanic corrosion by using a fastener with about the same electrical potential as the base metal. Galvanic corrosion can also be thwarted if the relative surface area of the anode is much larger than that of the cathode, which is the case when the fastener is the cathode. There is also the issue of the corrosion resistance of the fastener itself in the service environment, independent of the metal being joined. In light of these considerations, here's how the various materials stack up for use in connecting aluminum parts:

Aluminum Fasteners   Higher strength aluminum alloys 2024 and 7075 are less corrosion resistant than 6061, but they are acceptable for most applications. Sometimes, higher strength alloys are anodized for additional protection. Aluminum *Specification* Section 5.1 specifies a 0.2 mil [0.005 mm] thick anodized coating for 2024 bolts when specified by the designer, and it's a good idea to do so if they will be exposed to moisture. Because the electrical-potential difference between aluminum alloys is small, aluminum bolts don't suffer significant galvanic corrosion when used in aluminum members. The full list of aluminum alloys produced in wire and rod suitable for manufacturing fasteners is given in Table 8.4.

Stainless Steel Fasteners   Whoever came up with the name "stainless steel" was a marketing genius—the alliteration and the image it conjures up are masterful. The problem is, stainless steel stains sometimes. While it's is often referred to as if it were one material, the American Iron and Steel Institute (AISI) actually recognizes dozens of stainless steel alloys. Furthermore, most alloys are available in different conditions that have different strengths. Three broad categories of stainless steel exist. From lowest to highest regarding corrosion resistance, they are: *martensitic, ferritic,* and *austenitic;* about all they have in common is that they all have at least 10.5% chromium. Chromium is what differentiates stainless steel from rust; nickel helps, too, and both are the reason stainless steel is so much more expensive than plain steel. Carbon increases strength, but it decreases the corrosion resistance effect of chromium.

The martensitic stainless steels, the least corrosion resistant of the group, are iron-chromium alloys with 12% to 17% chromium. They also have enough carbon to be heat-treatable, so they can be provided in high strengths, but that's also why they aren't very corrosion resistant. Type 410 is a popular example; with only 12.5% nominal chromium content, it's one of the least expensive stainless steels. The ferritic stainless steels are also iron-chromium alloys, but without carbon. They are not heat-treatable, so they don't come in high strengths. Type 430 is the most popular ferritic stainless steel alloy; it has a nominal chromium content of 16%. The austenitic stainless steels (200 and 300 series) have about 18% chromium and 8% nickel; as such, they are very corrosion resistant. They're not heat-treatable, but they are strengthened by cold-working.

The austenitic stainless steels are the most suitable for use with aluminum and constitute about 80% of all stainless steel fasteners made. If you don't specify which austenitic stainless steel alloy you require, so-called 18-8 is typically furnished. A number of alloys fit this description, so any one of them might be supplied. For example, type 303 is among the least expensive and machines well, but it has the lowest corrosion resistance among the 300 series stainless alloys. Type 304 is more corrosion resistant, and type 316 is better still (at a 50% cost premium), especially in marine environments. Which alloy do you suppose will be supplied on your job if you don't specify one?

The point is that you should specify exactly which alloy you want. Type 304 works well in most industrial atmospheres; the corrosion resistance of type 316 in marine environments can justify its cost premium. Any material submerged in liquid should be subject to special investigation and then tested in that environment. Once the alloy is selected, its condition, which establishes its strength, should be chosen. We'll discuss this further under the topic of fastener strengths below.

It's only a matter of time before plated steel and ferritic and martensitic stainless steel screws rust when exposed to the elements. The heads of through-fastened screws used to attach the cladding of a structure *are* exposed to the weather, so they rust right out there for everyone to see. Fortunately, plated steel and ferritic and martensitic stainless steels are magnetic, while the preferred aluminum and austenitic stainless steel screws are not. Thus, a magnet offers an easy way to identify an offending fastener. For locations not exposed to the elements, the proposed North American Fenestration standard, which is similar to AAMA/NWWDA 101, allows the use of both non-austenitic stainless steel with a nominal content of 16% or more chromium and coated carbon steel fasteners. This allows 430 stainless but excludes 410 stainless.

Carbon Steel Fasteners   Plated or galvanized carbon steel bolts may be useful in less exposed or corrosive applications. Coated A307 and A325 bolts may be used, but A490 bolts are not allowed by the Aluminum *Specification* because they become brittle when galvanized (122). This is no great loss, however, since A490 bolts are most profitably used with high strength steel members, and aluminum alloy strengths this high are not generally used in construction. Because A449 bolts have less stringent quality control provisions than A325 bolts, A449 bolts are permitted only by the Steel *Specification* (38, 39) for diameters not available in A325, which is above $1\frac{1}{2}$ in. diameter. Bolts of this size are not often needed in either steel buildings or aluminum construction. Steel rivets are prohibited by the Aluminum *Specification* unless corrosion is of no concern.

8.1.4 Fastener Mechanical Properties

The *minimum mechanical properties* of concern (shear and tensile ultimate strengths) for aluminum bolts are listed in Aluminum *Specification* Table 5.1.1.1-1. Specifications for aluminum bolts are also contained in ASTM F468, *Nonferrous Bolts, Hex Cap Screws, and Studs for General Use* (80); and in ASTM F467, *Nonferrous Nuts for General Use* (79), for aluminum nuts. (However, see the clarifying note below in the paragraph regarding tensile loads on fasteners before using minimum tensile strengths from ASTM F468.) Each bolt manufactured to ASTM F468 must be marked with a code designating its alloy as tabulated in Table 8.3.

ASTM F901, *Aluminum Transmission Tower Bolts and Nuts*, addresses this special application and is limited to $\frac{5}{8}$, $\frac{3}{4}$, and $\frac{7}{8}$ in. diameter 2024-T4 bolts with 6061-T6 or 6262-T9 nuts.

Minimum strengths and allowable stresses for bolts and driven rivets are given in Aluminum *Specification* Tables 5.1.1.1-1 and 5.1.1.2-1. Table 8.4 also gives this information but is updated for current alloys available for making aluminum bolts and rivets (from ASTM B316, *Aluminum and Aluminum-Alloy Rivet and Cold-Heading Wire and Rods* (58)). The most available alloys are 2024-T4, 6061-T6, and 7073-T73; specify the others only after ascertaining their availability.

Minimum tensile strengths for stainless steel fasteners are given in ASTM F593, *Stainless Steel Bolts, Hex Cap Screws, and Studs* (81). (Strengths for full-size tests and machined specimen tests are given; preference should be given to full-size test strengths.) Tensile strengths vary by alloy; condition, such as cold-worked or strain-hardened; and diameter. Table 8.5 shows a few examples. Bolts conforming to ASTM F593 are easy to identify: each bolt must be marked "F593" and with a letter code designating the condition. An example of this is shown in Figure 8.8.

ASTM F594 addresses stainless steel nuts. Nuts should be of the same alloy group as the bolt. (As defined by ASTM F593, alloys 303 and 304 are in alloy group 1, and 316 is in alloy group 2).

Strengths for carbon steel bolts are given in ASTM specifications, such as A307 and A325.

Beware of fastener-manufacturer provided "typical strengths," which may be given for some proprietary fasteners. Metal design specifications require that safety factors (or resistance factors) for fasteners be applied to minimum strengths, not typical values. A good way to establish the minimum strength is to identify conformance of the material to an ASTM specification. If the material specification cannot be precisely identified, the alternative is testing. Tests conducted to establish minimum strength, however, must be conducted and interpreted in accordance with Aluminum *Specification* Section 8.3.1, Tests for Determining Mechanical Properties. This way, minimum strengths

TABLE 8.3   ASTM F468 Head Marking Code

Marking	Alloy
F468X	2024-T4
F468Y	6061-T6
F468Z	7075-T73

TABLE 8.4 Allowable Stresses for Aluminum Fasteners for Building-Type Structures

Alloy Temper	Minimum Shear Strength		Allowable Shear Stress on Effective Area		Minimum Tensile Strength		Allowable Tensile Stress on Effective Area	
	(ksi)	(MPa)	(ksi)	(MPa)	(ksi)	(MPa)	(ksi)	(MPa)
2017-T4	33	225	14	95	55	380	24	165
2024-T42	37	255	16	110	62	425	26	180
2117-T4	26	180	11	75	38	260	16	110
2219-T6	30	205	13	90	55	380	24	160
6053-T61	20	135	8.5	60	30	205	13	90
6061-T6	25	170	10.5	75	42	290	18	125
7050-T7	39	270	17	115	70	485	30	205
7075-T6	42	290	18	125	77	530	33	225
7075-T73	41	280	18	120	68	470	29	200
7178-T6	46	315	20	135	84	580	36	250
7277-T62	35	240	15	105	60	415	26	175

Notes

Strengths apply for 0.063 in. [1.60 mm] to 1.00 in. [25.00 mm] diameters, except 7277 diameters, which are 0.500 in. [12.50 mm] to 1.250 in. [32.00 mm].

Allowable stress is the minimum strength divided by 2.34.

All rivets are driven cold as received.

TABLE 8.5 Minimum Tensile and Shear Strengths of Stainless Steel Fastener Alloys 303, 304, and 316

ASTM F593 Condition		Nominal Diameter D (in.)	Tensile Strength (ksi)	Shear Strength (ksi)
A	Annealed	$\frac{1}{4}$ to $1\frac{1}{2}$	75	45
CW1	Cold-Worked	$\frac{1}{4}$ to $\frac{5}{8}$	100	60
CW2	Cold-Worked	$\frac{3}{4}$ to $1\frac{1}{2}$	85	51
SH1	Strain-Hardened	$\frac{1}{4}$ to $\frac{5}{8}$	120	72
SH2	Strain-Hardened	$\frac{3}{4}$ to 1	110	66

Diameters are inclusive.

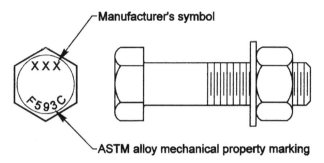

Figure 8.8   ASTM F 593 head markings.

established by testing are determined on a consistent basis with the minimum strengths for aluminum alloys elsewhere in the Aluminum *Specification*.

Yield strengths are not used in determining fastener shear strength because the effect of shear yielding of a fastener is typically imperceptible and does not limit the strength of the connection. For example, if a $\frac{3}{4}$ in. [20 mm] diameter rivet reaches yield in shear, the shear displacement (based on the 0.2% offset definition) is $0.002 \times \frac{3}{4} = 0.0015$ in. [0.038 mm]. This displacement is too small to have any effect. Each hole in a bolted connection may be oversized by as much as 0.0625 in. [1.6 mm], so shear yield displacement is even less important in bolted connections than riveted connections. For this reason, the Aluminum *Specification*—nor the AISC (38, 39) or AISI (40) steel specifications—does not limit fastener strength for yield considerations.

8.1.5 Types of Loads on Fasteners

Fasteners in aluminum structures may be subjected to shear, tension, or combined shear and tension. While each of the fasteners discussed above is suitable in shear, not all fasteners are acceptable for tensile loads. Table 8.6 shows the functions each fastener performs.

TABLE 8.6 Fastener Loads

Fastener	Resists Shear?	Resists Tension?	Holds Connected Parts Together?
Rivet	Yes	No	Yes
Bolt	Yes	Yes	Yes
Pin	Yes	No	No
Screw	Yes	Yes	Yes

8.1.6 Types of Bolted Connections

Rivets were the first kind of fastener devised for both steel and aluminum structures. They were labor-intensive to install and tolerated almost no hole mismatch, but were simple and permanent. Another drawback, though, was that they were unsuitable for any appreciable tensile loads. Bolts, too, were unreliable under cyclic tensile loads because the nuts could loosen, and so were confined to temporary use during erection. The introduction of high-strength steel bolts in the 1950s overcame this problem in steel construction. They were strong enough to be tightened sufficiently that there was no danger of the nuts coming off. These bolts could, if needed, use friction between the *faying* surfaces to prevent the connected parts from slipping in relation to each other, although at a reduced design strength. In the AISC Steel *Specification* (38, 39), the connections designed for no slippage are called *slip-critical connections* (formerly called *friction connections*).

Aluminum Friction Connections Friction connections (or slip-critical connections; call them what you will), however, are a much more recent development in aluminum structures, being introduced to the Aluminum *Specification* in Section 5.1.17 only in 2000. Even so, the circumstances under which they can be used are relatively limited: the aluminum faying surfaces must be abrasion-blasted with coal slag to an average substrate profile of 2.0 mils [0.05 mm], for which a slip coefficient of 0.50 is given (121). Such aluminum surfaces may be used in contact with similarly treated aluminum surfaces or a steel surface with a 4 mil [0.1 mm] maximum thickness zinc-rich paint. Only zinc-coated A325 bolts may be used; this is to prevent galvanic corrosion between the steel bolt and the aluminum parts. The connected aluminum parts must have a tensile yield strength of at least 15 ksi [105 MPa] to keep the bolt from crushing the connected parts. Given these limitations and the fact that the smallest A325 bolt diameter is $\frac{1}{2}$ in. (and $\frac{3}{4}$ in. is much more common), aluminum slip-critical connections are only for the stronger aluminum alloys in heavy duty structural uses.

Because aluminum and steel have different coefficients of thermal expansion, some have speculated that aluminum friction connections could loosen if the temperature changed from the installation temperature. For example, if the temperature drops after tightening, the aluminum parts in the grip of the bolt should contract more than the steel bolt. This could cause relaxation of the bolt tension. Logical as this sounds, tests conducted in the U.S. and Canada (104, 123) have not been able to consistently demonstrate such an effect, and no one has been able to explain why.

Tests (124) were also conducted to determine the proper turn-of-nut tightening method for aluminum friction connections. Since aluminum has a lower modulus of elasticity than steel, investigators expected it might take more turns to achieve the same bolt tension with aluminum parts in the bolt grip versus steel parts. In the tests, however, it took nearly identical turns after the snug-tight condition was reached (defined as having all plies of the joint in firm but not necessarily continuous contact) to attain the prescribed preload with aluminum parts as for steel. The reason may be that aluminum connected parts can be brought to firmer contact, and that bolt tension is higher at the snug-tight point than for steel parts because aluminum is more flexible. Whatever the behavior, as a result of this investigation, the Aluminum *Specification* prescribes the same number of turns for aluminum friction connections as for steel friction connections, given in the Research Council on Structural Connections (RCSC) *Specification for Structural Joints Using ASTM A325 or A490 Bolts* (131).

Aluminum Bearing Connections Most aluminum bolted connections are designed under the assumption that the connected parts can slip in relation to each other. These kinds of bolted connections are called *bearing connections* because the bolt may bear on the side of the hole under design load. Is this a handicap? Not really, since other than a greater fatigue strength, the main benefit of slip-critical connections is that they allow larger hole oversizing. In steel, several kinds of holes for bolts are allowed, including oversized and short and long slotted holes. These are shown in Figure 8.9. Oversized holes and slotted holes where the load does not act perpendicular to the slot length are allowed in steel to facilitate fit-up where alignment is expected to be difficult, but only when the joint is designed using the lower strength of a slip-critical connection. In bearing connections, therefore, only standard holes and slots with the load acting perpendicular to the slot length are allowed. Consequently, aluminum fabrication must be accurate, or holes must be drilled with the parts held in alignment.

Using lockbolts (described in Section 8.1.2) doesn't guarantee a slip-proof joint, but they safely resist both shear and tension like high-strength steel bolts. And, as with high-strength steel bolts, there is little danger of a lockbolt disengaging, so it can be used on connections subject to vibration.

Because bearing connections cannot be relied on to prevent slippage of the connected parts in relation to each other, bolts cannot share loads with welds in aluminum bearing connections.

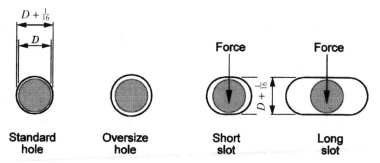

Figure 8.9 Hole size options for steel bolts.

8.1.7 Holes

Holes in aluminum parts must not be more than $\frac{1}{16}$ in. [1.6 mm] larger than the nominal bolt diameter (Aluminum *Specification* Section 6.4d) for the reasons just discussed. Hole diameters for cold-driven rivets are limited to no more than 4% larger than the nominal rivet diameter (*Specification* Section 6.4b). When blind rivets are used, hole sizes should be in accordance with the manufacturer's instructions.

The preferred method of hole fabrication is punching because it can be performed quickly. However, punching cannot be used when the metal thickness is greater than the hole diameter. This is because punching in such cases creates a ragged, enlarged hole (see Figure 8.10). In these cases, the hole must be drilled, or subpunched and then reamed to the final diameter. The diameter of the subpunched hole must be smaller than the diameter of the final hole by one-quarter the thickness or $\frac{1}{32}$ in. [0.8 mm], whichever is less (Aluminum *Specification* Section 6.4a). For purposes of calculating net area, the diameter of punched holes is taken as $\frac{1}{32}$ in. larger than the fabrication diameter to account for the effect of tear out at the back of the hole. The Steel *Specification,* on the other hand, requires that the size of all holes, regardless of method of fabrication, be taken as $\frac{1}{16}$ in. [1.6 mm] larger than the fabrication diameter.

Some oversizing of the hole is necessary to allow for tolerance on fabrication and temperature differences between parts. Because aluminum has a coefficient of thermal expansion about twice that of steel, temperature differences, if they occur, will cause a greater effect in aluminum structures. For aluminum members fabricated to $\frac{1}{32}$ in. [1 mm] tolerance, a hole sized $\frac{1}{32}$ in. greater than the fastener diameter permits ready fit-up.

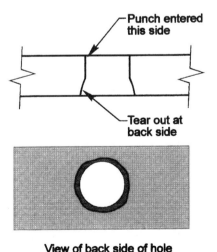

View of back side of hole

Figure 8.10 Tear out at back of punched hole.

Holes more than $\frac{1}{16}$ in. [1.6 mm] larger than the bolt diameter are often needed to align holes in base plates with anchor bolts embedded in concrete. (Embedded anchors tend not to be located very precisely.) You can avoid this situation by using drilled-in-place concrete expansion or adhesive anchors. The hole in the base plate can be used as a template to locate anchors precisely enough to match holes only $\frac{1}{16}$ in. oversized. Concrete anchor bolts are available in galvanized carbon steel and 304 and 316 stainless steels, so they are acceptable for use with aluminum base plates. Double base plate details that permit oversized holes and adjustable embedded anchor systems may also be used.

Recommended hole sizes for aluminum sheet metal screws, which tap their own hole, are given in *Aluminum Design Manual*, Part VII, Table 5-14. Hole sizes are a function of screw size and type and material thickness. If the hole is too large, the threads in the connected part may strip; conversely, if the hole is too small, the screw head may break off during installation.

Sometimes holes are fabricated in the wrong place. Fabricators usually want to weld up the hole, but it's usually better to leave the hole in place since welding often introduces weld defects that are more harmful than the hole itself. Additionally, welding robs aluminum of strength derived from cold-work or precipitation-heat treatment; designers must account for this in the design.

8.1.8   Failure Modes for Mechanically Fastened Joints

In order to design joints, you have to look at the dark side of things, anticipating how they might fail. For example, bolts subjected to tension may break. Failure modes for bolted and riveted joints in shear are shown in Figure 8.11 and are: 1) shear failure of the fastener, 2) bearing failure, 3) tension failure of the connected parts, and 4) break out of the end of the connected part, also referred to as *shear rupture, block shear,* or *tear out.* Long connectors may also be subject to bending, so their strength is limited by Aluminum *Specification* Section 5.1.8, Grip of Rivets, Screws, and Bolts, which is similar to provisions for steel fasteners other than high-strength bolts. Screws can fail in yet other ways; we'll address those separately in Section 8.1.17 below. For each failure mode, we'll examine how the Aluminum *Specification* determines the strength.

Mechanically fastened connections may also fail in fatigue under repeated cycles of loading and unloading. The fatigue strength of connections is discussed in Section 9.2, Fatigue.

First, let's establish the ground rules. Structural design for fasteners of a given material are taken from the structural specification for the material used to make the fastener:

1) *Aluminum*—The Aluminum *Specification* requires a safety factor of 2.34—or a resistance factor of 0.65—to be applied to the minimum tensile

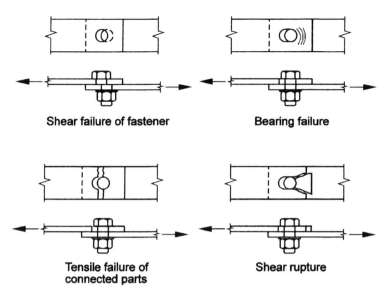

Figure 8.11   Failure modes of bolted connections.

and shear ultimate strengths of aluminum bolts and rivets for *building structures*. (See *Specification* Section 5.1.1.) For screws, a safety factor of 3.0, or a resistance factor ϕ of 0.50, is used.

2) *Stainless Steel*—ASCE 8-90, *Specification for the Design of Cold-Formed Stainless Steel Structural Members* (82), governs the design of stainless steel fasteners used in aluminum structures. In that specification, Table E requires factors of safety on minimum strengths of 3.0 for bolt shear and 3.0 for bolt tensile strength. (Minimum strengths are given in ASTM F593 as discussed above.) Table 6 of ASCE 8-90 gives resistance factors if LRFD is used. Because stainless steel material specifications do not provide a minimum shear strength (unlike the Aluminum *Specification*), ASCE 8-90 provides the relationship that shear strength is 60% of tensile strength. (See its Section 5.3.4 commentary.)

3) *Carbon Steel*—The AISC *Specification for Structural Steel Buildings* governs the design of carbon steel bolts in buildings. Strengths for design may be taken from Steel *Specification* Table J3.2, but only bearing connections may be used unless the faying surfaces are prepared in accordance with Aluminum *Specification* Section 5.1.17. As we discussed above, bearing connections have higher design strengths than slip-critical connections, so from a strength standpoint, this is not a disadvantage. The AAMA *Metal Curtain Wall Fasteners* report (34) offers guidance on design of steel fasteners not included in the AISC *Specification*.

8.1.9 Tensile Loads on Fasteners

Aluminum differs from steel when it comes to the definition of the effective area of a bolt in tension. For steel bolts, the tensile effective area is the tensile stress area:

$$\text{tensile stress area} = A_s = 0.7854 \left(D - \frac{0.9743}{n} \right)^2$$

where:

D = nominal size (basic major diameter)
n = number of threads per in.

The tensile stress area is an empirically determined area used for computing the tensile strength of externally threaded fasteners. It has been selected so that when the fastener strength (in units of force) is divided by the tensile stress area (in units of area), the resulting calculated stress equals the tensile strength of the material (in units of force per unit area). The diameter on which the tensile stress area is based is between the outside diameter of the fastener and the diameter at the root of the threads, and it accounts for the notch and helix effect of the threads. Tensile stress areas are tabulated in the Steel *Manual* (39), page 4-147, for coarse threads, and the same equation is used for stainless steel fasteners and tabulated in ASTM F593, Table 4, for both coarse and fine thread series. (Coarse threaded bolts are typically used in construction.)

The effective area of aluminum bolts in tension is defined as the *root area* in Aluminum *Specification* Section 5.1.1.1. The root area is not defined in the *Specification*, nor in the IFI *Fastener Standards* (112), but *Fastener Standards* defines the *root diameter* as the non-preferred term for the minor diameter of external thread for which it defines the area as:

$$A_r = 0.7854 \left(D - \frac{1.3}{n} \right)^2$$

Comparing the equations for A_s and A_r, you can see that the area used to calculate tensile stress on an aluminum bolt (A_r) is less than the area used to calculate tensile stress area on a steel bolt (A_s). Table 8.7 gives effective tensile areas for coarse-thread series for steel and aluminum bolts. The effective areas used for steel are 10% to 18% greater than the effective tensile areas used for aluminum. There is no physical difference in size for bolts of the two materials; their definitions of effective tensile area are simply different. Because ASTM F468 uses the steel definition for effective tensile area, minimum tensile strengths for aluminum alloys in ASTM F468 have been factored down to give the same tensile force as the Aluminum *Specification* when area is multiplied by strength.

TABLE 8.7 Cross-Section Areas for Bolts (UNC [Coarse] Threads)

Nominal Diameter D (in.)	Threads/in. n (1/in.)	Nominal Area A (in.^{2})	Tensile Stress Area A_s (in.^{2})	Root Area A_r (in.^{2})	A_s/A_r
0.216	24	0.0366	0.0242	0.0206	1.17
$\frac{1}{4}$	20	0.0491	0.0318	0.0269	1.18
$\frac{5}{16}$	18	0.0767	0.0524	0.0454	1.15
$\frac{3}{8}$	16	0.110	0.0775	0.0678	1.14
$\frac{7}{16}$	14	0.150	0.1063	0.0933	1.14
$\frac{1}{2}$	13	0.196	0.142	0.1257	1.13
$\frac{9}{16}$	12	0.249	0.182	0.162	1.12
$\frac{5}{8}$	11	0.307	0.226	0.202	1.12
$\frac{3}{4}$	10	0.442	0.334	0.302	1.11
$\frac{7}{8}$	9	0.601	0.462	0.419	1.10
1	8	0.7854	0.606	0.551	1.10

Example 8.1: What is the allowable tensile load on a $\frac{1}{2}$ in. diameter 7075-T73 aluminum bolt?

The allowable tensile load for aluminum bolts is the root area times the allowable tensile stress. The root area for a $\frac{1}{2}$ in. diameter bolt is taken from Table 8.7 as 0.1257 in.^{2}. The allowable tensile stress is the ultimate tensile strength divided by the safety factor for connections, 2.34. (Refer to Aluminum *Specification* Section 5.1.1.1.) The ultimate tensile strength is 68 ksi in Table 8.4, so the allowable tensile stress is (68 ksi)/2.34 = 29 ksi. (This value is also given in *Specification* Table 5.1.1.1-1.) The allowable tensile force is, then, (0.1257 in.^{2})(29 ksi) = 3.6 k.

Allowable tensile forces on aluminum bolts for the three alloys commonly used for bolts are given in Table 8.8.

Bolts designed to carry tensile loads should be installed with enough torque to prevent nuts from loosening in service. The RCSC *Specification for Structural Joints Using ASTM A325 or A490 Bolts* permits four ways to do this for such connections in steel: turn-of-nut, calibrated wrench, alternative fasteners, or tension-indicating devices. In all four methods, the bolt is first made snug-tight, which is defined as having all plies of the joint in firm but not necessarily continuous contact. Tension-indicating devices (also called *DTIs*, for *direct tension indicators*) include washers with bumps that flatten when bolts are properly tensioned. Alternative fasteners, including lockbolts, incorporate a feature, such as a twist-off end that fails when enough tension is applied. The calibrated-wrench method requires daily calibration and hardened washers, but does not, contrary to urban legend, allow the use of stan-

TABLE 8.8 Allowable Tensile Force in Aluminum Bolts

Nominal Diameter D (in.)	Effective Tensile Area A_r (in.²)	*Allowable Tensile Force* (k) (for building-type structures)		
		2024-T4 Allowable Stress = 26 ksi	6061-T6 Allowable Stress = 18 ksi	7075-T73 Allowable Stress = 29 ksi
$\frac{1}{4}$	0.0269	0.70	0.48	0.78
$\frac{5}{16}$	0.0454	1.2	0.82	1.3
$\frac{3}{8}$	0.0678	1.8	1.2	2.0
$\frac{7}{16}$	0.0933	2.4	1.7	2.7
$\frac{1}{2}$	0.1257	3.3	2.3	3.6
$\frac{9}{16}$	0.162	4.2	2.9	4.7
$\frac{5}{8}$	0.202	5.3	3.6	5.9
$\frac{3}{4}$	0.302	7.9	5.4	8.8
$\frac{7}{8}$	0.419	11	7.5	12
1	0.551	14	10	16

Note

Allowable tensile forces in Table 8.8 are calculated using the rounded off allowable stresses for aluminum alloys from Aluminum *Specification* Table 5.1.1.1-1 for building-type structures and aluminum stress areas from Table 8.7.

dard torques from tables or formulas. Turn-of-nut tightening is much more commonly used than any other. In this method, after the bolt is made snug-tight, the nut is turned an additional specified portion of a turn, usually with a long-handled wrench. As noted in Section 8.1.6 above, tests have shown that for A325 bolts in aluminum parts, using the same number of turns as for steel assemblies will reliably produce the proper bolt tension. It might seem that aluminum, with one-third the modulus of elasticity of steel, would require more turns than steel, but apparently aluminum's flexibility enables bolts to reach greater tension at the snug-tight condition than stiffer steel parts.

Tightening aluminum bolts is another matter. The Aluminum *Specification* offers no guidance on this, but since aluminum bolts have not been demonstrated as suitable for slip-critical connections, they would be applied only to bearing connections and, thus, require only snug-tight installation.

8.1.10 Shear Loads on Fasteners

Aluminum and steel specifications also differ in their accounting methods for fastener shear area. The steel specifications (39, 82) base all shear stress computations on the nominal area of the fastener and adjust allowable stresses downward when the threaded portion of the bolt extends into the shear plane. (For A307 steel bolts, code writers assumed the threads will be in the shear plane, so only one allowable shear stress is provided.) The two cases are illustrated in Figure 8.12.

The Aluminum *Specification,* on the other hand, calls for shear stress to be calculated on what is defined as the effective area of the bolt, which is the nominal area when threads are not in the shear plane, and the root area when threads are present in the shear plane. Consequently, only one allowable shear stress exists for an aluminum bolt, regardless of where the threads lie, but the areas used for shear stress computation differ for the two cases.

Example 8.2: What is the allowable shear load on a $\frac{3}{8}$ in. diameter 7075-T73 aluminum bolt with 1) no threads in the shear plane, and 2) threads in the shear plane?

1) The allowable shear load for aluminum bolts with no threads in the shear plane is the nominal area times the allowable shear stress. The nominal area for a $\frac{3}{8}$ in. diameter bolt is taken from Table 8.7 as 0.110 in.². The allowable shear stress is the ultimate shear strength divided by the safety factor for connections, 2.34. (Refer to Aluminum *Specification* Section 5.1.1.2.) The ultimate shear strength is 41 ksi in Table 8.4, so the allowable shear stress is (41 ksi)/2.34 = 18 ksi. (This value is also given in *Specification* Table 5.1.1.1-1.) The allowable shear force is, then, (0.110 in.²)(18 ksi) = 2.0 k.

2) The allowable shear load for aluminum bolts with threads in the shear plane is the root area times the allowable shear stress. The root area for a $\frac{3}{8}$ in. diameter bolt is taken from Table 8.7 as 0.0678 in.². The allowable shear stress is the same as for no threads in the shear plane calculated in (1) above (18 ksi). (This value is also given in *Specification* Table 5.1.1.1-1.) The allowable shear force is, then, (0.0678 in.²)(18 ksi) = 1.2 k.

Table 8.9 lists allowable shear loads for the commonly used aluminum bolt alloys. You can use the values in this table for threads excluded from the

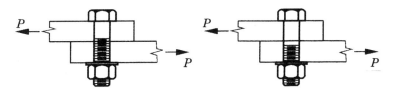

Threads in shear plane   No threads in shear plane

Figure 8.12 Bolt threads in and out of the shear plane.

TABLE 8.9 Allowable Shear Loads (k) on Aluminum Bolts

Nominal Diameter D (in.)	2024-T4 Allowable Stress = 16 ksi		6061-T6 Allowable Stress = 11 ksi		7075-T73 Allowable stress = 18 ksi	
	Threads in Shear Plane	No Threads in Shear Plane	Threads in Shear Plane	No Threads in Shear Plane	Threads in Shear Plane	No Threads in Shear Plane
$\frac{1}{4}$	0.43	0.79	0.30	0.54	0.48	0.88
$\frac{5}{16}$	0.73	1.2	0.50	0.84	0.82	1.4
$\frac{3}{8}$	1.1	1.8	0.75	1.2	1.2	2.0
$\frac{7}{16}$	1.5	2.4	1.0	1.7	1.7	2.7
$\frac{1}{2}$	2.0	3.1	1.4	2.2	2.3	3.5
$\frac{9}{16}$	2.6	4.0	1.8	2.7	2.9	4.5
$\frac{5}{8}$	3.2	4.9	2.2	3.4	3.6	5.5
$\frac{3}{4}$	4.8	7.1	3.3	4.9	5.4	8.0
$\frac{7}{8}$	6.7	9.6	4.6	6.6	7.5	11
1	8.8	13	6.1	8.6	9.9	14

Notes

Allowable shear loads shown in Table 8.9 are based on single shear using the rounded off values for allowable stress from Aluminum *Specification* Table 5.1.1-1, for building-type structures.

Threads in shear plane: area based on root diameter; see Table 8.7.

No threads in shear plane: area based on nominal diameter; see Table 8.7.

shear plane for aluminum lockbolts since their shank diameters in the shear plane are almost identical to the same nominal-size hex bolt.

Example 8.3: What is the allowable shear load on a $\frac{3}{8}$ in. diameter ASTM F593 type 304 condition CW stainless steel bolt with 1) no threads in the shear plane, and 2) threads in the shear plane?

1) The allowable shear load for stainless steel bolts with no threads in the shear plane is the nominal area times the allowable shear stress. The nominal area for a $\frac{3}{8}$ in. diameter bolt is taken from Table 8.7 as 0.110 in.^{2}. The allowable shear stress is the minimum shear strength divided by the safety factor for stainless steel bolts (3.0). The minimum shear strength is 60 ksi in Table 8.5, so the allowable shear stress is (60 ksi)/3.0 = 20 ksi. The allowable shear force is, then, (20 k/in.^{2})(0.110 in.^{2}) = 2.2 k.

2) The allowable shear load for stainless steel bolts with threads in the shear plane is the nominal area times the allowable shear stress times 0.75. The nominal area for a $\frac{3}{8}$ in. diameter bolt is taken from Table 8.7 as 0.110 in.^{2}. The allowable shear stress must be multiplied by 0.90, however, for bolts less than $\frac{1}{2}$ in. in diameter because a greater proportion of the diameter of smaller bolts is lost to threading than for larger bolts. (See ASCE 8-90, Table 6, footnote g.) The allowable shear force is, then, (0.110 in.^{2})(20 k/in.^{2})(0.75)(0.9) = 1.5 k.

Tables 8.10 and 8.11 list allowable shear loads for stainless steel bolts, calculated according to ASCE 8-90.

The Aluminum *Specification* does not prescribe the 20% reduction in shear capacity of bolts in long connections, as the Steel *Specification* (39) does (Table J3.2) for splices with a length greater than 50 in. [1300 mm]. Con-

TABLE 8.10   Allowable Shear Loads (k) on Stainless Steel Bolts, No Threads in Shear Plane

Nominal Diameter D (in.)	Minimum Tensile Strength (ksi)				
	75	85	100	110	120
$\frac{1}{4}$	0.74	0.83	0.98	1.1	1.2
$\frac{3}{8}$	1.7	1.9	2.2	2.4	2.7
$\frac{1}{2}$	2.9	3.3	3.9	4.3	4.7
$\frac{5}{8}$	4.6	5.2	6.1	6.7	7.4
$\frac{3}{4}$	6.6	7.5	8.8	9.7	11
$\frac{7}{8}$	9.0	10	12	13	14
1	12	13	16	17	19
$1\frac{1}{4}$	18	21	25	27	29

TABLE 8.11 **Allowable Shear Loads (k) on Stainless Steel Bolts, Threads in Shear Plane**

Nominal Diameter D (in.)	Minimum Tensile Strength (ksi)				
	75	85	100	110	120
$\frac{1}{4}$	0.50	0.56	0.66	0.73	0.80
$\frac{3}{8}$	1.1	1.3	1.5	1.6	1.8
$\frac{1}{2}$	2.2	2.5	2.9	3.2	3.5
$\frac{5}{8}$	3.5	3.9	4.6	5.1	5.5
$\frac{3}{4}$	5.0	5.6	6.6	7.3	8.0
$\frac{7}{8}$	6.8	7.7	9.0	9.9	11
1	8.8	10	12	13	14
$1\frac{1}{4}$	14	16	18	20	22

nections this long are rare in aluminum structures, but bolt strengths should also be reduced wherever they occur. The steel requirement is based on consideration of ductility and compatibility of strains through the joint (138) by which bolts at the ends of long connections attract higher shears and should apply to aluminum, as well as steel.

Aluminum Design Manual, Part VII, Table 5-2 gives reductions in shear strength of aluminum rivets used in thin sheets or shapes relative to the rivet diameter. This table is based on an article titled "Joining Aluminum Alloys," which was published in 1944 (110). The reason cited there for the reduction is the cutting action of very thin plates on the shank of aluminum rivets, which is not fully accounted for by bearing stress limitations. In single shear, this reduction occurs only if the thinner plate joined is thinner than $\frac{1}{3}$ the rivet diameter. The reduction is larger for double shear and is used when the middle plate is thinner than $\frac{2}{3}$ the rivet diameter.

8.1.11 Combined Shear and Tension on Bolts

The Aluminum *Specification* contains no provisions for combined shear and tension on bolts, but it would seem prudent to reduce the allowable shear stress on bolts in tension, and vice versa. Both the steel (39) and stainless steel (82) specifications have such provisions. These provisions can be used when fasteners of these materials are used in aluminum structures. If you use aluminum fasteners in both shear and tension, however, you're on your own. The steel specifications are based on an elliptical interaction curve approximated by three straight lines for simplicity. The elliptical interaction equation is:

$$\left(\frac{f_s}{F_s}\right)^2 + \left(\frac{f_t}{F_t}\right)^2 \leq 1.0$$

where:

f_s = shear stress in bolt
F_s = allowable shear stress in bolt
f_t = tensile stress in bolt
F_t = allowable tensile stress in bolt

This can also be expressed as:

$$f_t \leq \frac{F_t}{F_s} \sqrt{F_s^2 - f_s^2}$$

8.1.12 Bearing Strength and Edge Distance

The area used to calculate bearing stress is the length in bearing times the hole diameter for rivets and the nominal bolt diameter for bolts, even if threads occur in the bearing area. For countersunk rivets or bolts, one-half the depth of the countersink shall be deducted from the length (Aluminum *Specification* Section 5.1.4). These provisions are identical in the aluminum and steel (39) specifications. Countersinking the full thickness of the material is poor practice because the knife edge created may damage the fastener or be easily distorted.

The allowable bearing stress on rivets and bolts is given in Aluminum *Specification* Section 3.4.5 as $2F_{tu}/n_u$, when the distance from the center of the fastener to the edge of the part in the direction of the applied force (edge distance) is two fastener diameters (D) or more (Figure 8.13). Since n_u is 1.95, the allowable bearing stress is approximately equal to the tensile ultimate strength F_{tu}. Limiting the stress to $2F_{tu}$ limits hole elongation to about one-third of the bolt diameter (or less, when the material around the hole is confined by outer plies of connected parts or tightly installed nuts). The edge distance may be as small as $1.5D$, but if it's less than $2D$, the allowable bearing stress must be reduced by the ratio of edge distance to $2D$. The maximum reduction is, therefore,

$$25\% = 1 - 1.5/2.0.$$

The allowable bearing stress for a pin in a hole or a bolt in a slot is limited by *Aluminum Specification* Section 3.4.6 to $2F_{tu}/(1.5n_u)$ (approximately $\frac{2}{3}F_{tu}$), which is the allowable bearing stress for rivets and bolts in holes reduced by one-third. This reduction is arbitrary as opposed to rational.

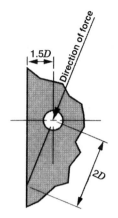

D = **Fastener diameter**

Figure 8.13   Bolt hole edge distance.

8.1.13   Tension Strength of Connected Parts

The tensile strength of connected parts is addressed in Section 5.1, Tension. An issue that applies only to bolted connections, however, is covered here.

The Aluminum *Specification* does not place a limit on the *net area*, which is the portion of the *gross area* of a bolted member that acts to resist tensile load. The Steel *Specification*, however, does limit the net area of bolted and riveted splice and gusset plates (as opposed to structural shapes) in tension to 85% of the gross area (Section B3). It is reasonable to apply this limitation to aluminum gusset plates as well. Different limitations on net area apply to structural shapes, as discussed in Section 5.1.

Example 8.4: What is the net area of the splice plate shown in Figure 8.14?

The net area is the lesser of the net width times the thickness and 0.85 times the gross area. Since the hole is drilled, the width of the hole may be taken as the drilled diameter. The net width times the thickness is

$$(6 - (13/16))(3/4) = 3.89 \text{ in.}^2.$$

The gross area times 0.85 is

$$(6)(3/4)(0.85) = 3.83 \text{ in.}^2, \text{ so the net area is } 3.83 \text{ in.}^2.$$

8.1.14   Shear Rupture

As shown in Figure 8.15, sometimes the strength of a connection is governed by tear out of the material outside the bolt holes of a part. The material may

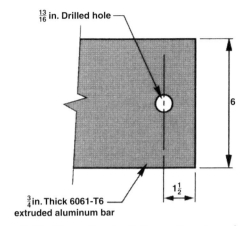

Figure 8.14 Example of a bolt near the edge of a bar.

fail in tension along part of the line of fracture and shear elsewhere, so the term used in the Aluminum and AISC specifications (*block shear rupture*) is somewhat misleading, and some engineers prefer to call the behavior *tear out* or *break out*. Aluminum *Specification* Section 5.4 (added in the 7th edition) (125) addresses the issue in the same manner as the AISC *Specification*. First, the failure path is determined; this determines how the strength is calculated. For wide, shallow bolt patterns, the strength is the sum of the tensile ultimate strength on the net tensile area plus the shear yield strength on the gross shear area; for narrow, deep patterns, the strength is the sum of the tensile yield strength on the gross tensile area plus the shear ultimate strength on the net

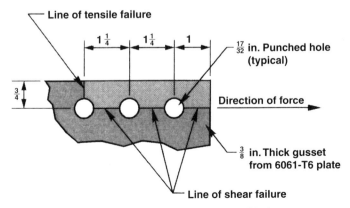

Figure 8.15 Example of a row of bolts.

shear area. The Aluminum *Specification* tear out provisions also apply to welded connections.

Example 8.5: What is the allowable tear out force on the 6061-T6 gusset plate in the direction shown in Figure 8.15?

Since the hole is punched, the effective width of the hole is the nominal hole diameter plus $\frac{1}{32}$ in. $= \frac{17}{32} + \frac{1}{32} = \frac{9}{16}$ in.

gross tensile area $= A_{gt} = (\frac{3}{4})(\frac{3}{8}) = 0.281$ in.^{2}

net tensile area $= A_{nt} = (\frac{3}{4} - (\frac{1}{2})\frac{9}{16})(\frac{3}{8}) = 0.176$ in.^{2}

gross shear area $= A_{gv} = (1\frac{1}{4} + 1\frac{1}{4} + 1)(\frac{3}{8}) = 1.312$ in.^{2}

net shear area $= A_{nv} = (1\frac{1}{4} + 1\frac{1}{4} + 1 - 2.5(\frac{9}{16}))(\frac{3}{8}) = 0.785$ in.^{2}

$$F_{tu} A_{nt} = (42 \text{ k/in.}^2)(0.176 \text{ in.}^2) = 7.4 \text{ k} < F_{su} A_{nv}$$

$$= (27 \text{ k/in.}^2)(0.785 \text{ in.}^2) = 21.2 \text{ k, so}$$

allowable tear out force

$$= P_{sr} = (F_{su} A_{nv} + F_{ty} A_{gt})/n_u$$

$$= [((27 \text{ k/in.}^2)(0.785 \text{ in.}^2) + (35 \text{ k/in.}^2)(0.281 \text{ in.}^2)]/1.95$$

$$= 15.9 \text{ k}$$

For a single bolt connection (Figure 8.11), the *Specification* break out strength is always greater than the *Specification* bearing strength, so you don't need to check tear out. This can be demonstrated by calculating bearing and tear out strengths for an edge distance of 1.5D and 2D:
For an edge distance of 1.5D, the bearing strength is:

$$F_{brg} = (1.5D/2D)2F_{tu}Dt = 1.5F_{tu}Dt$$

and the tear out strength is:

$$F_{bo} = 2(1.5D)0.6F_{tu}t = 1.8F_{tu}Dt$$

For an edge distance of 2D, the bearing strength is:

$$F_{brg} = 2F_{tu}Dt$$

and the tear out strength is:

$$F_{bo} = 2(2D)0.6F_{tu}t = 2.4F_{tu}Dt$$

For greater edge distances, the tear out strength increases while the bearing strength remains $2F_{tu}Dt$.

8.1.15 Minimum Spacing and Edge Distance

For spacing and edge distances, all distances are measured from the center of the hole in the Aluminum *Specification*. The *Specification* allows a minimum spacing between bolts of $2\frac{1}{2}$ times the bolt diameter and 3 diameters for rivets (see Section 5.1.9), compared to $2\frac{2}{3}$ or 3 diameters that the AISC requires, depending on how bearing stresses are limited. While the steel provisions are more precise and relate such factors as edge distance and spacing to bearing strength, the Aluminum *Specification* gives fairly similar results to steel and are simple to apply. Using the Aluminum Association spacing of $2\frac{1}{2}$ diameters does not place bolts so close together that adjacent bolts hamper installation. As discussed above with respect to bearing, the minimum edge distance for aluminum is two fastener diameters unless allowable bearing stresses are reduced. Bolt and rivet edge distances and spacings are given in Table 8.12 for common fastener diameters.

Some specifications refer to distances from the fastener to the edge of the member in a direction perpendicular to the force direction as edge distance, and to distances from the fastener to the edge of the member in the direction of force as end distance (Figure 8.16).

Example 8.6: What is the allowable bearing stress on the rightmost bolt shown in Figure 8.15?

The allowable bearing stress for bolts is

$$2F_{tu}/n_u = 2 \ (42 \ \text{ksi})/1.95 = 43 \ \text{ksi}.$$

This value need not be reduced for edge distance since the edge distance is

$$1.0 \ \text{in.} = 2 \ (\tfrac{1}{2} \ \text{in.}) \geq 2 \ (\text{bolt diameter}).$$

8.1.16 Maximum Edge Distance and Spacing

In built-up compression members, the Aluminum *Specification* dictates maximum fastener spacing to prevent buckling between fasteners (see Figure 8.17). Material between fasteners should be treated as a flat plate supported along both edges. As a result, it can be checked for local buckling using the

TABLE 8.12 Minimum Spacing and Edge Distances for Bolts and Rivets in Aluminum Members

Nominal Bolt Diameter D (in.)	Minimum Edge Distance $1.5D$ (in.)	Minimum Edge Distance for Full Bearing Allowable $2D$ (in.)	Minimum Bolt Spacing $2.5D$ (in.)	Minimum Rivet Spacing $3D$ (in.)
$\frac{1}{4}$	$\frac{3}{8}$	$\frac{1}{2}$	$\frac{5}{8}$	$\frac{3}{4}$
$\frac{3}{8}$	$\frac{9}{16}$	$\frac{3}{4}$	$\frac{15}{16}$	$1\frac{1}{8}$
$\frac{1}{2}$	$\frac{3}{4}$	1	$1\frac{1}{4}$	$1\frac{1}{2}$
$\frac{5}{8}$	$\frac{15}{16}$	$1\frac{1}{4}$	$1\frac{9}{16}$	$1\frac{7}{8}$
$\frac{3}{4}$	$1\frac{1}{8}$	$1\frac{1}{2}$	$1\frac{7}{8}$	$2\frac{1}{4}$
$\frac{7}{8}$	$1\frac{5}{16}$	$1\frac{3}{4}$	$2\frac{3}{16}$	$2\frac{5}{8}$
1	$1\frac{1}{2}$	2	$2\frac{1}{2}$	3

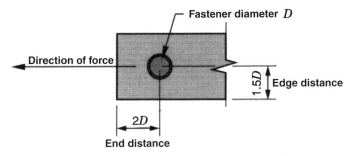

Figure 8.16   Bolt hole end distance and edge distance.

provisions of Aluminum *Specification* Section 3.4.9, and for overall buckling using Section 3.4.7. *Specification* Section 5.1.9 allows the plate width b (for local buckling) to be taken as 80% of the gauge distance—that is, perpendicular to the axis of the member—between fasteners and the plate length (for overall buckling) as one-half the spacing between fasteners.

There should also be limits on spacing and edge distances for bolted connections that are to be caulked leak-tight. A good guideline is that spacing should not exceed $4D$ nor $10t$, and edge distance should not exceed $1.5D$ nor $4t$, where D is the fastener diameter and t is the plate thickness (1).

Aluminum *Specification* Section 5.1.10 limits the spacing of fasteners in built-up tension members to 3 in. $+ 20t$ where $t =$ the thickness of the outer ply (in.) [76 mm $+ 20t$ where $t =$ the thickness of the outer ply (mm)].

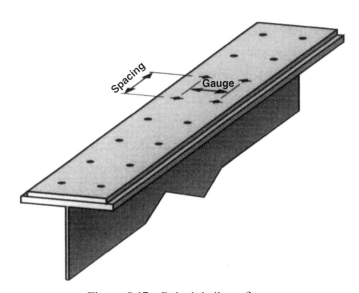

Figure 8.17   Bolted, built-up flange.

No maximum edge distance requirements exist for aluminum members, unlike steel, for which a 12*t* or 6 in. maximum edge distance is prescribed (39). This is partly because aluminum does not suffer corrosion to the extent steel does, so the need to avoid crevices that can trap dirt and moisture is less urgent. However, you should consider this effect, especially for the less corrosion-resistant alloys, such as 2014-T6.

8.1.17 Screw Connections

The Aluminum *Specification* addresses tapping screw connections in Section 5.3. You have additional failure modes to consider for screws beyond those considered for bolts: (1) pull-out in tension, (2) pull-over in tension, and (3) tilting in shear. These modes are illustrated in Figure 8.18 and are included in the calculation procedures given below.

Screw Connections in Tension The Aluminum *Specification* procedure for calculating the allowable strength of screw connections in tension is:

1) *Screw Tensile Strength:* Determine the force required to fail the screw itself in tension (let's call this P_{st}, since the Aluminum *Specification* doesn't assign a name to it). This can be done by testing or can be calculated as the screw material's minimum tensile ultimate strength times the root area of the screw.

2) *Pull-Out Strength:* Determine the force to pull the screw out of its threaded engagement with the bottom connected part, which is called the *pull-out strength* of the screw connection [P_{not}]) from Section 5.3.2.1 of the *Specification*:

(1) for UNC threads (screw thread types C, D, F, G, and T):

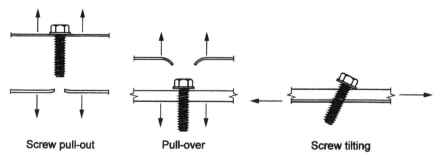

Screw pull-out   Pull-over   Screw tilting

Figure 8.18   Failure modes of sheet metal screws.

a) for 0.060 in. $\leq t_c \leq$ 0.125 in. [1.5 mm $\leq t_c \leq$ 3 mm]

$$P_{not} = K_s D t_c F_{ty2}$$

where:

K_s = 1.01 for 0.060 in. $\leq t_c <$ 0.080 in. [1.5 mm $\leq t_c <$ 2 mm]
K_s = 1.20 for 0.080 in. $\leq t_c \leq$ 0.125 in. [2 mm $\leq t_c \leq$ 3 mm]

b) for 0.125 in. $< t_c <$ 0.25 in. [3 mm $< t_c <$ 6.3 mm]

$$P_{not} = 1.2 D F_{ty2}(0.25 - t_c) + 1.16 A_{sn} F_{tu2}(t_c - 0.125)$$

c) for 0.25 in. $\leq t_c \leq$ 0.375 in. [6.3 mm $\leq t_c \leq$ 10 mm]

$$P_{not} = 0.58 A_{sn} t_c F_{tu2}$$

(2) for spaced threads (screw thread types AB, B, BP, BF, and BT):
   a) for 0.038 in. $\leq t_c \leq 2/n$ [1 mm $\leq t_c \leq 2/n$]

$$P_{not} = K_s D t_c F_{ty2}$$

where:

K_s = 1.01 for 0.038 in. $\leq t_c <$ 0.080 in. [1 mm $\leq t_c <$ 2 mm]
K_s = 1.20 for 0.080 in. $\leq t_c \leq 2/n$ [2 mm $\leq t_c \leq 2/n$]

b) for $2/n < t_c < 4/n$

$$P_{not} = 1.2 D F_{ty2}(4/n - t_c) + 3.26 D F_{tu2}(t_c - 2/n)$$

c) for $4/n \leq t_c \leq$ 0.375 in. [$4/n \leq t_c \leq$ 8 mm]

$$P_{not} = 1.63 D t_c F_{tu2}$$

where:

A_{sn} = thread stripping area of internal thread per unit length of engagement (see Table 8.13)
D = nominal screw diameter
F_{ty2} = tensile yield strength of member not in contact with the screw head
F_{tu2} = tensile ultimate strength of the member not in contact with the screw head
K_s = coefficient that depends on member thickness
n = number of threads per unit length for a screw

TABLE 8.13 Internal Thread Stripping Area For Class 2B UNC Threads

Nominal Size-Threads per in.	A_{sn} Internal Thread Stripping Area (in.^{2}) per in. of Engagement
8–32	0.334
10–24	0.401
12–24	0.458
$\frac{1}{4}$–20	0.539
$\frac{5}{16}$–18	0.682
$\frac{3}{8}$–16	0.828

t_c = the length of thread engagement in the member not in contact with the screw head

Equations 1(a) and 2(a) model circumferential yielding of the material around the hole that tends to occur in thin material, releasing the screw without stripping the threads. Equations 1(c) and 2(c) represent stripping of the threads in thicker tapped parts. Equations 1(b) and 2(b) are a combination of the yielding and stripping behavior, which occurs in intermediate thicknesses.

The pull-out force method given above was introduced in the 7th edition of the *Specification*. A simpler but slightly more conservative pull-out force from the prior edition is given by:

$$P_{not} = 0.85t_c DF_{tu2}$$

where:

t_c = length of thread engagement in the member not in contact with the screw head

D = nominal screw diameter

F_{tu2} = tensile ultimate strength of the member not in contact with the screw head.

3) *Pull-Over Strength:* Determine the force required to pull the top connected part over the head of the screw (called the *pull-over strength* of the screw connection [P_{nov}]) from Equation 5.3.2.2-1 of the *Specification*:

$$P_{nov} = Ct_1 F_{tu1}(D_{ws} - D_h)$$

where:

C = 0.7 for fasteners with a gap between the joined parts (such as through the crown of corrugated roofing) and 1.0 for all others

t_1 = thickness of the member in contact with the screw head

F_{tu1} = tensile ultimate strength of the member in contact with the screw head

D_{ws} = larger of the screw head diameter or washer diameter, but not greater than $\frac{1}{2}$ in. [13 mm]

D_h = nominal hole diameter.

4) Determine the nominal tensile strength of the screw connection (P_{nt}) as the minimum of $P_{st}/1.25$, P_{not}, and P_{nov}.

5) Determine the allowable tensile force on the screw (P_{at}) as:

$$P_{at} = P_{nt}/n_s$$

where n_s = safety factor for screws = 3.0.

For LRFD, rather than allowable tensile force, you determine the design strength of the screw connection $\varphi P_{nt} = 0.5P_{nt}$. Screws are allotted a higher safety margin than bolts and rivets because screws are more subject to the vagaries of installation.

Screw Connections in Shear   The Aluminum *Specification* procedure for calculating the allowable strength of screw connections in shear is:

1) *Screw Shear Strength:* Determine the shear strength of the screw itself (let's call this P_{ss}). This can be done by testing or can be calculated as the screw material's minimum tensile ultimate strength times 0.6 times the root area of the screw ($P_{ss} = 0.6F_{tu}A_r$).

2) *Bearing Strength:* Determine the bearing strength of the screw connection (let's call this P_{bs}) as the minimum of the results from Equations 5.3.1.1-2 and 5.3.1.1-3 of the *Specification*, which we'll call P_{bs1} and P_{bs2} and define next.

The allowable bearing force of the member in contact with the screw head (P_{bs1}) is:

$$P_{bs1} = 2F_{tu1}Dt_1 \frac{n_s}{n_u}$$

The allowable bearing force of the member not in contact with the screw head (P_{bs2}) is:

$$P_{bs2} = 2F_{tu2}Dt_2 \frac{n_s}{n_u}$$

where:

n_s = safety factor for screw connections = 3.0
n_u = safety factor for fracture = 1.95
D = nominal screw diameter
F_{tu1} = tension ultimate strength of the member in contact with the screw head
F_{tu2} = tension ultimate strength of the member not in contact with the screw head
t_1 = thickness of the member in contact with the screw head
t_2 = thickness of the member not in contact with the screw head

3) *Screw Tilting:* If the thicker of the two members joined is in contact with the screw head—in other words, $t_1 > t_2$, meaning the top material is thicker than the bottom material, an unlikely situation if you're screwing sheet to a structural member—determine the tilting strength of the screw connection (let's call this P_{ts}) from Equation 5.3.1.1-4 of the *Specification*:

$$P_{ts} = 4.2(t_2^3 D)^{1/2} F_{tu2}$$

where F_{tu2} = tensile ultimate strength of the member not in contact with the screw head.

4) Determine the nominal shear strength of the screw connection (P_{ns}) as the minimum of $P_{ss}/1.25$, P_{bs}, and P_{ts}.

5) Determine the allowable shear force on the screw (P_{as}) as:

$$P_{as} = P_{ns}/n_s$$

where n_s = safety factor for screws = 3.0.

For LRFD, rather than allowable shear force, determine the design strength of the screw connection, which is

$$\varphi P_{ns} = 0.5 P_{ns}.$$

8.1.18 Minimum Requirements for Connections

The Aluminum *Specification*, unlike the Steel *Specification*, does not require minimum connection strengths. (See the AISC Section J1.6 (38), which requires that connections carrying calculated stresses be designed to support not less than 6 k.) Because the Aluminum *Specification* covers a wide variety of structures, some of which might be designed for only light loads, this is not altogether surprising. However, when aluminum structures are designed for building-type structures, it's appropriate to consider certain size connections as a minimum, even if smaller fasteners could be demonstrated to work on paper. As a practical matter, the smaller and fewer the fasteners, the more serious the consequences when a fastener is omitted or improperly installed. Also, there is a minimum connection size that looks robust enough to inspire confidence in laypersons. Last, unanticipated loads, which often occur during shipping or construction, for example, may overload a connection designed strictly for the specified loads.

Table 8.14 offers guidelines for minimum connection sizes. Single fastener connections have no redundancy, so they are suggested only for secondary members. The minimum strength guidelines would also apply to welded connections.

8.2 WELDED CONNECTIONS

Unlike mechanical connections, designing *welded* connections is more often in the realm of specialists than in the hands of structural engineers. Metallurgical considerations involved in welding may loom as large as structural issues. For example, selection of the welding process, joint detailing, and *filler* alloy selection can demand the skills of welding engineers whose decisions can affect the structural design. This section is not intended to cover all the issues involved in weld design, but instead to give structural engineers some background in the Aluminum *Specification* provisions for welded connections.

8.2.1 Aluminum Welding Processes

Welding is the process of uniting pieces by fusion with heat. Many aluminum welding processes are available, some with exotic names, like electron beam welding; some with nifty acronyms, like shielded metal arc welding (SMAW); and yet others that sound slightly suspect, like electroslag welding. Forget all these. The welding processes identified in the Aluminum *Specification* and used for most construction applications are *gas metal arc welding* (GMAW),

TABLE 8.14 Suggested Minimum Connection Sizes for Building Structures

Type of Connection	Minimum Allowable Strength (k)	Minimum Allowable Strength (N)	Minimum Number of Fasteners
Primary member	6	25	2
Secondary member, area > 1.0 in.² [600 mm²]	3.5	15	1
Secondary member, area ≤ 1.0 in.² [600 mm²]	1.8	8	1

Note

Minimum fastener size for members other than building sheathing should be $\frac{3}{8}$ in. [10 mm].

commonly called MIG (for metal inert gas); and *gas tungsten arc welding* (GTAW), referred to as TIG (tungsten inert gas). These are the processes for which minimum welded mechanical properties are given in the *Specification* Table 3.3-2. For other processes, you're on your own—and that can be a lonely place.

As a structural engineer, you don't need to know how MIG and TIG welds are made. You can leave that to the fabricator's welding engineer, but it might be helpful to have a rough idea of what these weld processes are. MIG welding uses an electric arc within a shield of inert gas between the base metal and an electrode filler wire. The electrode wire is pulled from a spool by a wire-feed mechanism and delivered to the arc through a gun. In TIG welding, the base metal and, if used, filler metal are melted by an arc between the base metal and a nonconsumable tungsten electrode in a holder. Tungsten is used because it has the highest melting point of any metal (6,170°F [3,410°C]) and reasonably good conductivity: about one-third that of copper. Inert gas that flows from a nozzle on the holder shields the weld metal and the electrode. In both MIG and TIG welding, the inert gas (usually argon, sometimes helium, or some of each) protects the molten metal from oxidation, allowing coalescence of the base and filler metals.

TIG welding was developed before MIG and was originally used for all metal thicknesses. TIG welding, however, is slower and doesn't penetrate as well as MIG welding. With the development of MIG welding, TIG has been generally limited to welding thin material (up to about $\frac{1}{4}$ in. [6 mm] thick). In MIG welding, the welding machine controls the electrode wire speed, and once it is adjusted to a particular welding procedure, doesn't require readjustment. As a result, even manual MIG welding is considered to be semiautomatic. MIG welding is suitable for all aluminum material thicknesses and costs about $5 /ft [$16/m] once all labor, consumable, and material costs are included.

Before 1983, the ASME *Boiler and Pressure Vessel Code, Section IX, Welding and Brazing Qualifications* was the only widely available standard for aluminum welding. Many aluminum structures other than pressure vessels were welded in accordance with the provisions of the *Boiler and Pressure Vessel Code* because no other standards existed. In 1983, the American Welding Society's (AWS) D1.2 *Structural Welding Code—Aluminum* (91) was introduced as a standard for general-purpose structural MIG and TIG welding (in other words, everything except aerospace or pressure vessel applications). In addition to rules for qualifying aluminum welders and weld procedures, D1.2 includes design, fabrication, and inspection requirements. The requirements in D1.2 and the *Boiler and Pressure Vessel Code* are similar, but D1.2's format roughly parallels the AWS structural steel welding code (D1.1) for the convenience of users since most of them are familiar with steel welding. Unlike D1.1, though, D1.2 does not include design stresses for aluminum welds; for that, it refers to the Aluminum *Specification*.

8.2.2 Selecting a Filler Alloy

The material specification for MIG and TIG filler alloys is AWS A5.10, *Specification for Bare Aluminum and Aluminum-Alloy Welding Electrodes and Rods* (88). The term *bare* differentiates these fillers from shielded metal arc-welding fillers that come with a crusty flux cover and were used before inert-gas shielding methods were developed in the 1940s.

The structural design of welded connections is a function of the filler wire used in the weld. Table 7.2-1 of the Aluminum *Specification* dictates the filler alloy based on the parent metal alloy(s). In many cases, several filler alloys may be used. You have many factors to consider in addition to strength, including ductility, color match after anodizing, corrosion resistance, hot-cracking tendency, MIG wire feedability, and elevated temperature performance. In spite of this, the vast majority of applications use 4043 or 5356 filler, which costs $2.50 to $3.25/lb [$5.50 to $7.00/kg]. (4043 is slightly less expensive than 5356, but not enough to override other factors in deciding which to use.) Other fillers can cost as much as $10 to $12/lb [$22 to $26/kg] and are not as available, so you need a good reason before using them.

Filler alloys 5356, 5183, and 5556 were developed to weld the 5xxx series alloys, but they have also become useful for welding 6xxx and 7xxx alloys. Alloy 5356 is the most commonly used filler due to its good strength, compatibility with many base metals, and good MIG electrode wire feedability. Alloy 5356 also is used to weld 6xxx series alloys because it provides a better color match with the base metal than 4043 when anodized. (4043 tends to turn black when anodized, a look the architect might not have had in mind.) Alloy 5183 has slightly higher strength than 5356, and 5556 higher still. The most notable example of where the higher strength of 5556 is needed is for welding 5083 or 5456. Because 5356, 5183, and 5556 contain more than 3% magnesium and are not heat-treatable, however, they are not suitable for elevated temperature service or post weld heat treating. Alloy 5554 was developed to weld alloy 5454; both contain less than 3% magnesium so as to be suitable for service more than 150°F [66°C].

Alloy 5654 was developed as a high-purity, corrosion-resistant alloy for welding 5652, 5154, and 5254 components used for hydrogen peroxide service. Its magnesium content exceeds 3% so it is not used at elevated temperatures.

Alloy 4043 was developed for welding the heat-treatable alloys, especially those of the 6xxx series. Its has a lower melting point than the 5xxx fillers, so it flows better and is less sensitive to cracking. Alloy 4643 is for welding 6xxx base-metal parts more than 0.375 in. [10 mm] to 0.5 in. [13 mm] thick that will be heat-treated after welding. Although filler alloys 4047 and 4145 are listed for some parent metal combinations in *Specification* Table 7.2-1, they are primarily used for brazing. *Brazing* is the process of joining metals by fusion using filler metals with a melting point above 840°F [450°C], but lower than the melting point of the base metals being joined. Brazing is not the same process as welding, and we won't be discuss it here. Filler 4047 has more silicon than 4043, so it flows better, making it better suited to producing seal welds. 4145 is used for welding 2xxx alloys, and 4047 is used instead of 4043 in some instances to minimize hot cracking and increase fillet weld strengths.

Alloy 2319 is used for welding 2219; it's heat treatable and has higher strength and ductility than 4043 when used to weld 2xxx alloys that are post weld heat treated.

Pure aluminum alloy fillers are often needed in electrical or chemical industry applications for conductivity or corrosion resistance. Alloy 1100 is usually satisfactory, but for even better corrosion resistance (due to its lower copper level) 1188 may be used. These alloys are soft and sometimes have difficulty feeding through MIG conduit.

The filler alloys used to weld castings are castings themselves (C355.0, A356.0, 357.0, and A357.0), usually $\frac{1}{4}$ in. [6 mm] rod used for TIG welding. They are mainly used to repair casting defects. More recently, wrought versions of C355.0 (4009), A356.0 (4010), and A357.0 (4011) have been produced, so that they can be produced as MIG electrode wire. (Alloy 4011 is available only as rod for GTAW, however, since its beryllium content produces fumes too dangerous for MIG welding.) Like 4643, 4010 can be used for post weld heat treated 6xxx weldments.

The mechanical property given for aluminum filler alloys in the Aluminum *Specification* is not the minimum tensile ultimate strength as for steel electrodes, but rather minimum shear strength. Aluminum *Specification* Table 7.2-2 shear strengths are shown in Table 8.15. The values for 1100, 4043, 5356, 5554, and 5556 are the longitudinal shear strengths given in the *Welding Research Supplement* February 1966 article titled "Shear Strengths of Aluminum Alloy Fillet Welds." (127) The values for 5183 and 5654 were added after the 1st edition of the *Specification* was issued, and no published documentation of their strengths exists.

TABLE 8.15 Minimum Shear Strengths for Aluminum Filler Alloys

Filler Alloy	Minimum Shear Strength (ksi)	Minimum Shear Strength (MPa)
1100	7.5	50
4043	11.5	80
5183	18.5	130
5356	17	115
5554	17	115
5556	20	140
5654	12	85

Heat-treatable base metals and fillers (a good example is 6061 welded with 4043 filler) will undergo natural aging after welding—as opposed to post weld heat-treatment artificial aging, a controlled way to increase strength—by which the strength increases over time. Most of the strength increase from natural aging occurs within two to three months after welding. The welded strengths given in Aluminum *Specification* Tables 3.3-2 and 7.2-2 already take this increase into account.

Once welded, the strongest filler alloy is not necessarily the one with the highest shear strength as listed in Table 8.15, but often the one that can be welded with the fewest flaws. A higher-strength filler does not make a higher-strength weldment if it is prone to cracking or other defects. Therefore, the filler alloy selection process should include welding fabricators for their input.

8.2.3   Types of Welds

To identify welds, the AWS standard weld symbols (AWS A2.4 (86)) are used for aluminum, as for steel. Two categories of welds are commonly used for structural members: *groove welds* and *fillet welds*. Plug and slot welds and stud welds are also used.

Groove Welds   Groove welds (also called *butt welds*) are used at *butt joints* where the edge of one part butts against another part. Some examples of the various kinds of groove welds, such as square, V, bevel, U, and J, are shown in Figure 8.19. The purpose of the bevel is to allow welders full access to the joint and to permit enough filler volume to achieve the proper alloying of the weldment. The tensile ultimate and yield strengths across a groove weld are given in Aluminum *Specification* Table 3.3-2, when the base metal-filler alloy pairings as given in *Specification* Table 7.2-1 are used.

The stress on a groove weld is calculated on its effective area, which is the product of its size and its *effective length*. The effective length is the width of the narrower part joined, perpendicular to the direction of stress. The size of a groove weld depends on whether it's a complete joint penetration (CJP) weld or a partial joint penetration (PJP) weld. For CJP welds, the weld size is the thickness of the thinner part joined. For PJP welds, the weld size is usually the depth of preparation of the joint, meaning the depth of the V, bevel, U, or J, but the weld size depends on the included angle, type of preparation, welding method, and welding position (see AWS D1.2).

How do you know if a groove weld has complete or only partial penetration? Follow the D1.2 rules: a groove weld is full penetration if it's:

(1) welded from both sides with the root of the first weld backgouged to sound metal before welding the second side

(2) welded from one side using a permanent or temporary backing

(3) welded from one side using a AC-GTAW root pass without backing

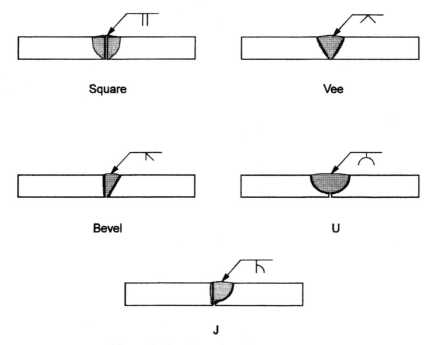

Figure 8.19 Examples of groove welds.

(4) welded from one side using plasma arc welding with variable polarity (PAW-VP) in the keyhole mode

All other groove welds are partial penetration welds. From the above, it's clear that you have to know how the weld will be fabricated before you can design it. Unfortunately, you may need to know its size before you know how it will be fabricated, so this can be an iterative process. The lion's share of CJP welding is done with methods (1) and (2); which is used boils down to whether it's more economical for the fabricator to backgouge or to use backing. Permanent backing has very low fatigue strength: the joint between the backer and the part is essentially a pre-existing crack. Temporary backing, usually part of a fixture, tape, or anodized bar, can be expensive. Option (3) (AC-GTAW) is used nearly exclusively for welding pipe because it's not usually possible to gain access to the back side of a weld inside a pipe. If you aren't welding an exotic aerospace application, you probably won't be able to afford the welding equipment for option (4) (PAW-VP).

The Aluminum *Specification* prescribes a safety factor of 1.95 ($= n_u$) on the tensile ultimate strength for groove welds. As discussed in the section on welded members, tensile ultimate strengths for welded aluminum alloys are additionally factored by 0.9 to account for defects that visual inspection may not detect. The net safety factor on fracture is, thus,

$$1.95/0.9 = 2.17.$$

Example 8.7: What is the allowable tensile force across a groove weld between two 0.5 in. thick, 4 in. wide 5086-H112 plates welded from one side using a temporary backing with 5356 filler alloy?

Since this groove weld is made from one side with a temporary backing, it is a full penetration weld (see [2] above), so the weld size is the thickness of the thinner part joined (0.5 in.). The weld's effective length is the width of the narrower part joined, perpendicular to the direction of stress, which is 4 in., so the weld area is

$$(0.5 \text{ in.})(4 \text{ in.}) = 2 \text{ in.}^2.$$

From *Specification* Table 3.3-2 for welded 5086-H112, the tensile ultimate strength (F_{tuw}) is 35 ksi. The allowable tensile stress is

$$0.9 \, F_{tuw}/n_u = \frac{(0.9)(35 \text{ ksi})}{1.95} = 16.2 \text{ ksi.}$$

The allowable tensile force is, then,

$$(16.2 \text{ k/in.}^2)(2 \text{ in.}^2) = 32.4 \text{ k.}$$

Let's take a moment to compare the allowable tensile force we just calculated to the allowable tensile force of the bar away from the weld. *Specification* Table 3.3-1 gives the tensile ultimate strength for 5086-H112 plate 0.5 in. thick as 35 ksi, which is the same as the welded strength. The only difference in the design strengths is the 0.9 factor placed on welded strengths, so the design strength across the weld is only 10% less than the design strength of the base metal. Before you start welding, though, realize that this situation occurs only with certain 5xxx alloys. This is because they are not heat-treatable, and in certain tempers (like H112) have little strength increase from cold work, so they have little to lose by welding. Their strength comes from their alloying elements (primarily magnesium), and welding doesn't affect that, since they're welded with aluminum-magnesium alloy fillers.

Example 8.8: What is the allowable tensile force across a groove weld between two 0.5 in. thick, 4 in. wide 5086-H112 plates with a $\frac{7}{16}$ in. V groove, welded from one side without a backing with 5356 filler alloy?

Since this groove weld is made from one side without temporary or permanent backing, it is a partial penetration weld and the weld size is the depth of the groove ($\frac{7}{16}$ in.). The weld length is 4 in., so the weld area is

$$(\tfrac{7}{16} \text{ in.})(4 \text{ in.}) = 1.75 \text{ in.}^2.$$

As in Example 8.7, the allowable tensile stress is

$$0.9F_{tuw}/n_u = (0.9)(35 \text{ ksi})/1.95 = 16.2 \text{ ksi.}$$

The allowable tensile force is, then,

$$(16.2 \text{ k/in.}^2)(1.75 \text{ in.}^2) = 28.3 \text{ k.}$$

Fillet Welds   A fillet weld is shown in Figure 8.20. Fillet welds are roughly triangular in cross section and the size specified by giving the side dimension, which is taken to be equal for both legs unless noted otherwise. In aluminum, as for steel, the stress on the weld is computed as shear stress on the effective area of the weld, which is the product of the *effective length* and the *effective throat*. The effective throat is the shortest distance from the root to the face of the weld. For equal length leg fillet welds, this distance is 0.707 (= cos 45°) times the leg length. The effective length is the overall length of the full size weld, including end returns. The effective length must be at least four times the weld size, however, or the weld size shall be taken as $\tfrac{1}{4}$ the effective length.

Smaller fillet welds are more cost-effective than large ones. This is because the strength is proportional to the fillet weld size, while the volume of filler—and hence the cost—is proportional to the square of the fillet weld size. Keeping the weld small also minimizes distortions that the residual stresses induced by welding cause. Maximum fillet weld sizes are established by the Aluminum *Specification* since it requires conformance with (in *Specification* Section 7.3) AWS D1.2. The maximum size of fillet welds prescribed in D1.2 is given in Table 8.16.

On the other hand, welds that are much smaller than the thickness of the metal joined may induce large residual stresses. This is because the amount

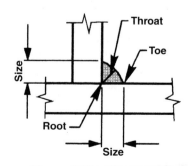

Figure 8.20   Fillet weld.

TABLE 8.16 Maximum Size of Fillet Welds Along Edges

Base-Metal Thickness t		Maximum Fillet Weld Size	
(in.)	(mm)	(in.)	(mm)
$t < \frac{1}{4}$	$t < 6$	t	t
$\frac{1}{4} \leq t$	$6 \leq t$	$(t - \frac{1}{16})*$	$(t - 2)*$

*Unless the weld size is designated on the drawing or the weld is designated on the drawing to be built out to obtain full throat thickness.

of heat needed to lay a small weld bead is insufficient to expand the base metal. As the weld metal cools, it is restrained against contraction by the already cool base metal. For this reason, the Aluminum *Specification* (by reference to AWS D1.2) requires minimum fillet weld sizes, shown in Table 8.17.

Aluminum fillet welds, like those in steel, are stronger when loaded transversely than longitudinally (133). (See Figure 8.21.) You can rationalize this by imagining the transverse failure as a tensile failure and the longitudinal failure as a failure in shear. But rather than attempt to make this distinction, the Aluminum *Specification* conservatively uses the lower longitudinal shear strengths. (If you're curious, read the *Welding Research Supplement* article mentioned in Section 8.2.2 above, which provides transversely loaded strengths.) This prevents complications because even at the same joint, some of the welds may be loaded transversely and others longitudinally. Furthermore, for design purposes, the strength of the filler alloy is taken as the shearing strength, regardless of the direction of the load on the weld. This is conservative because the shear strength of the weld metal is less than its tensile strength.

TABLE 8.17 Minimum Size of Fillet Welds For Statically Loaded Structures

Base-Metal Thickness (t) of Thicker Part Joined		Minimum Size of Fillet Weld	
(in.)	(mm)	(in.)	(mm)
$t \leq \frac{1}{4}$	$t \leq 6$	$\frac{1}{8}$	3
$\frac{1}{4} < t \leq \frac{1}{2}$	$6 < t \leq 13$	$\frac{3}{16}$	5
$\frac{1}{2} < t$	$13 < t$	$\frac{1}{4}$	6

Notes

Minimum fillet welds must be made in one pass (one progression of the electrode down the length of the weld).

Minimum size need not exceed the thickness of the thinner part joined. In such cases, care should be taken to provide sufficient preheat to ensure weld soundness.

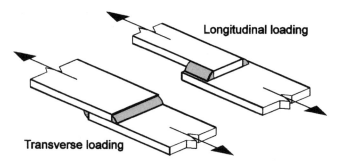

Figure 8.21 Transverse and longitudinal loads on fillet welds.

A fillet weld may fail in one of two ways: (1) by shear or tensile failure through the narrowest part of the fillet, called the *throat;* or (2) by shear or tensile failure of the underlying, weld-affected base metal (Figure 8.22). Unlike steel, both the base metal and the filler alloy must be assumed to be potential sources of failure of the joint, using reduced strengths for the weld-affected base metal. The reduced strengths were discussed in the chapters on material properties and welded members, and are summarized in Appendix C. When two different base metal alloys are used, for example, 6061 welded to 6063, the as-welded strength of the weaker of the two must be used for the base-metal strength.

The Aluminum *Specification* prescribes a factor of safety on ultimate strength of 2.34 for fillet welds. This is the safety factor on the ultimate strength of 1.95 used elsewhere in the *Specification*, multiplied by 1.2. (For LRFD, the resistance factor is 0.65.) This is consistent with the philosophy of designing connections with greater margin on failure than the members they join. As for mechanical connections, the *Specification* does not require

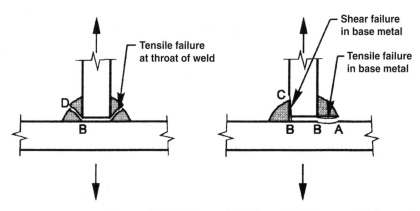

Figure 8.22 Fillet weld failure modes.

a check on yielding at the welded connection since it does not typically constitute a limit state. The small amount of elongation that would take place due to yielding at the weld is insignificant compared to the length of the welded member, except for very short members.

Example 8.9: What is the allowable stress on the fillet weld shown in Figure 8.22 and made with 4043 filler alloy on 6061-T6 base metal?

First, let's consider the shear strength of the filler alloy. The failure plane is along line BD. From Table 8.15, the minimum shear strength of the 4043 filler is 11.5 ksi. The allowable shear stress in the filler alloy is, then,

$$(11.5 \text{ ksi})/2.34 = 4.9 \text{ ksi}.$$

Next, consider the base metal. One mode of failure is that the base metal can fail in tension along line AB, which is 1.41 times the length of line BD. The minimum welded tensile ultimate strength of the 6061-T6 base metal is found in Aluminum *Specification* Table 3.3-2 as

$$F_{tuw} = 24 \text{ ksi}.$$

The allowable stress on the base metal, adjusting for the longer failure line compared to the throat, is

$$1.41(F_{tuw})(0.9)/1.2n_u k_t = 1.41(24 \text{ ksi})(0.9)/(2.34)(1.0) = 13 \text{ ksi}.$$

(Note that for all welded alloys, k_t is 1.0 per *Specification* Table 3.4-2.)
The base metal can also fail in shear along line BC, which also is 1.41 times longer than line BD. The minimum welded shear ultimate strength of the 6061-T6 base metal is found in Aluminum *Specification* Table 3.3-2 as

$$F_{suw} = 15 \text{ ksi}.$$

The allowable stress is

$$1.41(F_{suw})/1.2n_u = 1.41(15 \text{ ksi})/2.34 = 9.0 \text{ ksi}.$$

For equal leg fillets, shear failure in the base metal (failure on line BC) will always govern over tension failure in the base metal (line AB) since the shear strength is less than the tensile strength.

The least of 4.9 ksi, 13 ksi, and 9.0 ksi is 4.9 ksi, which is rounded to 5 ksi, as used for the allowable shear stress for 4043 filler and 6061 base metal in Aluminum *Specification* Table 7.2-2, Allowable Shear Stresses in Fillet Welds for Building Type Structures. For most filler-base metal combinations, the filler alloy shear strength is the weak link and, thus, governs the strength of the fillet weld.

TABLE 8.18   Minimum Diameter of Hole for Plug Welds or Width of Slot for Slot Welds

Material Thickness (t)		Minimum Hole Diameter or Slot Width
(in.)	(mm)	
$t < \frac{1}{8}$	$t < 3$	$3t$
$t \geq \frac{1}{8}$	$t \geq 3$	$2.5t$

Plug and Slot Welds   Plug and slot welds are used to transmit shear between lapped parts and made by fabricating a hole or slot in the top part and filling the hole with filler metal. An example is attaching a cover plate to a beam flange. The effective area of plug and slot welds is the nominal area of the plug or slot in the plane of the faying surfaces. Minimum sizes for plug and slot welds are a function of the material thickness in which they're placed and are given in Table 8.18.

Stud Welds   *Stud welding* (SW) is a process used to attach externally or internally threaded rod or wire (studs) to a part. Two methods are used for aluminum: arc stud welding, which uses a conventional welding arc over a timed interval, and capacitor-discharge stud welding, which uses an energy discharge from a capacitor. Arc stud welding is used to attach studs ranging from $\frac{56}{1}$ in. [8 mm] to $\frac{1}{2}$ in. [13 mm] in diameter, while capacitor-discharge stud welding uses studs $\frac{1}{16}$ in. [1.6 mm] to $\frac{5}{16}$ in. [8 mm] in diameter. (Neither method produces studs as large as those for steel.) Capacitor-discharge stud welding is very effective for thin sheet (as thin as 0.040 in. [1.0 mm]), because it uses much less heat than arc stud welding and, thus, is much less likely to mar the appearance of the sheet on the opposite side from the stud. Studs are inspected using the bend, torque, or tension test requirements given in D1.2. Alloys available for stud welding are given in Table 8.19. Minimum tensile strengths are available only for 5183, 5356, and 5556 alloy studs and are given in Table 8.20. The D1.2 minimum base-metal thicknesses for stud welds are given in Table 8.21. Stud-welding requirements are included in AWS D1.2 Section 6. Headed studs, such as those used to embed steel studs in concrete, are not available in aluminum, but you can thread nuts onto aluminum studs

TABLE 8.19   Stud Welding Alloys

Arc	Capacitor Discharge
	1100
2319	2319
4043	4043
5183	5183
5356	5356
5556	5556
	6061
	6063

TABLE 8.20 Minimum Tensile Strengths for 5183, 5356, and 5556 Studs

Stud Size	Arc		Capacitor Discharge	
	(lb)	(N)	(lb)	(N)
6–32			375	1,670
8–32			635	2,820
10–24	770	3,420	770	3,420
$\frac{1}{4}$–20	1,360	6,050	1,360	6,050
$\frac{5}{16}$–18	2,300	10,200	2,300	10,200
$\frac{3}{8}$–16	3,250	14,500		
$\frac{7}{16}$–14	4,400	19,600		
$\frac{1}{2}$–13	5,950	26,500		

for this purpose. More common applications include threaded studs welded to the back of aluminum signs.

8.2.4 Comparing Aluminum and Steel Fillet Weld Safety Factors

Design of aluminum welds is different from steel in the manner in which the allowable shear stress for filler wire is determined. In the AISC steel *Specification*, the *tensile* strength of the filler alloy is multiplied by 0.3; if the shear strength is 60% of the tensile strength, this is effectively a safety factor of 2.0 (= 0.6/0.3). In aluminum the ultimate *shear* strength of the filler is divided by a safety factor of 2.34. The difference in the two is almost entirely due to the additional 1.2 factor the Aluminum *Specification* includes in the safety factor on connections.

Other structural design codes for aluminum may require higher safety factors for shear stresses on fillet welds. ASME B96.1 *Welded Aluminum-Alloy Storage Tanks*, for example, requires a factor of safety of 4.0, rather than 2.34, on fillet welds.

8.2.5 Weld Fabrication

Aluminum *Specification* Section 7.3 requires that aluminum welding comply with AWS D1.2. AWS D1.2 divides welded aluminum members into four categories shown in Table 8.22 and assigns different requirements to each.

TABLE 8.21 Minimum Base-Metal Thickness (t) for Stud Welds

Arc	Capacitor
$t \geq 0.5$ (stud diameter)	$t \geq 0.25$ (stud base diameter)*

*Capacitor studs can be flanged or nonflanged. If flanged, the stud base diameter = flange diameter. If nonflanged, the stud base diameter = stud diameter.

TABLE 8.22   AWS D1.2 Welded Member Categories

	Type of Member	
Type of Loading	Statically Loaded Non-tubular	Statically Loaded Tubular (Class I)
	Cyclically Loaded Non-tubular	Cyclically Loaded Tubular (Class II)

D1.2 includes workmanship requirements regarding:

1) conditions, such as temperature and moisture, under which welding may be performed
2) preparation of the base metal
3) assembly
4) welding technique
5) control of distortion and shrinkage
6) weld termination
7) dimensional tolerances of welded members
8) weld profiles
9) repairs

Distortion   Aluminum welding fabrication issues are a book in themselves, but a few comments might be useful here for structural engineers. Welding can cause significant distortions to structural assemblies because aluminum, like steel, shrinks upon cooling after welding. Predicting the resulting distortion can be difficult, but transverse shrinkage at a groove weld is approximately:

$$\Delta_t = 0.06 \, A_w / t$$

where:

A_w = cross-sectional area of the weld
t = thickness of the parts welded

Expressed another way, the transverse shrinkage is about 6% of the average width of the weld. Steel shrinks more—about 10% of the weld width—because even though its thermal expansion coefficient is half of that of aluminum, it has a much higher melting point and so has to be heated to a much higher temperature to weld. Any prediction of shrinkage must be considered approximate since it's a function of the filler and base metals, welding speed, welding method, and other welding parameters.

Example 8.10: How much will the weld in Figure 8.23 shrink transversely?

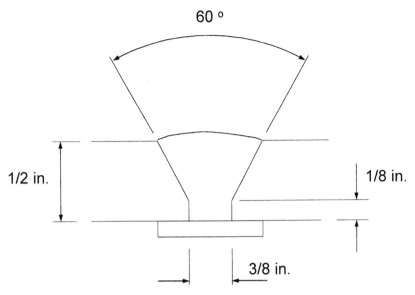

Figure 8.23   Transverse shrinkage of a groove weld.

The cross-sectional area of the weld is:

$$A_w = (\tfrac{3}{8} \text{ in.})(\tfrac{1}{2} \text{ in.}) + 2(\tfrac{1}{2})(\tfrac{3}{8} \text{ in.})(\tfrac{3}{8} \text{ in.})\tan 30 = 0.269 \text{ in.}^2$$

The transverse weld shrinkage is:

$$\Delta_t = 0.06 \, A_w/t = 0.06(0.269 \text{ in}^2)/(\tfrac{1}{2} \text{ in.}) = 0.032 \text{ in., or about } \tfrac{1}{32} \text{ in.}$$

While this isn't a big number, it can accumulate to a significant dimension in assemblies involving multiple welds. An example is a cylindrical pressure vessel with a series of circumferential welds that can cause considerable shortening of the vessel.

Weld shrinkage also produces curling in members with welds that are not balanced about the member's neutral axis. The member will curl toward the side where the welds exert the greater bending moment. The moment can be calculated by multiplying the cross-sectional area of the welds by the distance from the centroid of the weld to the member's neutral axis. If the amount of curling is too large to be called *camber*, the member must be straightened. You can do this by adding weld to the more lightly welded side of the neutral axis or by applying heat with a torch on the side you want to shrink. You can use heat crayons to monitor the temperature to prevent heating so much that strengths are affected by more than a specified amount. (Maximum cumulative times at elevated temperatures for some common alloys shown in Table 8.23 will result in no more than a 5% loss in strength.) Because welding on the opposite side will affect strengths, you must account for this in the design. You can perform either of these methods while loading or constraining the member.

8.2.6 Weld Quality Assurance

Long before quality became the latest business fad, quality assurance was an important issue in welding. It still is. This is because the strength of weldments is subject to such great variability depending on the care taken in their production. In this sense, weldments are like concrete. The strength of concrete in place can vary greatly depending on how it was installed. For example, too much water added to the mix can render the best concrete design flawed. Mill-produced shapes of steel and aluminum, on the other hand, typically exceed minimum strength requirements quite consistently. Furthermore, while careful visual inspection of weldments is useful, visual inspection alone is very limited in the kinds of defects (euphemistically called *discontinuities*) that it detects.

AWS D1.2 requires, for each type of weld, that the fabricator:

1) Prepare a written welding procedure specification.
2) Qualify this procedure.
3) Qualify the welders by testing their work before they are allowed to do any production welding.
4) Maintain records of each of these steps.

Aluminum and steel welding differ significantly in that AWS D1.2 does not allow prequalified joint details and welding procedures for aluminum. This is because of the wide range of aluminum welding conditions and alloys allowed under the D1.2 code. Therefore, all aluminum joint details and welding procedures must be individually qualified. AWS D1.2 Annex B includes some recommended aluminum joint details for partial and complete joint penetration groove welds. D1.2 Annex E includes sample forms for welding procedure specifications (WPS), procedure qualification records (PQR), and manufacturers' records of welder qualification tests. An example of a WPS is AWS B2.1.015, a standard welding procedure specification for TIG welding 10 through 18 gauge aluminum sheet (89).

Inspection Methods and Acceptance Criteria   It may surprise you to learn that neither the Aluminum *Specification* nor the AWS D1.2 code it references requires any inspection of production welds other than visual. Other than the requirements on procedure and welder qualification, workmanship, and visual

TABLE 8.23   Temperature Exposure Limits for Artificially Aged Tempers of 6005, 6061, and 6063

Temperature*		Time	Time
		(no more than 5% loss in	(per AWS
°F	°C	strength)	D1.2)
450	232	5 min	5 min
425	218	15 min	15 min
400	204	30 min	30 min
375	191	2 hr	1 to 2 hr
350	177	10 hr	8 to 10 hr
325	163	100 hr	50 hr
300	149	1,000 hr	50 hr
212	100	100,000 hr	

*Interpolate time (t) for other temperatures (T) using:

$$\log t = \log t_2 + \frac{\log(T_2/T)}{\log(T_2/T_1)}(\log t_1/t_2)$$

where:

T_1 = next lower temperature in Table 8.23 than T
T_2 = next higher temperature in Table 8.23 than T
t_1 = time corresponding to T_1
t_2 = time corresponding to T_2

inspection, no other mandatory quality assurance requirements for welding are required by the *Specification*. If you want any other inspection by the fabricator or verification inspection by another party, you must specify so elsewhere, such as in contract documents.

Two types of inspection are available: *nondestructive testing*, sometimes referred to by the acronym NDT, and *destructive testing* (also referred to as *mechanical testing*). Both refer to the condition of the inspected part after testing.

Nondestructive testing includes:

1) *Visual inspection* detects incorrect weld sizes and shapes, such as excessive concavity of fillet welds, inadequate penetration on butt welds made from one side; undercutting; overlapping; and surface cracks in the weld or base metal.

2) *Radiographic inspection* (X-ray photographs) (RT) is limited to groove welds in butt joints. This is because radiographic techniques are generally best applied to a joint of fairly constant thickness. Thus, fillet welds tend to be unsuitable for radiographic inspection. Radiography can detect defects as

small as 2% of the thickness of the weldment, including porosity, internal cracks, lack of fusion, inadequate penetration, and inclusions.

3) *Dye-penetrant inspection* (PT) may be performed on all types of welds and joints. The procedure is to apply a penetrating dye, wait 5 to 10 minutes, and then clean the surface. A developer is then applied, and the surface is inspected for evidence of discontinuities revealed by die that has penetrated them. This method will show only discontinuities that have access to the surface of the weld. Care should be taken to fully remove any dye or developer before repair welding.

4) *Ultrasonic inspection* (UT) uses high-frequency sound waves to detect similar flaws, but it is expensive and requires trained personnel to interpret the results. Its advantage over radiography is that it is better suited to detecting thin planar defects parallel to the X-ray beam. Ultrasonic inspection is limited to groove welds, and D1.2 does not include ultrasonic inspection methods or acceptance criteria. Therefore, when ultrasonic testing is required, the procedure, such as ASTM E164, *Standard Recommended Practice for Ultrasonic Contact Examination of Weldments*, and acceptance criteria must be specified.

Destructive testing, such as bend tests, tension tests, and fracture tests, are usually reserved for qualifying welders or weld procedures. They can also be used on production weldments if you don't mind making a few extra.

9 Special Topics

9.1 WELDED MEMBERS

Welding aluminum members introduces a consideration that doesn't get much thought for steel: the effect of welding on the strength of the base metal. (In hindsight, after the cracking of welded steel frames in the Northridge earthquake, we realized we probably should have been giving this more thought.) Sometimes it's just too easy to call for joints to be welded to the hilt to avoid the need to design and detail bolted connections. But the "less is more" maxim applies to aluminum welding. Not only does welding tend to cause distortions, it significantly reduces the strength of the aluminum base metal.

The prerequisite for this section is an understanding of the behavior of aluminum that's *not* welded. Once you're comfortable with that, you're ready to see what welding changes. This section covers only the design of welded aluminum members. Section 8.2 addresses the structural design of welded aluminum connections.

9.1.1 What Welding Does to Aluminum

The heat of welding reduces yield and ultimate strengths of aluminum alloys that get their strength from *artificial aging* and those that derive their strength from *cold-working*. (That just about covers all of the tempers you're ever likely to use.) In other words, welding offsets the strength the producers worked so hard, by one means or the other, to get into the metal. How much the strength is reduced depends on whether the alloy was artificially aged or cold-worked. Cold-worked alloys—(those with an -H in the suffix)—are reduced to the *annealed condition* (-O temper), their weakest state. Heat-treated alloys—(those with a -T in the suffix)—fare a little better, but not by much: their post-welded strength is slightly greater than their solution-heat-treated strength (-T4 temper). The effect is illustrated in Figure 9.1 for one alloy from each of these two groups. On the other hand, welding does not affect the *modulus of elasticity* (*E*).

This reduction in strength is apart from any effect from the strength of the *filler weld alloy;* even a filler stronger than the base metal, as is common in steel, doesn't help. That's because the heat of welding is the culprit. Of course, if the filler metal in the weld is weaker than the reduced strength base metal, the strength of the weld will be reduced even further. But if you select your filler from the ones in Aluminum *Specification* Table 7.2-1 (and not an

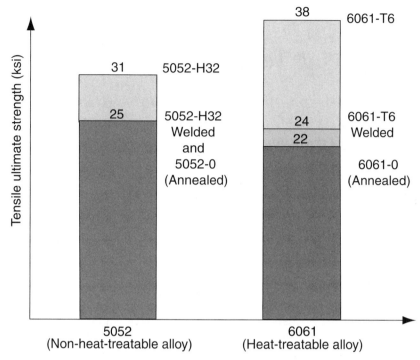

Figure 9.1   Effect of welding on tensile ultimate strength of alloys 5052 and 6061.

alternate filler given in the notes to this table), the filler metal should never be the weak link.

The damage is summarized in Aluminum *Specification* Table 3.3-2, Minimum Mechanical Properties for Welded Aluminum Alloys (included here as Appendix C and discussed in Section 4.5). While it's not a pretty sight, at least we know what we're dealing with. Actually, even this statement should be qualified. The values in Table 3.3-2 for ultimate tensile strengths (F_{tuw}) must be multiplied by 0.9 before they are used for structural design. This is to account for the fact that welds may get only visual inspection and might not reliably attain the strengths in Table 3.3-2 due to undetected defects. This 0.9 factor is not applied to minimum properties other than ultimate tensile strength; here, as elsewhere in the *Specification*, the margin of safety against tensile fracture is the greatest concern.

A picture of the effect of welding on strength is given in Figure 9.2. This shows that the heat-affected zone (HAZ) is limited to about 3 in. [75 mm] to either side of the weld centerline, and is most pronounced within 1 in. [25 mm] of the weld. This graph is important for two reasons. First, it's the reason the Aluminum *Specification* treats the HAZ as that area of the cross section within 1 in. of the centerline of the weld. Second, as we'll explain, it is why

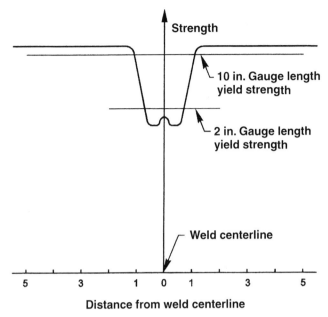

Figure 9.2 Strength distribution with distance from weld.

the gauge length affects the value of the yield strength across the weld so dramatically.

The gauge length for *yield strength* measurements is the length over which length measurements are taken during load testing. The yield strength is the average yield strength over the gauge length. A gauge length of 2 in. [50 mm] is typically used to report elongation at tensile failure for unwelded aluminum alloys.

The minimum yield strengths in Table 3.3-2 for welded alloys, however, are based on a 10 in. [250 mm] gauge length. The 10 in. length includes roughly 8 in. that are not weld-affected at all, however, and a shorter gauge length gives a lower yield strength because a greater proportion of the material in the gauge length is heat-affected. Using a 2 in. gauge length (over which almost all of the material is heat-affected) gives a yield strength of about 75% of yield strengths for a 10 in. gauge length. This is illustrated in Figure 9.2. The 2 in. gauge length value for tensile and compressive yield strengths is effectively required by the Aluminum *Specification* for longitudinal welds since all the material over the length of such a weld is weld-affected. The way the *Specification* requires the 2 in. gauge length values is by prescribing a 0.75 factor on the 10 in. gauge length yield strengths.

The minimum mechanical properties (strengths) for welded aluminum alloys are summarized in Appendix C. In Appendix C, the tensile ultimate strengths have been factored by 0.9 from those given in Aluminum *Specification* Table 3.3-2, and the tensile and compressive yield strengths have been factored by 0.75 from those given in Table 3.3-2 for the reasons cited above so that they may be used directly for structural calculations. One more subtlety: if the welded compressive yield strength (F_{cyw}) exceeds 90% of the unwelded compressive yield strength (F_{cy}), then the unwelded compressive yield strength may be used to compute allowable compressive stresses in the weld-affected area. In this case, the welded material has properties close enough to the unwelded material that the *Specification* allows you to ignore the slight difference in strength. Other than annealed tempers, this rule winds up applying only to 5086-H111 and 5454-H111 extrusions (which are rarely used), so it doesn't warrant a lot of attention. If you're starting with annealed (-O temper) base metal before welding, of course, then the welded strengths are the same as the unwelded strengths, and welding has no effect on design. Aluminum structures that are going to be heavily welded, such as pressure vessels, are often made of annealed material because for design purposes, the material is going to wind up annealed at the end of fabrication anyway.

9.1.2 Types of Welded Members

The effect of welding on member strength depends, in part, on the orientation of the weld with respect to the length of the member. Welds, therefore, are divided into two types for the purpose of divining their influence on member strength: transverse welds and longitudinal welds.

Transverse welds are those lying in a plane transverse to the long axis of the member. Examples of transverse welds are a *groove weld* at a splice in the length of a member, and the weld attaching a base plate to a column (Figure 9.3). Welds made at connections are often transverse welds.

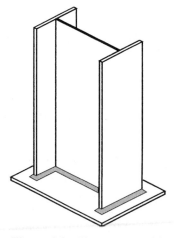

Figure 9.3 Transverse weld.

Longitudinal welds are those that run parallel to the length of the member. An example of a longitudinal weld is the weld between the web and the flange of a plate girder. Longitudinal welds are typical of *built-up members* (Figure 9.4).

Another factor in determining the effect of welding on strength is whether the member is in tension or compression. Therefore, we'll address welded tension members and welded compression members separately below.

9.1.3   Welded Tension Members

Transverse Welds in Tension Members   A tensile member with a transverse weld across its full cross section is a good example of the old adage that a chain is only as strong as its weakest link. Where's the weak link? Here are the possibilities:

1) *The weld could fracture.* What about yielding there? Yielding at the transverse weld would cause very little elongation of the member in relation to its overall length because the weld-affected length typically constitutes such a small portion of the total length of the member. Consequently, yielding at the transverse weld is not a limit state, and the welded yield strength does not affect the tension strength of a member with a transverse weld. The fracture strength of the weld is given in Aluminum *Specification* Table 3.3-2 as F_{tuw}.

2) *The member may yield or fracture at a cross section away from the weld.* Since the length of the member over which the member's properties are not weld-affected comprises most of the member's length, you do have to check yielding of the non-weld-affected cross section. You can rule out

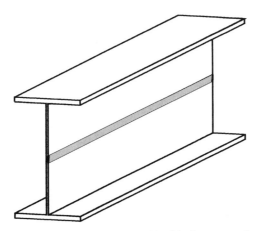

Figure 9.4   Longitudinal weld of built-up member.

fracture away from the weld, however, since the weld-affected metal is always weaker than the non-weld-affected metal. (In other words, $F_{tuw} \leq F_{tu}$ is always true.)

Therefore, the allowable tensile force T_a on a member with a transverse weld with a cross-sectional area A_w affecting the full cross section of area A is:

$$T_a = \min \left(\frac{0.9\, F_{tuw}\, A_w}{n_u}, \frac{F_{ty}\, A}{n_y} \right)$$

Some observations: the cross-section area at the weld is limited to the cross-sectional area of the member away from the weld by AWS D1.2 (91), so $A_w \leq A$. This is because, for design purposes, the effective thickness of a weld can't exceed the welded member's thickness, and the effective width of the weld can't exceed the welded member's width. If the weld is a partial joint penetration weld, the weld's cross-sectional area is less than the member's cross-sectional area. (See Section 8.2.3 for how to determine if a weld is a partial joint penetration weld—it may be more often than you think.) Also, the tension coefficient k_t, which accounts for notch sensitivity in certain alloy tempers, is always 1.0 for welded material since that material is essentially annealed and, hence, very ductile—about the only advantage of a weld.

Example 9.1: What is the allowable tensile stress on a 6061-T6 member with a full penetration transverse groove weld across the full width?

In this case, the area of the weld equals the area of the section ($A_w = A$) and

$$F_w = \min \left[\frac{0.9\, F_{tuw}}{n_u}, \frac{F_{ty}}{n_y} \right] = \min \left[\frac{0.9(24\ \text{ksi})}{1.95}, \frac{(35\ \text{ksi})}{1.65} \right]$$

$$= \min[11.1, 21.2] = 11.1\ \text{ksi}$$

Sometimes a transverse weld extends across only part of the cross section. For example, improbable as it seems, someone might drill a hole in the wrong place and weld a plug back in to cover it up. In this case, the fracture strength at the cross section that's partially weld-affected is the sum of the fracture strength of the weld-affected part and the fracture strength of the part that's not weld-affected, so the allowable tensile force T_a is:

$$T_a = \min \left(\frac{0.9\, F_{tuw}\, A_w + (A - A_w)\, F_{tu}/k_t}{n_u}, \frac{F_{ty}\, A}{n_y} \right)$$

Transverse welds in tension are used in aluminum structures, for example, the vertical welds on cylindrical tanks storing liquids, but such structures are usually subject to special inspection methods as discussed briefly in Section 3.3. Because the integrity of these welded structures are so dependent on the reliability of these welds, special standards have typically been developed in each industry that uses this type of construction. ASME Standard B96.1, *Welded Aluminum-Alloy Storage Tanks* (85), as one example, requires a joint factor less than 1.0 be applied to such welds that have not been fully *radiographed*. Additional requirements apply to the certification of the welder and the weld and weld repair procedures. You'll need to consult the appropriate codes for additional design considerations for structures that rely on transverse welds in tension.

Longitudinal Welds in Tension Members Longitudinal welds in tensile members (in axial tension or bending tension) are treated with a weighted-average approach. The formula for allowable stress is, therefore,

$$F_{pw} = F_n - (A_w/A)(F_n - F_w) \tag{9.1}$$

where: F_{pw} = allowable stress on a cross section, part of whose area lies within 1.0 in. [25 mm] of a weld

F_n = allowable stress for the cross section if none of it were weld-affected

F_w = allowable stress for the cross section if the entire cross section were weld-affected. For ultimate tensile strength, use $0.9F_{tuw}$ and for tensile yield strength use $0.75F_{tyw}$ to calculate F_w.

A_w = portion of the area of the cross section lying within 1.0 in. [25 mm] of the centerline of a groove weld or the heel of a fillet weld

A = net area of the cross section (for axial tension) or the tension *flange* (for beams). For this purpose, the flange is that portion of the section farther than $2c/3$ from the neutral axis, where c = the distance from the neutral axis to the extreme fiber.

The weighted-average approach means that the strengths of the unaffected portion of the section and the weld-affected portion of the section are weighted in proportion to their share of the full cross section. One simplification is used: If the weld-affected portion of the cross section is less than 15%, the strength reduction from welding is deemed too small to worry about, and the effect of welding can be neglected. (See Aluminum *Specification* Section 7.1.2.) The inaccuracy introduced by this simplification is on the order of 5%.

Example 9.2: What is the allowable axial tensile stress for the member with the cross section shown in Figure 9.5? The member is 6061-T6 welded with 4043 filler alloy.

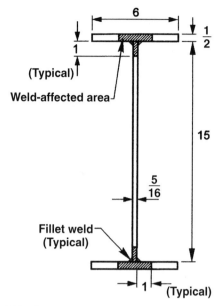

Figure 9.5   Cross section of a welded built-up member.

The net area of the cross section is A:

$$A = (0.5 \text{ in.})(2)(6 \text{ in.}) + (0.3125 \text{ in.})(15 \text{ in.}) = 10.7 \text{ in.}^2$$

The weld-affected area is A_w:

$$A_w = (0.3125 + 2)(\text{in.})(0.5 \text{ in.})(2) + (0.3125 \text{ in.})(1 \text{ in.})(2) = 2.94 \text{ in.}^2$$

The weld-affected area of the cross section is more than 15% of the cross section, so you must use Equation 9.1 to determine the strength of the section. You can obtain the unwelded tensile allowable stress from *Aluminum Design Manual*, Part VII, Table 2-21, as 19 ksi for *Specification* Section 3.4.1, Axial Tension.

The welded allowable tensile stress is:

$$F_w = \min[0.9F_{tuw}/(k_t \, n_u) \text{ and } 0.75F_{tyw}/n_y]$$

where:

F_{tuw} = welded tensile ultimate strength (from Aluminum *Specification* Table 3.3-2)

F_{tyw} = welded tensile yield strength (from *Specification* Table 3.3-2)

k_t = coefficient for tensile members = 1.0 in the weld-affected zone (*Specification* Table 3.4-2)

n_u = safety factor on ultimate = 1.95
n_y = safety factor on yield = 1.65.

Note that F_{tyw} is 15 ksi rather than 20 ksi because the flange plate thickness is greater than 0.375 in. and 4043 filler alloy is used. This gives:

$$F_w = \min[0.9(24 \text{ ksi})/[(1.0)(1.95)] \text{ and } 0.75(15 \text{ ksi})/1.65]$$

$$F_w = \min[11.1 \text{ ksi}, 6.8 \text{ ksi}] = 6.8 \text{ ksi}$$

Calculating the allowable stress for the partially weld-affected section (F_{pw}) using Equation 9.1:

$$F_{pw} = 19 - (2.94/10.7)(19 - 6.8) = 15.6 \text{ ksi}$$

9.1.4 Welded Compression Members

Transverse Welds in Compression Members For beams or columns supported at both ends and with a transverse weld within 5% of the length of the member from an end, Aluminum *Specification* Section 7.1.1 allows you to neglect the effect of welding on the member as long as you consider the member to be pin-ended. An example is a column in a building frame with a welded base plate.

For all other beams and columns with transverse welds, the Aluminum *Specification* requires that the member be designed as if the entire member were weld-affected, even if the transverse weld is only at one point along the member length. You calculate the strength by substituting $0.75F_{cyw}$ for F_{cy} in the equations for strength (in Aluminum *Specification* Table 3.4-3), and in the equations for the buckling constants (B, D, and C). As discussed above, the equations to be used for buckling constants are those for non-artificially aged alloys (*Specification* Table 3.3-3, *not* Table 3.3-4), even if the welded alloy in question was artificially aged before welding.

Example 9.3: What is the allowable compressive stress for the cross section of Example 9.2 as an axially loaded column with a transverse weld at its midheight and a slenderness ratio (kL/r) of 30?

In accordance with Aluminum *Specification* Section 7.1.3, since the weld is farther than 5% of the member length from the member ends, the column is to be treated as if the entire column were weld-affected. The compressive yield strength is 0.75(15 ksi) = 11 ksi (see Table 3.3-2). The allowable stress for columns with a slenderness ratio less than slenderness limit S_1 is:

$$F_w = F_{cy}/n_y = (11 \text{ ksi})/1.65 = 6.7 \text{ ksi} \tag{9.2}$$

The allowable stress for columns with a slenderness ratio between slenderness limits S_1 and S_2 is:

$$F_w = (B_c - D_c(kL/r))/n_u \tag{9.3}$$

where B_c, D_c = buckling constants calculated from Table 3.3-3 using 11 ksi for F_{cy}.

Calculation of the buckling constants for the welded column requires the use of Table 3.3-3, which is for unwelded tempers -T1 through -T4 and all welded alloys, rather than Table 3.3-4, which is for unwelded tempers -T5 through -T9, even though the alloy is of -T6 temper:

$$B_c = F_{cy}\left[1 + \left(\frac{F_{cy}}{1000}\right)^{1/2}\right] = 11\left[1 + \left(\frac{11}{1000}\right)^{1/2}\right] = 12.2 \text{ ksi}$$

$$D_c = \frac{B_c}{20}\left(\frac{6B_c}{E}\right)^{1/2} = \frac{12.2}{20}\left[\frac{6(12.2)}{10,100}\right]^{1/2} = 0.052 \text{ ksi}$$

So, the allowable stress is:

$$F_w = 12.2/1.95 - (0.052/1.95)(kL/r) = 6.3 - 0.027 \; (kL/r)$$

The allowable stress for columns with a slenderness ratio greater than S_2 is:

$$F_w = \pi^2 E/[(n_u)(kL/r)^2] = \pi^2(10,100)/(1.95)(kL/r)^2 = 51,100/(kL/r)^2 \tag{9.4}$$

To determine which of Equations 9.2, 9.3, and 9.4 to use for this member, we need to determine the slenderness limit S_2:

$$S_2 = C_c = \frac{2B_c}{3D_c} = \frac{2(12.2)}{3(0.052)} = 157$$

Since the slenderness ratio is 30, which is less than $S_2 = 157$, the allowable stress is given by the inelastic buckling equation (Equation 9.3):

$$F_w = 6.3 - 0.027(30) = 5.5 \text{ ksi}.$$

This example illustrates how tedious it is to calculate the strength of welded compression members. Most of the work lies in calculating the welded buckling constants. For non-welded members, the buckling constants are given in *Aluminum Design Manual*, Part VII, Table 2-1, but buckling constants for welded aluminum alloys aren't given in the *Manual*. Therefore, we provide the welded buckling constants in Appendix K.

Longitudinal Welds in Compression Members You calculate the strength of longitudinal welds in columns and beams by the weighted-average approach as for tension members with longitudinal welds. One difference, however, is that the strength of a weld-affected compression member involves calculating buckling constants, whereas tension strengths do not. The formula for allowable stress is:

$$F_{pw} = F_n - (A_w/A)(F_n - F_w) \tag{9.5}$$

where:

F_{pw} = allowable stress on a cross section, part of whose area lies within 1.0 in. [25 mm] of a weld

F_n = allowable stress for the cross section if none of it were weld-affected

F_w = allowable stress on the cross section if the entire cross section were weld-affected. For compressive yield strength, use $0.75F_{cyw}$ to calculate F_w. For buckling constants, use Table 3.3-3, regardless of temper before welding.

A_w = portion of the area of the cross section lying within 1.0 in. [25 mm] of the centerline of a groove weld or the heel of a fillet weld

A = gross area of the cross section (for columns) or compression flange (for beams). For this purpose, the flange is that portion of the section farther than $2c/3$ from the neutral axis, where c = the distance from the neutral axis to the extreme fiber.

For beams, the 1983 Canadian aluminum standard (96) uses I_w/I in place of A_w/A in Equation 9.5 above, where I_w is the moment of inertia of the heat-affected zone (HAZ) about the axis of bending and I is the moment of inertia of the whole section. This accounts for the location of the weld in the cross section but complicates the determination of strength.

Example 9.4: What is the allowable axial compressive stress for overall member buckling for the member in Example 9.2 as a column with a slenderness ratio (kL/r) of 30?

The allowable stress for the cross section if there were no welds is F_n, which is determined from *Aluminum Design Manual*, Part VII, Table 2-21, Section 3.4.7:

$$F_n = 20.2 - 0.126(30) = 16.4 \text{ ksi}$$

The compressive yield stress in the HAZ is:

$$F_{cy} = 0.75F_{cyw} = (0.75)(15 \text{ ksi}) = 11 \text{ ksi}$$

The buckling constants are the same as those in Example 9.3 because the yield strength is the same and in both cases the buckling constant formulas are taken from Table 3.3-3, so:

$$B_c = 12.2 \text{ ksi}, \ D_c = 0.052 \text{ ksi, and } C_c = 157$$

Since the slenderness = $30 < S_2 = 157$, the allowable stress is:

$$F_w = 12.2/1.95 - (0.052/1.95)(kL/r)$$

$$F_w = 6.3 - 0.027(kL/r) = 6.3 - 0.027(30) = 5.5 \text{ ksi}$$

The allowable stress on the cross section is, then:

$$F_{pw} = 16.4 - (2.94/10.7)(16.4 - 5.5) = 13.4 \text{ ksi}$$

What about local buckling checks in welded compression members? Aluminum *Specification* Section 7.1.1 addresses only the effect of welding on local buckling when the welds are located on the supported edge of flat elements. In such cases, the *Specification* allows you to neglect the effect of welding on the strength, so in Example 9.4 above, the local buckling checks would be performed as if there were no welds. The rationale behind this is that welding along a supported edge reduces the support to that of a pinned edge, and since the local buckling strengths of flat elements are already based on this assumption, welding does no further harm.

Where welds are not located at the supported edge of an element—for example, as in Figure 9.4, where the weld is at the midheight of the web—the Aluminum *Specification* is silent. It's not even obvious that treating the element as if it were entirely of heat-affected material is conservative since welding could introduce residual stresses or distortions that could reduce the strength even more than if the material were simply all of reduced strength. Also, the Aluminum *Specification* provisions for the compressive strength of transversely welded curved elements (Sections 3.4.10, 3.4.12, and 3.4.16.1) are limited to relatively stocky elements, those with a slenderness ratio $R_b/t \le 20$. This is because tests have indicated that the *Specification* is unconservative in such cases; fortunately, all schedule 40 pipe through 12 in. nominal diameter have $R_b/t \le 20$. We hope that future editions of the *Specification* will plug these gaps.

9.1.5 Post-Weld Heat Treatment

Designers familiar with steel think of post-weld heat treatment as stress-relief treatment—something done to a weldment to relieve residual stresses induced by differential cooling after welding. In aluminum, however, post-weld heat treatment is more often used to regain some of the strength lost by welding. If you think about it, the solidified weld metal is essentially annealed material, and if it's a heat-treatable alloy, it could benefit from heat treatment (see Chapter 2). A couple of problems occur with this, though. First, only heat-treatable base metals welded with heat-treatable filler alloys can benefit from heat treatment. The only heat-treatable fillers are 4xxx fillers, which boils down to 4043 most of the time; heat treatment won't work with any of the 5xxx series fillers (like 5356). Second, unless the base metal wasn't already solution-heat-treated or precipitation-heat-treated before welding or you can limit the heat treatment to only the weld HAZ, applying heat treatment after welding will overage the material away from the weld, thereby reducing its strength and ductility.

The post-weld heat treatment can be a solution heat treatment and artificial aging (precipitation heat treatment) or just artificial aging. While solution heat treating and aging will recover more strength than aging alone, the rapid quenching required in solution heat treating can cause distortion of the weldment because of the residual stresses that are introduced.

The size of the weld bead laid in one pass must also be limited if post-weld artificial aging is to be performed. A limit of $\frac{5}{16}$ in. [8 mm] on the size of a stringer-bead fillet weld for all welding positions is used. The purpose of the bead size limit is to restrict the amount of heat that welding puts into the base metal.

With these caveats in mind, you'll find that in some instances, post-weld heat treatment of aluminum weldments is practical. One such case is the weld at the base of aluminum light poles, mentioned in Aluminum *Specification* Section 7.4. For example, 6063-T4 poles—the T4 temper means that they've been solution-heat-treated, but not precipitation-heat-treated—are welded with 4043 filler and then precipitation-heat-treated to the T6 temper. This process improves the welded strength from 40% (10 ksi/25 ksi) of the T6 temper to 85% of the T6 temper; this is enough to be worthwhile, especially since the stresses in the pole are highest at the base. The strength of these post-weld heat-treated weldments was determined by testing, and testing would be required to establish post-weld heat-treated strengths of other weldments. Filler 4643 is also heat-treatable and used in post-weld heat treatments, especially for weldments more than about $\frac{1}{2}$ in. [13 mm] thick.

A final note: the strength of such heat-treatable filler alloys as 4043 and 4643 increases by natural aging even without any post-weld heat treatment (see Section 8.2.2 for more on this).

9.2 FATIGUE

It's quite possible to spend a career designing aluminum structures without entering the imperfectly understood netherworld called "fatigue." Even if you're already familiar with steel's fatigue performance, some surprises may be in store. The purpose of this chapter is to help you identify when you should consider fatigue and to offer a brief primer to get you through some simple fatigue designs. And in the land of the blind, the one-eyed man is king.

9.2.1   Fatigue—What Is It Again?

Fatigue is cracking caused by the repeated application of stresses. These stresses must be tensile or a cycling back and forth between tension and compression, referred to as *stress reversal.* Stresses that vary but always remain compressive don't concern fatigue designers, because fatigue cracks don't initiate without tensile stress. A stress variation with a constant amplitude is illustrated in Figure 9.6. In reality, stress is rarely such a simple function of time, but you may idealize it this way. When a stress reversal occurs, the *stress amplitude,* or range, is the sum of the compressive and tensile stress amplitudes, which can be thought of as the algebraic sum of the maximum stresses. Tensile stresses are arbitrarily called *positive,* and compressive stresses are called *negative.*

Cracking caused by fatigue occurs at stresses less than that required to break the member with just one application of load. The more times the load is applied, the lower the stress required to cause failure. The relationship between the number of cycles of stress and the fatigue strength is shown in Figure 9.7, which is called an *S-N curve* (Stress vs. Number of cycles). When the number of constant amplitude cycles is between about 10^5 (100,000) and 5×10^6 (5 million), the relationship is exponential, so it appears as a straight line when the scales are shown logarithmically, as in Figure 9.8. The number of cycles before fatigue begins to rear its ugly head varies by alloy. Some higher strength alloys suffer the onset of fatigue at fewer cycles of load.

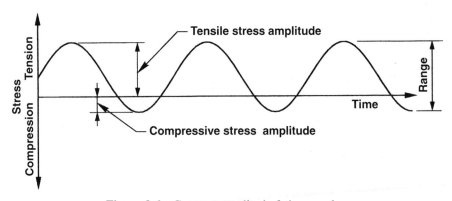

Figure 9.6   Constant-amplitude fatigue cycles.

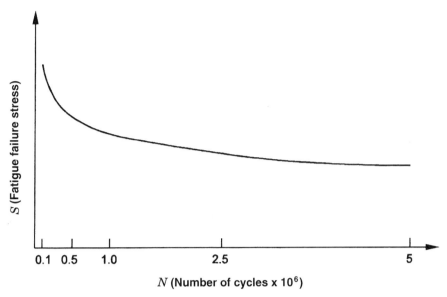

Figure 9.7   S-N curve for fatigue.

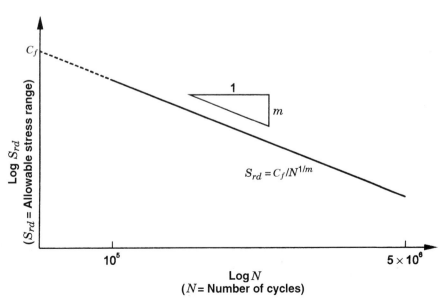

Figure 9.8   Log of S-N curve for fatigue.

For aluminum alloys, the *endurance limit* is defined as the fatigue strength of plain specimens for a very large number of cycles (5×10^8, or 500 million) of stress reversals, measured by a standard rotating beam test. Endurance limits for a number of alloys are given in *Aluminum Design Manual*, Part V, Material Properties, Table 5. These are typical values and are *not* to be used for design. The idea behind the endurance limit was that hardly anyone anticipated a structure being subjected to this many stress cycles, so the endurance limit would be a lower bound on the fatigue strength. Forget about the endurance limit; it's no longer considered a very useful indicator of fatigue behavior, partly because the standard test specimen was too small to be indicative of the performance of real structural parts. Fatigue behavior is a function of size because the larger the member, the more likely it is that a flaw will be present, and fatigue cracks tend to initiate at defects.

9.2.2   Fatigue Design: The Ground Rules

Fatigue strength determined in accordance with Aluminum *Specification* Section 4.8 is stated as an allowable stress range (S_{rd}) and is a function of two parameters:

1) the number of cycles of load (N)
2) the configuration of the detail under consideration.

In building and bridge structures, the goal is to size members and connections so that the applied stress range (S_{ra}) does not exceed the allowable stress range (S_{rd}) for the life of the structure. To start playing this fatigue design game, you must bring along the number of load cycles you expect will occur over the life of the structure and the type of detail you hope to use. Other approaches are available to assess fatigue, but they are more often used in such applications as aircraft design rather than construction. Designers using these other techniques often have different goals, such as predicting how much life is left in a member after a certain period of service.

For constant amplitude stress, once the number of cycles and the detail are known, you can apply Aluminum *Specification* Equation 4.8.1-2 to calculate the allowable-stress range:

$$S_{rd} = C_f N^{-1/m} = \frac{C_f}{N^{1/m}} \tag{9.6}$$

where:

S_{rd} = allowable stress range
C_f = constant (coefficient) from Aluminum *Specification* Table 4.8.1-1
m = constant (exponent) from *Specification* Table 4.8.1-1
N = number of constant amplitude cycles.

The constants C_f and m are a function of the configuration of the detail being designed, which we'll discuss more below. For constant amplitude loading, the number of cycles (N) used to calculate the allowable stress range need not be greater than 5×10^6, even if the actual number of cycles is greater than this. The allowable stress range at 5×10^6 cycles is the *constant amplitude fatigue limit* (CAFL), called the fatigue limit in the *Specification* and given in *Specification* Table 4.8.1-1 for each of the six different detail categories identified in *Specification* Table 4.8-1. For constant amplitude loading, then, the allowable stress range is never less than the fatigue limit.

Equation 9.6 replaced a table of values in the 1986 edition of the Aluminum *Specifications*, and the equation was also adjusted to take advantage of more extensive fatigue experience gained since the early 1980s, when the previous table was developed. Table 9.1 provides tabular values of fatigue strength based on *Specification* Equation 4.8.1-2.

How many cycles of stress must be anticipated before you worry about fatigue? The Aluminum *Specification* doesn't offer fatigue strengths for components subjected to less than 100,000 cycles of stress. Some designers begin

TABLE 9.1 Allowable Stress Range For Fatigue (ksi) (based on Aluminum *Specification* Equation 4.8.1-2)

Stress Category	Examples	Number of Cycles			
		100,000	500,000	2,000,000	5,000,000 (CAFL)*
A	Plain metal	18.0	14.2	11.6	10.2
B	Members with groove welds parallel to the direction of stress	12.0	8.6	6.5	5.4
C	Groove welded transverse attachments with transition radius 24 in. > R ≥ 6 in. [610 mm > R ≥ 150 mm]	11.8	7.6	5.2	4.0
D	Groove welded transverse attachments with transition radius 6 in. > R ≥ 2 in. [150 mm > R ≥ 50 mm]	7.2	4.7	3.2	2.5
E	Base metal at fillet welds	5.7	3.6	2.4	1.8
F	Fillet weld metal	6.0	3.8	2.5	1.9

* Constant Amplitude Fatigue Limit

their consideration of fatigue at 20,000 cycles for welded aluminum structures because welds may introduce significant residual stresses and abrupt changes in geometry, both of which shorten fatigue life. (The Canadian code (96) and earlier editions of the Aluminum *Specification* required fatigue design for welded structures subjected to more than 20,000 cycles.) The AISC Steel *Specification* suggests 20,000 cycles as the starting point for consideration of fatigue. In some cases, such as weldments and details with significant stress risers, it might be worthwhile to consider fatigue for aluminum components subjected to 20,000 to 100,000 cycles, based on the Aluminum *Specification* design fatigue strength for 100,000 cycles.

Numbers of these magnitudes are difficult to put in perspective, so you may find Table 9.2 useful. It shows the total number of cycles in terms of the number of cycles per day for various lifetimes. In buildings subjected to wind, seismic, and live loads from occupants, the number of load cycles is not enough to warrant fatigue consideration. However, vessels or piping subjected to frequent thermal-stress variation and crane runways are examples where the number of load cycles may justify a fatigue investigation.

The other parameter in fatigue is the configuration of the detail that is a measure of the severity of the local stress increase at the detail. The Aluminum *Specification* gives fatigue strengths for 6 different types of details called *stress categories*. (The AISC Steel *Specification* offers similar stress categories, but it has added two categories for a total of eight. AWS D1.1 (90) has even more categories). The categories range from A, which has the highest fatigue strength, to F. For example, stress category A includes plain-material base metal, such as a rolled or extruded shape, exposed to cycles of axial load or bending moment. This category is a close match for stress category A for steel. The stress categories are defined in the Aluminum *Specification,* and figures are also provided illustrating the categories (similar to the AISC Steel *Specification*). If your situation doesn't match one of these categories, you'll have to test to determine an allowable stress range. (If you don't have that much time or money, you could try to match your detail to one of the

TABLE 9.2   Approximate Total Number of Cycles for Given Rates/Day

Cycles/Day	Lifetime (Yr)	Total Number of Cycles (N)
2	25	20,000
10	25	100,000
50	25	500,000
200	25	2,000,000
1	50	20,000
5	50	100,000
25	50	500,000
100	50	2,000,000

steel details in the AISC Spec or AWS D1.1 and see what category is used for that.)

Example 9.5: Determine the fatigue strength for 6063-T5 extrusions subjected to 100,000 cycles of load. The detail under consideration is a lap-bolted splice connection, shown in Figure 9.9.

This detail matches Aluminum *Specification,* Figure 4.8-1, Case 8, which is identified as stress category E in *Specification* Table 4.8-1. When you use *Specification* Section 4.8.1 and look up the values for the variables in *Specification* Equation 4.8.1-2 from *Specification* Table 4.8.1-1, the allowable stress range is:

$$S_{rd} = C_f N^{-1/m} = \frac{160}{100,000^{1/3.45}} = 5.7 \text{ ksi} \tag{9.7}$$

You also could determine this value by looking it up in Table 9.1 above. If the connection were a bearing connection, this stress range would be compared to the stress range calculated on the *net section* of the extrusions. If the connection were a friction connection, the stress range would be compared to the stress range calculated on the *gross section* of the extrusions. This illustrates what is probably the main benefit of friction connections.

In static design, the *maximum stress* in the part must be limited to the design stress; in fatigue design, the *stress range* must be limited to the design stress range. The difference is easy to overlook, especially for designers used to static design. In fatigue, the *difference* between the maximum stress and the minimum stress is what matters. Since dead load acts all the time, only the difference in the live loads counts.

That's a basic primer. As with most provisions in the Aluminum *Specification,* you need to be aware of a few nuances. For one, fatigue design strength in accordance with the Aluminum *Specification* is not a function of

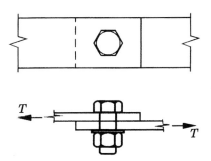

Figure 9.9 Bolted splice connection.

the aluminum alloy used. So, whether you use scrap metal from beer cans or an alloy from the space shuttle, if the material is a wrought aluminum alloy and all other things are equal, the fatigue design strength is the same. You can probably see where this is leading. If fatigue governs design, alloy selection might more properly be based on considerations other than strength, such as cost, corrosion resistance, and ease of fabrication. (The exceptions are cast alloys; their fatigue strengths can be established only by testing.) Also, the allowable fatigue stress range can't exceed the static design strength allowed without consideration of fatigue. This is an issue with only some of the weaker alloys loaded at relatively few cycles and in the less severe details.

One other important subtlety: you need not apply safety factors to fatigue design strength values provided by the Aluminum *Specification;* use the allowable stress ranges just as produced by Equation 4.8.1-2. The actual fatigue strengths have a 97.7% probability of survival with a confidence level of 95%. (This is a slightly lower probability than the likelihood that the minimum mechanical properties will be exceeded, which is expected to occur 99% of the time with a 95 % confidence level). Also, there is no difference between the *allowable stress design* and the *load and resistance factor design* (*LRFD*) *Specification* Sections 4.8 on fatigue. Both use *nominal loads* without applying any load factors, and neither factors the allowable stress range by a *safety factor* or a *resistance factor*.

9.2.3   Variable Amplitude Fatigue Design

Aluminum *Specification* Section 4.8.2 covers variable amplitude fatigue loading. Here, the stress versus timeline is not a neat sine wave of constant amplitude and frequency (Figure 9.10), but rather a mess (like Figure 9.11). When stress is not such a simple function of time, more complicated techniques are required to count cycles. For example, for the variable amplitude stress ranges shown in Figure 9.11, how many cycles have occurred? Methods

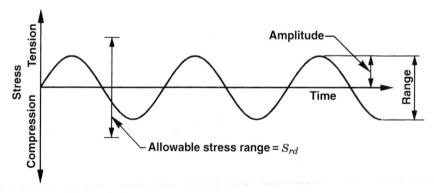

Figure 9.10   Example of constant-amplitude fatigue load.

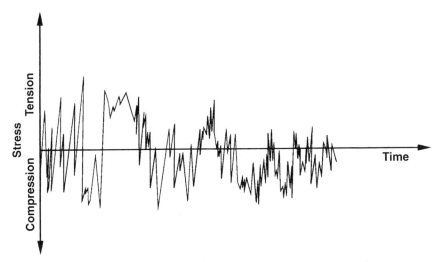

Figure 9.11 Example of variable-amplitude fatigue load.

of counting cycles with exotic monikers like "rainflow" are often used in these cases. This, however, is beyond the limited scope of this discussion. Additional information is available in other references (134).

These counting methods divide the stress amplitudes into several ranges (the Canadian Aluminum Code (96) calls these subdivisions *semi-ranges*) and count the number of cycles occurring in each range. Then Equation 4.8.2-2, which is called *Miner's Rule,* is applied. This rule says that the damage done by the cycling in each stress range is, in effect, cumulative. The form of the equation as used in the Aluminum *Specification* determines an equivalent stress range for the variable amplitude loading (S_{re}), which may be compared to the allowable stress range for constant amplitude load.

For variable amplitude loads, the Aluminum *Specification* does not limit how low the allowable stress range may go. In other words, no fatigue limit exists for variable amplitude loads. There is some debate over whether such an abrupt change in behavior between constant amplitude loads and variable amplitude loads is appropriate. If just one cycle of load has a different amplitude than all the others, the loading is variable, and no fatigue limit is allowed, whereas if that one cycle were the same as all the others, then a fatigue limit *would* be allowed. Since it only matters when the number of cycles is greater than 5×10^6, this doesn't come up often.

9.2.4 Aluminum Versus Steel in Fatigue

People often ask how aluminum's fatigue strength compares to that of steel. Table 9.3 shows that aluminum has about one-quarter to one-half the fatigue strength of steel. Design provisions for the two materials are otherwise very

TABLE 9.3   Ratio of Aluminum to Steel Fatigue Strengths

Stress Category	Number of Cycles			
	100,000	500,000	2,000,000	5,000,000
A	0.29	0.38	0.48	0.42
B	0.25	0.30	0.36	0.34
C	0.34	0.36	0.40	0.40
D	0.26	0.29	0.32	0.36
E	0.26	0.27	0.30	0.37
F	0.40	0.31	0.28	0.24

similar. Specifications for both materials consider fatigue strength to be independent of minimum material strength and use the detail classification method to assign fatigue strength categories.

9.2.5   Other Factors in Fatigue

Additional factors—other than number of cycles, stress range, and type of detail—affect fatigue strength. These factors include service temperature, the corrosiveness of the environment, and weld quality (133). At extremely low temperatures (−320°F [−200°C]), aluminum's fatigue strength is on the order of one to two times the fatigue strength at room temperature. Conversely, at elevated temperatures, aluminum's fatigue strength is diminished. Corrosive environments, such as marine atmospheres, also decrease the fatigue strength by as much as two-thirds, depending on alloy. Corrosion causes flaws, such as pits, that act as crack-initiation sites, thereby decreasing fatigue strength. Anodizing also decreases fatigue strength because it creates a brittle surface that is more prone to cracking.

These factors are not included in the Aluminum *Specification* fatigue provisions. Its predictions of fatigue strength are founded on an assumption of ambient temperature service and a normal environment, so different conditions should be considered when they are present. Enough information is not available to formulate design rules that include these factors.

Fatigue problems can be encountered in structures with resonant frequencies close to the frequency of a varying load, most often the gusting effect of wind. This effect is called *aerodynamic* or *aeroelastic response* and can be important in the design of flexible structures subjected to wind, such as light poles and signs. You can best address this situation not by proportioning the members large enough to resist the stresses caused by such loads, but rather by damping the structure to change its resonant frequency. This topic is beyond our scope here, so we refer you to other sources (134).

9.2.6   A Final Word

Even if you never use fatigue design provisions, it's wise to consider their lessons. Stress risers and uninspected welds that may have undetected flaws can be weak points in a structure. Avoid these situations whenever possible. Even relatively benign looking scratches can serve as crack-initiation sites, so careful workmanship and handling not only improve appearance, but also enhance fatigue life.

9.3   RECENT DEVELOPMENTS IN ALUMINUM STRUCTURES

Although it may seem the industry moves so slowly you have to line it up with a post to see if it's moving, sometimes change does occur. While not all the topics discussed below will ultimately affect non-aerospace structural applications, it's useful to be aware of them so you can decide which are worth monitoring. Go ahead—you deserve some light reading after wading through all the hard parts of the book.

9.3.1   Friction Stir Welding

One of aluminum's main disadvantages is the limitation imposed by its welding technology compared to that of steel. Several reasons for this exist: aluminum requires more joint preparation prior to welding than steel, and the heat of welding reduces the strength of aluminum while it does not do so for steel. Traditional arc welding used for joints made between butting aluminum parts usually requires that material first be removed from the joint, thereby creating a groove into which special filler material is redeposited under an inert shielding gas with an electric arc. This is especially inefficient for aluminum because aluminum production is so energy-intensive and arc welding requires most of the material in the joint be produced twice.

Welding limitations are also particularly disadvantageous for aluminum since the largest commonly produced aluminum structural shapes are smaller (12 in. [300 mm]) than those for carbon steel (44 in. [1,100 mm]). When larger structural shapes are needed, built-up shapes must be made by welding together plates or smaller shapes.

A new process, however, shows promise in helping to overcome these hurdles. The story goes that a technician at The Welding Institute (TWI) in England was cutting plastic pipe with a dull saber saw and noticed that the pipe tended to melt back together behind the blade due to friction-generated heat. True or not, in 1991 TWI filed a patent for friction stir welding (FSW), which uses frictional heat produced by a rotating, nonconsumable tool that is plunged into the interface between two closely abutted parts. The heat plasticizes the material and bonds it, producing welds with low heat input and, hence, little distortion and good properties in the heat affected zone. The process appears to be better suited to aluminum than any other material and has been used in the space shuttle's external fuel tanks (see Section 9.3.3 on aluminum-lithium alloys), other aerospace applications, and ships. It has been applied to 2xxx, 5xxx, 6xxx, and 7xxx alloys, in thicknesses up to 1 in. [25 mm]. No grooving operation, filler material, shielding gas, or electric arc are needed, so FSW offers considerable material, environmental, and safety advantages over conventional arc welding. Since FSW doesn't add any material to the weldment, it can't produce fillet welds, but it can be very useful for longitudinal butt welds between extrusions or plates. Friction stir welding also has been used to join 2xxx and 7xxx alloys that otherwise can't be welded.

FSW is not widely used yet, however. This is partly due to the cost to license the process from TWI and a natural delay in bringing a new technology to market, but it is also driven by a lack of technical standards. No design strengths and widths of heat affected zones have been established for FSW. Also, the FSW tool applies a high bearing force to the workpiece and tends to push apart the two parts to be joined, so only parts that can be properly supported from behind and on either side of the weld can be friction stir welded.

9.3.2   Alloy 6082

Registered in 1972, 6082 isn't exactly a new development, but it has only recently seen much use in the U.S. Its composition is compared to 6061 in Table 9.4. 6082 became popular in Europe because it has higher strengths than 6061 (see Table 9.5, which shows allowable tensile strengths are 10% to 18% higher, depending on thickness) and better general corrosion-resistance (due to lower copper content) than 6061, and about equal density (0.098 lb/in^3 [2,710 kg/m^3]), extrudability, and anodizing response. Although 6082 is a registered Aluminum Association alloy, it is not included in ASTM material specifications used in the U.S. because the necessary statistical data (100 tests from at least 10 different lots) required to establish U.S. minimum mechanical properties have not yet been collected and made available.

9.3.3   Aluminum-Lithium Alloys

Lithium (Li) is the lightest metallic element, and since the density of an alloy is the weighted average of the density of its constituents, lithium is attractive

TABLE 9.4   6061 and 6082 Chemical Compositions Percent by Weight

	Si	Fe	Cu	Mn	Mg	Cr	Zn	Ti
6061	0.40–0.8	0.7	0.15–0.40	0.15	0.8–1.2	0.04–0.35	0.25	0.15
6082	0.7–1.3	0.50	0.10	0.40–1.0	0.6–1.2	0.25	0.20	0.10

TABLE 9.5   6082-T6 Extrusions Versus 6061-T6 Extrusions

Alloy Temper	Thickness (in.)	Minimum Tensile Yield Strength (ksi)	Minimum Tensile Ultimate Strength (ksi)	Minimum Elongation (%)	Allowable Tensile Stress on Net Section (ksi)
6082-T6*	$t \leq 0.197$	36	42	8	21.5
	$0.197 < t \leq 0.984$	38	45	10	23.1
6061-T6	$t < 0.25$	35	38	8	19.5
	$t \geq 0.25$	35	38	10	

* 6082-T6 strengths are from EC9 (102) Table 3.2b [9].

as an alloying element if light weight is needed. But lithium has other benefits. In addition to a 3% decrease in density for every 1% of lithium added (up to the solubility limit of 4.2%), the elastic modulus increases by 5% to 6%. Aluminum-lithium alloys are also heat-treatable. These advantages are offset by the reactivity of lithium, which necessitates the use of an inert gas atmosphere when adding the liquid metal to the alloy during primary metal production, thereby driving up cost.

Al-Li alloys are often alloyed with copper, magnesium, zirconium, or other elements to improve properties. Since there is no aluminum-lithium alloy series, when lithium is the greatest alloying element, the designation number is 8xxx. When other alloying elements are in greater proportion than lithium, the designation number is based on the element in greatest proportion, such as 2195, which contains 4% copper and 1% lithium.

The Germans developed the first aluminum-lithium alloy in the 1920s, but the first Al-Li alloys to win commercial application were those developed for aircraft between the 1950s and 1970s. Alloy 2020 was used for compression wing skins of the RA5C Vigilante, but its registration was discontinued in 1974. Low ductility and fracture toughness hampered applications.

The second phase of Al-Li alloy development, which occurred in the 1980s, used relatively high levels of lithium (more than 2%) in order to maximize property improvements. Alloys 2090 and 8090, typical of this development phase, had some success but were limited by anisotropic behavior and relatively low corrosion-resistance. Finally, in the late 1980s and 1990s, work done at Martin Marietta yielded the Weldalite Al-Li alloys, which appear destined to achieve significant success in aerospace and aircraft applications. These alloys are weldable, as the name implies, and use copper as the primary alloy, with modest amounts of lithium (slightly more than 1%), and about 0.4% magnesium and 0.4% silver.

The most promising application for experimental, extremely light and strong materials is space launch vehicles, where the cost of attaining low earth orbit is about $8,000 per kg and the number of reuses is limited. The U.S. Space Shuttle external fuel tank is a good example. The first application of the Weldalite type alloys was the use of 2195 to replace 2219, a weldable aluminum-copper alloy, for the shuttle's liquid hydrogen and liquid oxygen tanks, producing a weight savings of 3,500 kg. Alloy 2197 is now being used to refurbish F-16 fighter-jet bulkheads, improving the range and performance of the aircraft. Commercial aircraft applications are anticipated next. As consumption has increased, Al-Li alloy material costs have fallen from a premium of 20 times that of common alloys to less than 4 times. It's unlikely, however, that Al-Li alloys will see any significant use outside aerospace applications.

9.3.4 The New Aluminum Automotive Alloys

The need to reduce emissions while enhancing performance and adding features has driven manufacturers to use more aluminum in automobiles and light trucks. This effort has been accompanied by the development of new aluminum alloys specifically tailored for these applications. Many of these alloys are too new to be listed in ASTM specifications or *Aluminum Standards and Data*, so detailed information is given here.

Since automobiles and light trucks undergo a paint-bake cycle at temperatures high enough to affect the temper of both heat-treatable and non-heat-treatable aluminum alloys, the automotive alloys are provided in the -T4 (solution-heat-treated) and -O (annealed) tempers, respectively. Both have the best formability in these tempers for the cold-working they undergo in the process of being formed into body panels (6). The forming operation increases strengths through cold-working. The subsequent paint bake artificially ages the heat-treatable alloys, which can additionally increase their strength, but re-anneals the non-heat-treatable alloys, erasing any strength increase due to cold work. High strength is not necessarily important in this application, however.

The automotive alloys fall in three groups:

- 2xxx series (aluminum-copper alloys), including 2008, 2010, and 2036. Alloys 2008 and 2010 were developed to provide improved formability over 2036. Alloy 2036 has more copper than 2008 and 2010, giving it about 40% higher strength but less corrosion resistance. These alloys are heat-treatable.

- 5xxx series (aluminum-magnesium alloys), including 5182 and 5754. 5182 was developed for the ends of beverage cans. It has a high magnesium content, providing high strength but also sensitivity to corrosion when exposed to temperatures above 150°F for extended periods. 5754 is a variant on 5454, with slightly more magnesium (3.1% versus 2.7%) and lower strength, but better formability.

- 6xxx series (aluminum-magnesium-silicon alloys), including 6009, 6111, and 6022. These alloys are heat-treatable and can attain fairly high strengths during the paint-bake cycle. The newest of these alloys, 6022, is used in the Plymouth Prowler body panels.

Extrusions have not seen significant automotive use, but some alloys, such as 7029, have been used in bumpers for some time (3).

9.3.5 Aluminum Metal Matrix Composites

A relatively new product, aluminum metal matrix composites (MMCs) consist of an aluminum alloy matrix with carbon, metallic, or, most commonly, ceramic reinforcement. Of all metals, aluminum is the most commonly used matrix material in MMCs. MMCs combine the low density of aluminum with the benefits of ceramics, such as strength, stiffness (by increasing the modulus of elasticity), wear resistance, and high-temperature properties. MMCs can be formed from both solid and molten states into forgings, extrusions, sheet and plate, and castings. Disadvantages include decreased ductility and higher cost; MMCs cost about three times more than conventional aluminum alloys. Yet even though they're still being developed, MMCs have been applied in such automotive parts as diesel engine pistons, cylinder liners, and drive shafts, and such brake components as rotors.

Reinforcements are characterized as continuous or discontinuous, depending on their shape and make up 10% to 70% of the composite by volume. Continuous fiber or filament reinforcements (designated f) include graphite, silicon carbide (SiC), boron, and aluminum oxide (Al_2O_3). Discontinuous reinforcements include SiC whiskers (designated w), SiC, or Al_2O_3 particles (designated p), or short or chopped (designated c) Al_2O_3 or graphite fibers. The Aluminum Association standard designation system for aluminum MMCs identifies each as:

matrix material/reinforcement material/reinforcement volume %, form

For example, 2124/SiC/25w is aluminum alloy 2124 reinforced with 25% by volume of silicon carbide whiskers; 6061/Al_2O_3/10p is aluminum alloy 6061 reinforced with 10% by volume of aluminum oxide particles.

PART IV
Design of Structural Systems

Aluminum domes. (Courtesy of Conservatek Industries, Inc.)

10 Structural Systems Built with Aluminum

In previous chapters, we discussed the design of various kinds of structural members, for example, columns, welded members, and beams, and failure modes, such as fracture and fatigue. In this chapter, we'll demonstrate some of these provisions in the design of aluminum structural systems. Section 10.1 deals with cold-formed sections; a latticed dome is presented in Section 10.2 to illustrate frame design; structural systems using aluminum with other materials (composite members) are addressed in Section 10.3; and Section 10.4 covers pressure piping.

10.1 COLD-FORMED ALUMINUM CONSTRUCTION

Aluminum *sheet* may be formed by various processes into structural members. We refer to these members as formed-sheet or *cold-formed construction,* and they have much in common with cold-formed steel. In fact, many of the issues in cold-formed metal design have been visited by steel researchers, as well as aluminum investigators. A popular kind of formed-sheet construction is building sheathing, such as corrugated siding, but sheet can also be used to fabricate structural frame members, such as channels and angles.

10.1.1 Building Sheathing

As we discussed in Section 3.1.5, many pre-formed aluminum products are useful in construction, especially building sheathing, which includes *roofing* and *siding.* Information on dimensions of profiles, load-span tables, and finishes is given in Section 3.1.5. You might have hoped that load-span tables would solve all your problems, but you knew it wasn't going to be that easy. Here are some additional considerations.

Thermal Expansion and Contraction The 5th edition of the Aluminum *Specifications* recommended (in Appendix B, Aluminum Formed-Sheet Building Sheathing Design Guide) that sheets with lengths in excess of 30 ft [10 m] should not be used unless provision was made to accommodate expansion and contraction. Changes in temperature caused the expansion and contraction of concern in this recommendation. Exposure to sunlight can produce maximum temperatures of about 140°F [60°C] for bare aluminum and 180°F [82°C] for dark painted metal. At night, metal temperatures may fall 10°F to 15°F [6°C to 8°C] below that of ambient air (21). Expansion or contraction can only be restrained by fastening with sufficient capacity and frequency, which may then result in unsightly buckling of the sheets. Allowing for movement may be more appropriate; if you go this route, movement may be calculated based on a *coefficient of thermal expansion* and contraction of approximately 13×10^{-6} /°F [23×10^{-6}/°C]. (See Section 4.9 for more.) This thermal coefficient is about twice that of steel.

Example 10.1: How much expansion allowance should be given for a 30 ft long mill-finished aluminum sheet? Assuming the design low temperature at the site is 0°F, an approximate temperature range for the material is about 150°F (minimum metal temperature of -10°F and maximum of 140°F). The amount of expansion is then 30 ft $\times$ 12 in./ft $\times$ 150°F $\times$ 13 $\times$ 10^{-6}/°F = 0.70 in., or about $\frac{3}{4}$ in.

Attaching and Sealing Building Sheathing   Aluminum *Specification* Section 5.5 requirements for spacing of through fasteners at sidelaps and endlaps for building sheathing are illustrated in Figure 10.1. Endlaps are usually located where the sheet is supported from below by a *purlin*. For more information on the fasteners used for the attachment of building sheathing to the building frame and at sheet-to-sheet laps, see Section 8.1, Mechanical Connections.

Some building-sheathing systems are not attached with through fasteners, but rather are held by clips that allow longitudinal movement of the sheathing while still providing hold-down against uplift (Figure 10.2). This prevents thermal forces from working the sheeting and elongating fastener holes, which may cause leaks. It can also permit the use of longer pieces, thereby eliminating some endlaps. These benefits must be weighed against the fact that clip-attached sheathing may not provide lateral bracing to the underlying frame members, especially under uplift.

Underwriters Laboratories (UL) publishes Standard UL 580, *Tests for Wind Uplift Resistance of Roof Assemblies*. This standard identifies three classes of uplift resistance for all types of roofing and its attachment to the building frame. In order of increasing resistance, they are: UL 30, 60, and 90. Manufacturers of building-sheathing systems may have their products rated by UL after testing in accordance with this standard. Some engineers and manufacturers have favored testing metal roofing panels with static air pressure, in a manner similar to ASTM Standard Test Method E330 (for exterior windows, curtain walls, and doors). There have, however, been conflicting views on how to apply this test to roofing panels. ASTM E1592 (78), *Standard Test Method for Structural Performance of Sheet Metal Roof and Siding Systems*

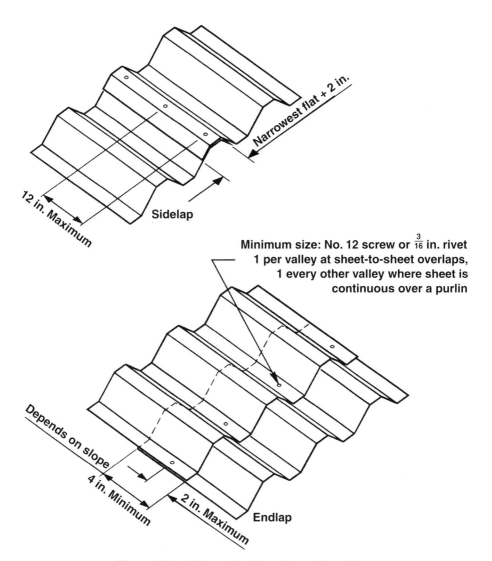

Figure 10.1 Side and end laps in metal roofing.

by Uniform Air Pressure Difference, became available in 1994 to specifically address the uplift resistance of sheet-metal roofing.

Elastomeric closure strips are available from building-sheathing suppliers to seal the profiled ends of common roofing and siding products (Figure 10.3). These closures match the profile of the particular sheeting profile used, such as V-beam or corrugated sheet. The closure strips are typically manufactured

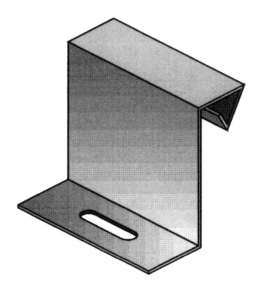

Attachment clip

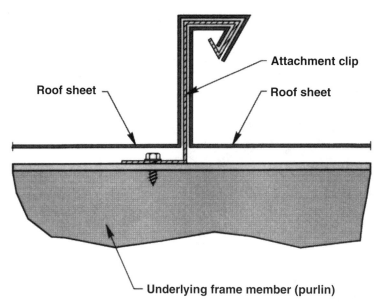

Figure 10.2   Clip-attached metal roofing panels.

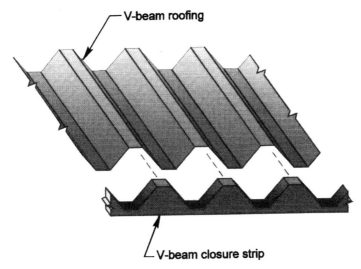

V-beam roofing

V-beam closure strip

Figure 10.3   Preformed elastomeric closure strip.

as a sponge gasket, and are made of EPDM (ethylene-propylene) or other rubber-like material.

Formed-Sheet Roof Slopes   The slope of a roof affects many aspects of roof design, including snow loads, weather-tightness, and endlap requirements. Minimizing the roof slope may offer economies by reducing building volume and height, or it may simply satisfy someone's sense of aesthetics. Reducing the roof slope can also increase design snow loads and make it more difficult to prevent wind-driven rain from penetrating the endlaps. The question is: how low can you go?

The Aluminum *Specification* does not dictate a minimum roof slope. *Specification* Section 5.5.1 offers endlap requirements based on roof slope; longer endlaps are required for lower slopes. For roof slopes less than 2 on 12 (9.5°), you're off the chart. For a slope of 2 on 12, the required endlap is 9 in. [230 mm], for lower slopes endlap requirements would presumably exceed 9 in., and caulk may be required to seal the laps.

Another issue for low-slope roofs is *ponding*. It is important to ensure positive slope at every point on the roof under design loads so that water will not pond, which can cause leakage or progressive structural failure. Since aluminum roofing is not typically used to support brittle materials, such as ceiling plaster, Aluminum *Specification* Section 8.4.4 and some building codes (Uniform Building Code [115] Section 2803 (c)) allow deflections of aluminum members to be as large as $L/60$, where L is the span. Within this requirement, a uniformly loaded simple span's deflected slope can be as large as about $\frac{3}{4}$ on 12 (3.6°)(Figure 10.4), suggesting this as the minimum slope for this case.

Shear Diaphragms   Two general classes of loads acting on buildings are:

- gravity loads, such as live load, dead load, or snow load, all of which act vertically
- lateral loads, such as wind and seismic loads, which have horizontal components.

The building frame may be designed to resist both types of loads, with lateral loads resisted by a rigid frame or with x-bracing (Figure 10.5a). However, the building cladding also tends to stiffen the structure against lateral deflection. The cladding acts like a *shear diaphragm,* a flat element that resists shear loads acting in its own plane (Figure 10.5b). How much shear corrugated panels can resist is a function of the sheet profile, thickness, span between purlins, end connections, and sidelap seam connections. Consequently, research has been directed toward evaluating the ability of roofing and siding to resist shear loads on metal buildings.

While more research has been done on steel diaphragms than aluminum, Sharp (133) presented shear test results for corrugated and V-beam aluminum sheeting. The test set-up is shown in Figure 10.6. The panels tested had sidelap fasteners in accordance with the Aluminum *Specification* Section 5.5.3 (No. 12 screws on 12 in. [300 mm] spacing). The corrugated sheet had No. 14 endlap fasteners in every fourth valley, and the V-beam sheet had No. 14 endlap fasteners at every valley. The results are summarized in Table 10.1.

Since some of these panels failed at the fasteners, a case could be made for a safety factor of 2.34 or more on the results. Also, only a few tests were run, so additional testing might be needed for design purposes. Special care must also be taken to ensure that the collection of parts (individual corrugated sheets and purlins) that make up the shear diaphragm are assembled so as to act in a manner equal to that of the tested assembly.

Another consideration in the design of shear diaphragms is that deflection must not exceed what the building can withstand. For example, when masonry walls are connected to the structure that resists lateral loads with a diaphragm,

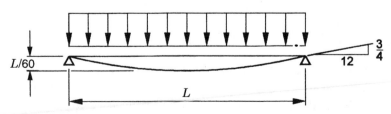

Figure 10.4   Slope of a simple span beam with a maximum deflection of $L/60$.

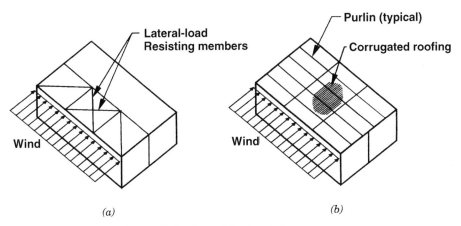

Figure 10.5   Lateral-load-resisting systems.

some building codes provide limits on deflection to prevent damage to the walls. The stiffness of a diaphragm is a function of its depth in the direction of the shear load, and other parameters. Sharp (page 173, (133)) provides a formula for calculating diaphragm deflection, as well as test results for deflection of the panels in Table 10.1 above.

10.1.2   Cold-Formed Aluminum Design

You can devise your own formed-sheet members rather than limiting yourself to the ones in the *Aluminum Design Manual*. Consider the example shown in

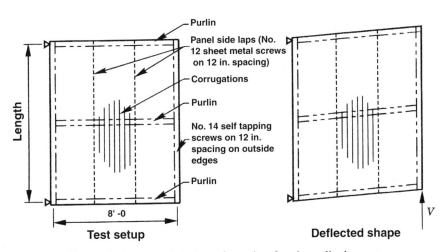

Figure 10.6   Sharp's test configuration for shear diaphragms.

TABLE 10.1   Shear Diaphragm Test Results

Product	Thickness (in.)	Average Maximum Shear (lb/ft)
Corrugated	0.024	254
Corrugated	0.032	377
V-Beam	0.040	570

Figure 10.8. In section 3.1.2, we determined the dimensions of this formed-sheet member. Now let's check its performance over a simple span of 20 ft 8 in. with a 0.00398 k/in. uniform live load, as shown in Figure 10.9. This produces a bending moment of:

$$M = wl^2/8 = (0.00398 \text{ k/in.})(248 \text{ in.})^2/8 = 30.6 \text{ in.-k.}$$

We'll show how to check this beam by first comparing stresses to allowable stresses using the methods discussed in Chapter 5, and then using the *effective width method* mentioned in Section 5.2.2 to calculate deflection.

First, calculate the section moduli to the top and bottom extreme fibers of the cross section (Figure 10.8.) Cold-formed-sheet designers often leave out the thickness in the area entries in Table 10.2 because all the elements of the section have the same thickness, but we'll include them here to avoid con-

Figure 10.7   Cold-formed sheet beams. (Courtesy of Temcor)

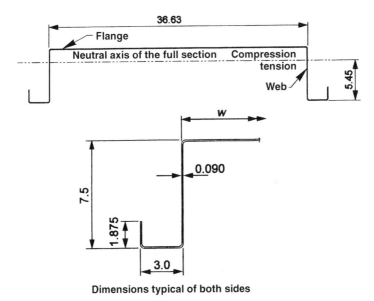

Figure 10.8   Section of a cold-formed sheet beam.

fusion. (This approach is called the *linear method* since it treats each element as a line.) Also, we'll treat the radii at the bends as square corners for simplicity.

From Table 10.2, the neutral axis is located

$$(29.86 \text{ in.}^3)/(5.476 \text{ in.}^2) = 5.45 \text{ in.}$$

from the bottom of the section.

The section modulus to the compression (top) side $= S_c$:

$$S_c = (45.22 \text{ in.}^4)/(7.5 - 5.45) \text{ in.} = 22.1 \text{ in.}^3$$

The section modulus to the tension (bottom) side $= S_t$:

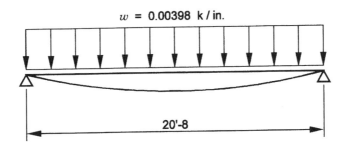

Figure 10.9   Load diagram for the cold-formed sheet beam example.

TABLE 10.2 Determining the Neutral Axis of a Formed-Sheet Member

Element	Area A (in.^{2})	Distance to Centroid y (in.)	Ay (in.^{3})
Bottom flange stiffeners	0.3213	0.9825	0.3157
Bottom flanges	0.54	0.045	0.0243
Webs	1.3176	3.75	4.941
Top flange	3.2967	7.455	24.577
	5.476 in.^{2}		29.86 in.^{3}

$$S_t = (45.22 \text{ in.}^4)/(5.45 \text{ in.}) = 8.29 \text{ in.}^3$$

The maximum compressive stress is

$$f_a = (30.6 \text{ in.-k})/(22.1 \text{ in.}^3) = 1.38 \text{ ksi.}$$

The maximum tensile stress is

$$f_t = (30.6 \text{ in.-k})/(8.29 \text{ in.}^3) = 3.69 \text{ ksi.}$$

Next, let's calculate the allowable stress in each of the *elements* of the beam. (Because the member is wider than it is tall, we know that it won't undergo lateral buckling, so only local buckling needs to be checked.)

Since there is no table of allowable stresses in *Aluminum Design Manual*, Part VII, for this alloy (5052-H36), we'll calculate them here.

The allowable bending compression stress in the top flange is determined in accordance with Aluminum *Specification* Section 3.4.16 since it is a flat plate supported along both edges and under uniform compression as a component of a beam. For this element,

$$b/t = (36.63 - 2[0.09])/0.09 = 405.$$

TABLE 10.3 Determining the Section Modulus of a Formed-Sheet Member

Element	A (in.^{2})	d (in.)	Ad^2 (in.^{4})	I (in.^{4})
Bottom flange stiffeners	0.3213	4.4704	6.421	0.0853
Bottom flanges	0.54	5.4079	15.793	0.0004
Webs	1.3176	1.7029	3.820	5.8833
Top flange	3.2967	2.0021	13.215	0.0022
		Moment of inertia =	39.25 +	5.97 = 45.22 in.^{4}

The

slenderness limit S_2 is $(k_1 B_p)/(1.6D_p) = [0.50(36.1)]/[1.6(0.263)] = 43$,

so the flange is well over the second slenderness limit S_2. The allowable compressive bending stress in the flange (F_{bf}) is therefore:

$$F_{bf} = \frac{k_2 \sqrt{B_p E}}{n_y(1.6b/t)} = \frac{2.04\sqrt{(36.1)(10,200)}}{1.65(1.6)(405)} = 1.16 \text{ ksi}$$

This is slightly less than the compressive stress of 1.38 ksi determined above. However, we can use the weighted-average allowable stress method (Aluminum *Specification* Section 4.7.2) to justify a higher allowable stress since the webs have a higher allowable stress than the flange.

The allowable stress for the webs is determined according to Aluminum *Specification* Section 3.4.18 since each web is a flat plate supported along both edges and under bending in its own plane as a component of a beam. The slenderness ratio is:

$$(7.5 - 2[0.09])/0.09 = 81.3,$$

which is less than S_2 since:

$$S_2 = \frac{k_1 B_{br}}{0.67 D_{br}} = \frac{0.50(48.1)}{(0.67(0.405)} = 88.6$$

So, the allowable compressive bending stress in the web (F_{bw}) is:

$$F_{bw} = \frac{B_{br} - 0.67 D_{br}h/t}{n_y} = \frac{48.1 - 0.67(0.405)(81.3)}{1.65} = 15.8 \text{ ksi}$$

For the weighted-average allowable stress, the flange is considered to consist of the 36.63 in. wide top flange and one-third of the distance from the extreme fiber to the neutral axis of the webs. The weighted-average allowable bending stress is, therefore:

$$F_b = \frac{(1.16)(36.63) + (15.8)(1/3)(7.5 - 5.45)(2)}{36.63 + (1/3)(7.5 - 5.45)(2)} = 1.69 \text{ ksi} > 1.38 \text{ ksi}$$

So, by using the weighted-average allowable stress method, we are able to justify the adequacy of the beam, whereas if we conservatively used the lower of the web and flange allowable stresses, we could not.

Effective Width Method for Calculation of Deflection   Now let's calculate deflection at the center of the beam using the effective width method, according to Aluminum *Specification* Section 4.7.6. To refresh your memory about our discussion in Section 5.2.2: elements in compression, such as the top flange of the beam we're considering here, may *buckle elastically* within allowable stresses if postbuckling strength is recognized. One way to treat this is to consider part of the flange to be buckled and, thus, ineffective in resisting load (see Figure 10.10). Only the parts of the top flange near the webs will remain straight and resist load.

An element in compression is fully effective if its slenderness ratio is less than the slenderness limit S_2 because the allowable stress for such slenderness ratios is less than the buckling stress. We just calculated the slenderness ratio of the web to be 81.3 and S_2 to be 88.6. Therefore, the webs are fully effective.

The top flange is another matter. Its slenderness ratio is so high (405 versus an S_2 of 43) that we expect that it will not be fully effective, even at relatively low stress. The criterion is whether its compressive stress exceeds the *local buckling stress* (F_{cr}) from Aluminum *Specification* Section 4.7.1 for an element supported along both edges (*Specification* Section 3.4.16):

$$F_{cr} = \frac{\pi^2 E}{(1.6b/t)^2} = \frac{\pi^2 (10,200)}{(1.6(405))^2} = 0.24 \text{ ksi}$$

Since

$$F_{cr} = 0.24 \text{ ksi} < 1.38 \text{ ksi} = f_a,$$

the flange, as suspected, is not fully effective. The portion of the flange that is effective is a function of the stress in the flange; the higher the stress, the smaller the effective width. We can calculate the effective width by using Aluminum *Specification* Equation 4.7.6-2:

$$b_e = b\sqrt{F_{cr}/f_a} = (36.63 - 2[0.09])\sqrt{0.24/1.38} = 15.2 \text{ in.}$$

Now that we know how much of the flange is effective, we can calculate the moment of inertia of the section, disregarding the portion of the top flange that is not effective, and then use this to calculate the deflection.

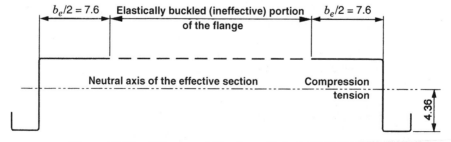

Figure 10.10   Effective width of a cold-formed sheet beam.

TABLE 10.4 Determining the Neutral Axis of a Formed-Sheet Member (Effective Width)

Element	Area A (in.^{2})	Distance to Centroid y (in.)	Ay (in.^{3})
Bottom flange stiffeners	0.3213	0.9825	0.3157
Bottom flanges	0.54	0.045	0.0243
Webs	1.3176	3.75	4.941
Top flange	1.368	7.455	10.198
	3.547 in.^{2}		15.48 in.^{3}

So, the neutral axis of the effective section is located

$$(15.48 \text{ in.}^3)/(3.547 \text{ in.}^2) = 4.36 \text{ in.}$$

from the bottom of the section.

Aluminum Design Manual, Part VII, Beam Formulas, Case 6 provides the equation for the deflection of a simply supported beam with a uniform load:

$$\delta = \frac{5wl^4}{384EI} = \frac{5(0.00398)(248)^4}{384(10,100)(33.28)} = 0.583 \text{ in.}$$

Notice that we would have underestimated the deflection by 36% had we used the moment of inertia we first calculated (45.22 in^4 in Table 10.3) instead of the moment of inertia of the effective section (33.28 in^4 in Table 10.5). Notice also that the modulus of elasticity used for deflection calculations is the average of the tensile and compressive moduli, or 100 ksi [700 MPa] lower than the compressive modulus given in Aluminum *Specification* Table 3.3-1.

It's important to keep in mind that the effective width we just calculated is valid only at the design load and, therefore, can't be used to determine the strength of the section. At failure, the stress in the top flange will be higher

TABLE 10.5 Determining the Section Modulus of a Formed-Sheet Member (Effective Width)

Element	A (in.^{2})	d (in.)	Ad^2 (in.^{4})	I (in.^{4})
Bottom flange stiffeners	0.3213	3.1086	3.1048	0.0853
Bottom flanges	0.54	4.0461	8.8402	0.0004
Webs	1.3176	0.3411	0.1533	5.8833
Top flange	1.368	3.0908	13.068	0.0009
		Moment of inertia = 27.31	+	5.97 = 33.28 in.^{4}

than the stress we used to calculate the effective width, so the effective width will be smaller than the 15.2 in. we calculated above. If you're familiar with the cold-formed steel *Specification* (40), you may realize that this is why the procedure used there for calculating load capacity differs from that used for determining deflection, even though the cold-formed steel *Specification* uses the effective width method for both.

10.1.3 Elastically Supported Flanges

Aluminum *Specification* Section 4.10 addresses *elastically supported flanges* of beams. If the heading of this section doesn't immediately tell you what this is about, you're probably not alone. The application of this provision may be clearer if you imagine the cross section of Figure 10.8 turned upside down, as shown in Figure 10.11. The two flanges of the beam are said to be elastically supported because in compression they would buckle laterally, but they are restrained from doing so by virtue of the elastic support that the tension flange and webs provide (see the dashed lines in Figure 10.12). Cross sections where this situation arises are more likely to be cold-formed than rolled or extruded.

This beam's buckling performance is somewhat like the single web beams treated under Aluminum *Specification* Section 3.4.11 since the compression flange in both cases buckles laterally and twists as the tension flange resists lateral movement. Consequently, the *Specification* allows the use of Section 3.4.11 provisions for the section shown in Figure 10.11 when you substitute an effective slenderness ratio for L_b/r_y. The only tricky step is the determination of β, a spring constant that measures the transverse force required to cause a unit lateral deflection of a unit length of the flange (Figure 10.12). *Aluminum Design Manual*, Part VIII, Example 23 illustrates such a calculation. Yu (141, Article 4.2.3.5) and Sharp (133, Section 8.3) also discuss this case in greater detail.

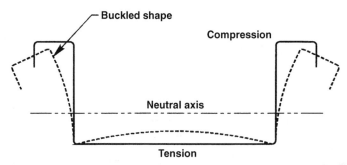

Figure 10.11 Buckling of an elastically supported flange.

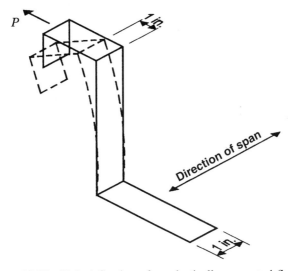

Figure 10.12   Unit deflection of an elastically supported flange.

10.2 ALUMINUM FRAMES

Undoubtedly, the most common form of clear-span aluminum structure is the triangulated dome, in variations on the geodesic framing principles that Buckminster Fuller popularized. Thousands of aluminum domes have been built in this country, ranging in function from highly visible aesthetic applications at Walt Disney World's EPCOT Center, to such utilitarian uses as covering tanks at sewage treatment plants. Let's walk through the design of an aluminum dome to illustrate the use of the Aluminum *Specification*.

10.2.1 System Description

We'll begin by describing the type of structure that we have in mind. You may want to refer to Figure 10.13, which shows a 200 ft. diameter dome under construction. The portion of the dome that appears relatively light in the picture has aluminum sheeting covering the frame. Where the triangles appear black, however, the sheeting has not yet been installed and the framing members are visible.

The structural frame of a typical aluminum dome is a single-layer triangulated lattice. The individual members are straight, but the lattice approximates a sphere by virtue of the connection nodes being located on a spherical surface. The resulting spaces are closed with light-gauge triangular aluminum sheets, which are called *panels*. These panels are held in place by batten bars

Figure 10.13 An aluminum dome under construction, partially sheeted. (Courtesy of Conservatek Industries, Inc.)

that clamp the panel edges to the top flange of the I-beam frame members. This clamping arrangement illustrates the flexibility in design that aluminum extrusions offer.

The I-beam and the batten bar are both custom extrusions, as we illustrated in Figure 3.8. While the details vary by manufacturer, they generally accommodate a preformed, shop-installed gasket to seal the panel edge. They also employ a screw chase in the top flange of the I-beam to receive the screws that secure the batten bar, thereby clamping the panel edge in place without piercing the panel material. The result is a connection that braces the flange of the I-beam yet eliminates the use of through fasteners. Through fasteners are avoided along the panel edges for the same reasons that they have fallen into disfavor in conventional metal roofing: thermal variations and wind fluctuations work the holes in the sheet metal, leading to leakage and the potential for panel failure.

In addition to I-beam members, batten bars, and panels, the dome requires some type of connector at the nodes. This is generally accomplished with dished gusset plates that sandwich the joint (Figure 8.1). This detail is preferred not only for simplicity, but also for its ability to achieve joint rigidity. As we will see when we work through the design, joint rigidity is critical to the stability of single-layer lattice frames. The gusset plate thickness and bolting pattern are designed to transmit moments through the joint, which generally implies that a minimum of two rows of fasteners is used in each I-beam flange, and that the ends of the I-beams are brought as close to the center of the joint as possible. The gusset plates should be designed for the actual stresses imposed and should be sufficiently stiff to develop the full moment capacity of the I-beams. A good rule of thumb is that the gusset plates be at least as thick as the I-beam flanges, particularly on larger domes.

The final component of the structural system is the support detail, which may employ a moment-relieving pin, as illustrated in Figure 8.6. Aluminum domes generally use a short outrigger to make the transition from the spherical surface to the horizontal plane of support. Since the reactions act at the end of this outrigger (Figure 10.14), this detail has a significant impact on the structural behavior.

10.2.2   Model for Analysis

We will model a 200 ft diameter dome, similar to the one pictured in Figure 10.13, with a slope at the perimeter of 33.72° from horizontal. Such a structure has nearly 1,000 members in the frame that, given an assumption of rigid joints, is highly indeterminate. Fortunately, however, you have more computing power on your desktop today than the entire U.S. atomic bomb program had during the Manhattan Project. The frame can be analyzed with any commercial stiffness analysis program, but we must be on our guard that we keep the model in touch with reality. As with almost any computer model, as-

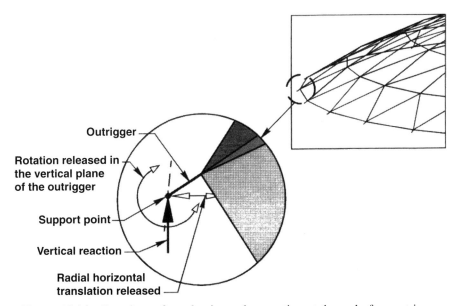

Figure 10.14   Reactions of an aluminum dome acting at the end of an outrigger.

sumptions and approximations are sure to be made in the creation of the model. While our emphasis is on the design process, let's review some of the analysis considerations that we must not overlook.

Load applied to the structure acts as pressure on the panels. The panels behave like membranes in tension to transmit the forces to the I-beam flanges. The I-beams, then, are subject to a continuous line load, as well as to the forces and moments resulting from the frame action. While there is a temptation to simplify the analysis model by applying the load only at the nodes, this would unconservatively ignore the bending induced by the panel attachment to the I-beam flange. The manner of load transmission to the frame must be accounted for in the analysis. You can do this either by appropriately applying the load continuously along the members in the computer model, or by adding the bending induced by the panel attachment to the results of a nodal analysis.

A simplification of the model that is permissible is to leave out the panels and batten bars. While the method of load transmission must be accounted for, the panel itself, generally, is not considered to be part of the primary frame. The contribution of the batten bar to the strength of the frame may also be neglected because the ability to transmit shear through the screw chase by which the battens are connected is not quantified.

We have, thus, included only beam elements as the members in our model, and we have been sure to include the outriggers at the supports. We'll release the supports horizontally in the radial direction to minimize the thrust that the dome imparts to the supporting structure (refer again to Figure 10.14). Let's assume that we have already developed the geometric arrangement of the framing, considering any fabrication and erection limitations. One such limitation is the width of coil material that is available for fabricating the panels. Each triangular space must have at least one altitude that is no greater than the width of the available material. Many such ancillary considerations exist in the production of a dome, but we will assume that these have been resolved.

Next we will input the geometry of the dome to the analysis program and begin to enter the material properties. We'll assume that we are using the old standby alloy, 6061-T6, and are now ready to enter the section properties of the members. How large should we guess that they need to be?

10.2.3   Getting Started

Aluminum's strength-to-weight ratio often allows domes spanning 200 ft [60 m] to be built with 6 in. to 8 in. [150 to 200 mm] deep I-beams. In order to pick member sizes for our first analysis run, however, it would be helpful to have some methods of approximation.

Let's take a moment to consider the overall behavior of the dome under load. A symmetric downward load, such as a uniform snow load, tends to flatten the dome. The dome is scrunched together as it is squeezed downward, and the perimeter starts to spread out (Figure 10.15). The base ring at the

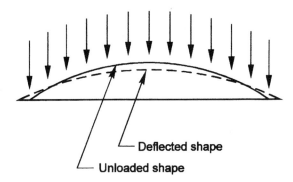

Figure 10.15 Behavior of a clear-span dome under uniform downward load.

perimeter of the dome acts as a big hoop in tension to keep the dome from flattening out. If we momentarily consider the joints to be pinned, we can derive an expression for the force in the base ring.

For a symmetrical load on a dome with evenly spaced supports, the vertical reaction (V) at each support is approximately:

$$V = P/n_s \tag{10.1}$$

where:

V = vertical reaction at a support
P = total vertical load applied to the dome
n_s = number of supports.

If we apply this vertical force to the nodes of the base ring, it must be carried by those members that are not in a horizontal plane (see Figure 10.16). Since these members lie in the slope of the dome at its perimeter, which we set at 33.72°, equilibrium requires that they also have a horizontal component of force. This horizontal component R is equal to:

$$R = V/\tan \beta \tag{10.2}$$

where:

R = radial horizontal force at a base-ring node
β = angle to horizontal of the dome at its base-ring node

Combining these two equations we have:

$$R = P/(n_s \tan \beta) \tag{10.3}$$

Now let's resolve these forces in the horizontal plane (Figure 10.17). The

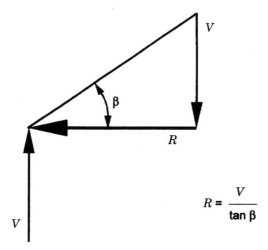

$$R = \frac{V}{\tan \beta}$$

Figure 10.16   Free-body diagram of forces acting in the vertical plane at a dome support.

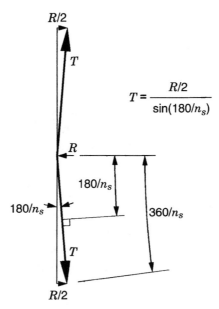

$$T = \frac{R/2}{\sin(180/n_s)}$$

Figure 10.17   Free-body diagram of forces acting in the horizontal plane at a dome support.

horizontal force component (R) from the nonhorizontal (inclined) members must be sustained by the two base-ring members framing into the joint. Each base-ring member, then, must have a force component in the radial direction of $R/2$. This force component is shown by geometry in Figure 10.17 to describe a force diagram having one angle equal to $180/n_s$. The resultant axial tension force (T) in the base ring, which is also called the *tension ring*, can then be expressed as:

$$T = R/(2 \sin [180/n_s])$$ (10.4)

Combining Equations 10.3 and 10.4 yields:

$$T = \frac{P}{2n_s \tan \beta \sin (180/n_s)}$$ (10.5)

For values of n_s greater than 6, $n_s \sin(180/n_s)$ is within 5% of π, so this expression can be simplified as:

$$T = \frac{P}{2\pi \tan \beta}$$ (10.6)

We can compare T to the allowable tensile stress for 6061-T6 aluminum to obtain a trial cross-sectional area.

Next, we will need a method of approximating the size of member required in the rest of the dome. We can use a stability check appropriate to this type of structure for this purpose. Numerous researchers have demonstrated that single-layer lattice domes are potentially subject to geometric nonlinearity (107). This is not the nonlinearity that results from material being stressed beyond yield, but rather is a function of the geometry of the structure. Linear behavior assumes that deflections of the frame are so small that any changes to the stiffness of the frame are negligible. When deflections are sufficiently large, however, the structure will behave differently in its deflected shape than in its original shape. If the stiffness of the structure changes appreciably under load, then the load-deflection relationship is no longer linear. Since aluminum is subject to larger deflections than steel by virtue of its lower modulus of elasticity, this concern is amplified in aluminum domes. When the onset of instability (buckling) occurs while the stress in the members is still in the elastic range, it is termed *elastic instability*.

This geometric nonlinearity introduces two modes of buckling behavior in addition to the buckling of individual members that we routinely address. These additional buckling modes are called *general buckling,* which is said to occur when a large area of the dome caves in, and *snap-through buckling,* which refers to the inverting of a single node. These instabilities are not predicted by linear, elastic finite element models. Several nonlinear finite element formulations are available, but when applied to this problem they have been shown to have a great deal of variability in their results (132). If the computer can't solve it, are we stumped?

Once upon a time, when engineers used slide rules and dinosaurs roamed the earth, the behavior of single-layer lattice domes was known to be similar to that of thin, solid shells. Interestingly, it's the validity of this very relationship that leads us to now model solid shells as a lattice of finite elements. In those days, however, the problem was turned around. Shell theory was well established, but computer-aided finite element methods of frame analysis were not available. By applying classical shell-buckling theory to spherical lattice frames, in 1965 Douglas Wright (139) resolved the general-buckling equation for the specific case of a single-layer triangulated lattice, resulting in the following:

$$p_{cr} = C(3\sqrt{2})\,\frac{AEr}{LR^2} \qquad (10.7)$$

where:

p_{cr} = critical pressure applied uniformly and normal to the shell
C = coefficient
A = cross-sectional area of the individual members in the lattice
E = modulus of elasticity
r = radius of gyration
L = typical length of the members
R = spherical radius of the dome

Kenneth Buchert conducted noteworthy investigation of the buckling of shell-like frames; he proposed a value of C equal to 0.365. This compares well with the value of 0.38 that Wright proposed (107). Multiplying these by the $(3\sqrt{2})$ factor yields coefficients of 1.55 and 1.6, respectively. We will refer to this as the "Wright-Buchert Equation." We can express the equation in terms of the member properties A and r and apply a suitable safety factor, in order to pick a minimum size of I-beam to use as a trial member in the analysis model.

Significant caution is warranted in the use of this equation. The premise that the lattice frame behaves in a manner similar to a solid shell relies on the assumption that the joints in the lattice frame are sufficiently rigid to enable the frame to act as a continuum. This was the basis for our earlier comment on the necessity of the joint being designed to develop the full-moment capacity of the I-beams. The equation is also sensitive to initial imperfections, including those due to fabrication and erection tolerances.

10.2.4   Analyzing the Dome

Let's use these methods to approximate the member sizes. The dome has the following dimensions (Figure 10.18):

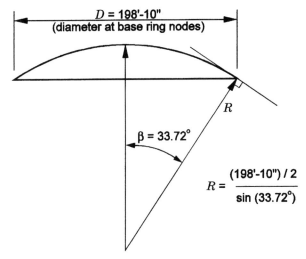

Figure 10.18 Geometric parameters defining the profile of a dome.

Dome diameter at the anchor bolts = 200 ft (2,400 in.)

Dome diameter at the base ring (D) = 198 ft 10 in. (2,386 in.)

Dome slope at the base (β) = 33.72 degrees

Dome spherical radius (R) = 179 ft 1 in. (2,149 in.).

The difference between the diameter to the anchor bolts and the diameter of the base ring is accounted for by the outrigger detail at the supports, as shown in Figure 10.19.

Let's apply a uniform downward load of 30 psf to the dome shown in Figure 10.20, and estimate a 3 psf dead load. In an actual design, all specified load conditions would have to be considered. The load combination, then, is:

Snow load = 30 psf

Dead load = 3 psf
_____

Total applied load = 33 psf.

The final parameter that we need is the average member length L. Remember that the limit on panel size is the width of the available coil material. We will assume that the panels are to be 0.050 in. thick 3003-H16 alloy, for which the greatest available width is 108 in. If 108 in. were the altitude of an equilateral triangle, the side lengths would be about 125 in. While some of the members will be longer and others shorter, we will use an approximate value of 130 in. for the average member length L.

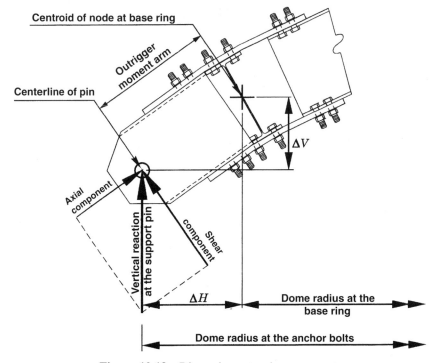

Figure 10.19   Dimensions at a dome support.

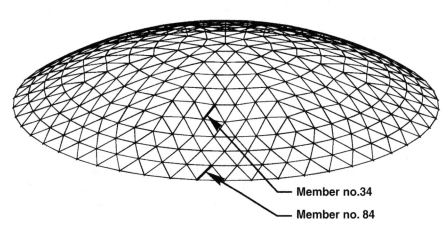

Figure 10.20   Geometry of the dome frame in the example.

We now have the information that we need to plug into Equations 10.6 and 10.7 in order to generate approximate member sizes. Let's start with the stability check, writing Equation 10.7 in terms of the product of the member properties A and r:

$$p_{cr} \frac{LR^2}{1.6E} (SF) = Ar \qquad (10.8)$$

where SF is the safety factor against buckling. Although our load is applied vertically, and p_{cr} is defined as a pressure normal to the surface, the difference is very small for a low profile dome. Substituting for the variables gives:

$$(33 \text{ psf}) \left(\frac{1 \text{ ft}^2}{144 \text{ in.}^2} \right) \frac{(130 \text{ in.})(2149 \text{ in.})^2}{1.6(10,100,000 \text{ psi})} (1.65) = 14 \text{ in.}^3$$

Thus, the trial member should have an area (A) and major axis radius of gyration (r) whose product is at least 14 in.^{3}. Let's look for an appropriate section among the Aluminum Association standard I-beams (refer to Appendix B). We see that the heaviest 6 in. deep I-beam has an area of 3.99 in.^{2} and an r of 2.53 in. for a product of 10.1 in.^{3}. This is less than the required 14 in.^{3}, so let's try the lightest 8 in. I-beam. It has an area of 5.26 in.^{2} and a radius of gyration equal to 3.37 in., resulting in a product of 17.7 in.^{3}. This is larger than 14 in.^{3}, so we'll select it.

Next, we need to approximate the cross-sectional area of the tension ring. We can readily compute the total vertical load (P) by multiplying the applied load of 33 psf by the horizontal projected area of the dome. The horizontal area of the 200-ft diameter dome is:

$$\pi d^2/4 = \pi 200^2/4 = 31,400 \text{ ft}^2$$

where d = diameter at the anchor bolts of the dome.

The total vertical load, then, equals:

$$P = 33 \text{ psf } (31,400 \text{ ft}^2) = 1,040,000 \text{ lb, or } 1,040 \text{ k.}$$

From Equation 10.6, we obtain the following:

$$T = \frac{P}{2\pi \tan \beta} = \frac{1,040 \text{ k}}{2\pi \tan 33.72} = 248 \text{ k}$$

Next, divide this force by the allowable tensile stress for 6061-T6 of 19 ksi (from Section 7.1.2) to determine an approximate tension ring cross-sectional area:

Required area (A) = 248 k/19 ksi = 13 in.^{2}

We won't select an I-beam this large; instead, we'll use a smaller I-beam and add reinforcement. To simplify the model, we can use the same size I-beam as that selected for the other members, but increase the area from 5.26 in.^{2} to 13 in.^{2}. We'll add the reinforcement symmetrically about the centroid of the tension ring, so that we don't have to model in any eccentricity.

10.2.5 Design Checks

Although the frame has 980 members, symmetry allows us to check fewer than 100 of them for this uniform vertical load case. That is still far more member checks than we will grind through in this example. We will select two members to illustrate the design steps. As shown in Figure 10.20, these members have been identified as no. 34 and no. 84. No. 34 is one of the more heavily loaded members other than those in the bottom bay or the tension ring. We'll then demonstrate with member no. 84 that the bottom bay of diagonal members is subject to moments that differ significantly from those in members farther from the supports.

Check Member No. 34   The forces and moments for member no. 34 are summarized in Table 10.6 below.

Member No. 34—Axial Compression   Referring to Section 7.2, we need to check the member as a column for both overall and local buckling. The first step is to determine the overall column slenderness ratio. The panels provide continuous bracing against minor axis buckling, but the full length of the member is unbraced for major axis buckling. Recalling Table 7.1 (Section 7.2.1) and obtaining r from the summary of section properties in Appendix B, we can summarize the flexural buckling slenderness ratios as follows and shown in Table 10.7.

Torsional buckling is not a concern because the continuous panel attachments brace the members against twisting. The overall slenderness ratio, then, is 40.1. The next step is to calculate the slenderness ratios of the cross-sectional elements, which we can skip because we already have them summarized in Appendix D for common shapes.

TABLE 10.6   Forces and Moments in Member No. 34

Member No. 34 L = 135 in.	Compressive Axial Force (k)	Major Axis Moment (in.-k)	Minor Axis Moment (in.-k)	Shear (k)
Top end	21.4	13.6	−0.19	0.55
Mid span	21.6	−5.49	0.09	0.01
Bottom end	21.8	11.7	0.37	−0.52

TABLE 10.7 Overall Column Slenderness Ratio Comparison for Member No. 34—Flexural Buckling

Axis	k	L	r	kL/r
Major axis (x)	1.0	135 in.	3.37 in.	40.1
Minor axis (y)	1.0	0 in.	1.18 in.	0

The overall slenderness ratio of 40.1 is between S_1 and S_2, so from Table 7.4 (Section 7.2.3) we determine the overall column allowable stress:

$$20.2 - 0.126 \, kL/r = 20.2 - 0.126 \, (40.1) = 15.1 \text{ ksi}$$

Comparing this value to the allowable stress for the elements from Appendix D:

Overall column allowable stress =	15.1 ksi
Element allowable stress =	17.6 ksi
Column allowable axial stress (F_a) =	15.1 ksi

Appendix D also indicates that none of the element slenderness ratios are greater than S_2, so we do not have to reduce the overall column allowable stress for the effect of local buckling.

We now compute the axial compressive stress (f_a). The maximum axial force is 21.8 k (Table 10.6), and the cross-sectional area is 5.26 in. (Appendix B). The axial compressive stress is:

$$f_a = 21.8 \text{ k}/5.26 \text{ in.} = 4.1 \text{ ksi}$$

Member No. 34—Major Axis Bending Having taken care of the axial load, we can check bending. Referring to Section 7.3, we'll begin by checking tension due to bending. For an I-beam bent about its strong axis, the allowable tensile bending stress (from Section 7.3.1) is 19 ksi.

As with all tension considerations, the calculated stresses are based on the net area (refer to Section 5.1 for calculating the net area). When tension is caused by bending, it occurs in only one flange while compression is induced in the other. The net section for tension bending, then, needs only to have the bolt hole deductions applied to the tension flange. To simplify the calculations, however, we will deduct holes from both flanges to compute a net section. Assuming two $\frac{1}{2}$ in. diameter bolts in each row of bolts, the I 8 × 6.18 trial member has a net section modulus of 11.8 in.³. The tensile bending stress (f_{btx}) is then:

$$f_{btx} = 13.6 \text{ in.-k}/11.8 \text{ in.}^3 = 1.15 \text{ ksi}$$

Note that this is less than the axial compressive stress that we calculated. The tensile bending stress will, thus, reduce the compressive stress in the tension flange, but it will not actually result in a state of tension.

Now let's switch to the other flange and consider compression due to bending. We need to check for both overall (lateral) buckling and local (element) buckling. The lateral buckling allowable stress is dependent on the unbraced length of the compression flange (L_b). The I-beams in the dome have their top flanges braced by the panels, so the compression flange is unbraced only when the bottom flange is in compression. To determine what length, if any, of the bottom flange is in compression, we need to consider the moment diagram for the member (Figure 10.21). (Our convention is that a positive moment results in compression in the bottom flange.) The bottom flange is in compression at the ends of the member, but the top flange is in compression in the middle. A conservative and simple method for determining the maximum length of unbraced compression flange for a moment diagram of this shape is to approximate the curves with straight lines. We can, then, solve by similar triangles for L_b as follows:

$$L_b = \frac{M'L}{2(M' + |M'|)}$$

where:

L_b = length of the member between points of lateral support of the compression flange (unbraced length)
M' = greater of the end moments
L = overall length of the member

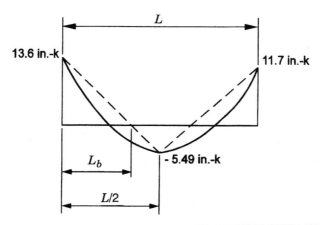

Figure 10.21   Moment diagram for member no. 34.

$|M''|$ = absolute value of the moment at the midspan of the member.

For member no. 34:

$$L_b = \frac{13.6 \text{ in.-k}(135 \text{ in.})}{2(13.6 \text{ in.-k} + 5.49 \text{ in.-k})} = 48.1 \text{ in.}$$

A cautionary note is in order here. In this example, the inflection point determines the unbraced length because one of the flanges is continuously braced and the other is not. If neither flange were braced, then the inflection point would not be deemed a brace point. It would indicate a change in which flange is subject to compression, but the member would still have an unbraced compression flange on either side of the inflection point.

The value of r_y for an I 8 × 6.18 I-beam is given in Appendix B as 1.18 in. The ratio L_b/r_y is, then:

$$L_b/r_y = 48.1 \text{ in.}/1.18 \text{ in.} - 40.8$$

From Table 7.9 (Section 7.3.2), we see that this is between S_1 and S_2, and the allowable stress for lateral buckling is:

$$23.9 - 0.124 \ (L_b/r_y) = 18.8 \text{ ksi}$$

Since this is 90% of the maximum allowable compressive stress (21 ksi from Table 7.9), we won't pursue a more precise value, even though the Aluminum *Specification* provides a way to do so (Section 5.3.2). Comparing to the allowable stress for the elements of beams from Appendix D:

Lateral buckling allowable stress =	18.8 ksi
Local buckling allowable stress =	22.7 ksi
Allowable compressive bending stress =	18.8 ksi = F_{bx}

We calculate the compressive bending stress (f_{bx}) by dividing the major axis moment from Table 10.6 by the gross section modulus from Appendix B:

$$f_{bx} = 13.6 \text{ in.-k}/14.9 \text{ in.}^3 = 0.91 \text{ ksi}$$

Member No. 34—Minor Axis Bending As shown in Table 10.6, member no. 34 has minor axis bending, as well as major axis bending. The allowable tensile bending stress for an I-beam bent about its weak axis is obtained from Figure 5.43 as 28 ksi.

To determine the tensile bending stress (f_{bty}), we must calculate the minor axis section modulus using the net section. Once again, we will deduct the bolt holes from both flanges as a conservative simplification, resulting in a minor axis net section modulus of 2.8 in.^{3}. Referring to Table 10.6, the greater minor axis end moment occurs at the opposite end of the member from the greater major axis end moment. Although we could check each end separately for combined stresses, the moments are relatively small and we don't want to have to check both ends. We'll conservatively use the larger value regardless of where it occurs. The minor axis tensile bending stress (f_{bty}) is, then:

$$f_{bty} = 0.37 \text{ in.-k}/2.8 \text{ in.}^3 = 0.13 \text{ ksi}$$

Combining this with the major axis tensile bending stress of 1.15 ksi produces a maximum bending tensile stress of 1.28 ksi. This is still less than the axial compressive stress of 4.1 ksi, so the entire cross section is in a state of (nonuniform) compression.

For minor axis bending compression, we note that a beam bent about its weak axis is not subject to lateral buckling. Furthermore, the web lies on the neutral axis, so it need not be checked for local buckling. Appendix D does not have values for minor axis bending allowable stresses, so we will have to calculate the allowable local buckling stress for the flanges. We can turn to Appendix D to find that the b/t ratio of the flanges is 5.96. From Figure 5.56, we see that this is less than S_1, and the allowable minor axis local buckling stress (F_{by}) is 28 ksi. Dividing the minor axis moment (Table 10.6) by the minor axis gross section modulus (from Appendix B), we calculate a minor axis bending stress (f_{by}):

$$f_{by} = 0.37 \text{ in.-k}/2.92 \text{ in.}^3 = 0.13 \text{ ksi}$$

Member No. 34—Shear   Finally, let's address shear stress. The allowable shear stress in the web is a function of its h/t ratio, which for shear in flat plate webs is based on the clear height between the flanges. From the dimensions provided in Appendix B, we can calculate h/t as follows:

$$h/t = [8 \text{ in.} - 2 (0.35 \text{ in.})]/0.23 \text{ in.} = 31.74$$

Note that this is slightly greater than the b/t ratio of 29.13 given in Appendix D for the web in compression, where the fillets were deducted from the height of the web. We see from Table 7.15 that 31.74 is less than S_1, and the allowable shear stress (F_s) is 12 ksi. The calculated shear stress is obtained from the approximation

$$f_s = \text{shear}/(\text{area of the web}) = 0.55 \text{ k}/[(7.3 \text{ in.})(0.23 \text{ in.})] = 0.33 \text{ ksi}$$

Member No. 34—Combined Stresses We've calculated a bunch of numbers, so let's summarize them before moving on.

Because the member has both axial force and bending, we must check it for combined stresses. From Section 7.4, we see that there are two cases to check for combined compression and bending, and another case to check for combined shear, compression, and bending. We'll begin with Equation 7.9:

$$\frac{f_a}{F_a} + \frac{C_{mx}f_{bx}}{F_{bx}(1 - f_a/F_{ex})} + \frac{C_{my}f_{by}}{F_{by}(1 - f_a/F_{ey})} \leq 1.0$$

Values for all of the variables except C_m and F_e are given in Table 10.8. All of the members that frame into a given node of a single-layer spherical lattice are within a few degrees of lying in the same plane, so we will consider the joint to be permitted to *sway* (i.e., allowed to translate in the plane of bending). The value of C_m is, thus, taken as 0.85, and all that remains to be calculated are F_{ex} and F_{ey}, given:

$$F_e = \frac{\pi^2 E}{n_u (kL/r)^2}$$

where:

E = modulus of elasticity (10,100 ksi)
n_u = safety factor on ultimate strength (1.95)
kL/r = slenderness ratio of the unsupported length in the plane of bending. Remember that this is the slenderness ratio for the member as a column.

Retrieving the values for (kL/r) from Table 10.7, we determine the following values for F_{ex} and F_{ey}:

$$F_{ex} = [\pi^2(10,100 \text{ ksi})]/[1.95\,(40.1)^2] = 31.8 \text{ ksi}$$

and

$$F_{ey} = [\pi^2\,(10,100 \text{ ksi})]/[1.95\,(0)^2] = (51,100 \text{ ksi}/0)$$

We've left F_{ey} expressed as a ratio, so we can follow the effect of the zero in the combined stress equation, which is now expressed as follows:

$$\frac{4.1}{15.1} + \frac{0.85(0.91)}{18.8(1 - 4.1/31.8)} + \frac{0.85(0.13)}{28(1 - 4.1(0/51,100)}$$

$$= \frac{4.1}{15.1} + \frac{0.85(0.91)}{18.8(0.87)} + \frac{0.85(0.13)}{28(1)}$$

$$= 0.27 + 0.047 + 0.004 = 0.32 < 1.0$$

The stresses in member no. 34 satisfy this first check on combined compression and bending, so let's perform the second check. Repeating Equation 7.10:

$$f_a/F_{ao} + f_{bx}/F_{bx} + f_{by}/F_{by} \leq 1.0$$

The only new variable in this equation is F_{ao}, which is the allowable compressive stress of the member considered as a short column. A short column would have an overall slenderness ratio less than S_1, for which the allowable overall column buckling stress is given in Table 7.4 as 19 ksi. The allowable compressive stress will, thus, be limited by the allowable buckling stress for elements of the column, which is given in Table 10.8 as 17.6 ksi. We can now express this combined stress equation as:

$$(4.1 \text{ ksi}/17.6 \text{ ksi}) + (0.91 \text{ ksi}/18.8 \text{ ksi}) + (0.13 \text{ ksi}/28 \text{ ksi}) = 0.29 < 1.0$$

Both of the equations for combined compression and bending are satisfied, but member no. 34 also has shear. The requirement for checking combined shear, compression, and bending is given in Equation 7.13:

$$(f_a/F_a) + (f_b/F_b)^2 + (f_s/F_s)^2 \leq 1.0$$

Because we have both major and minor axis bending, we will express this as:

$$(f_a/F_a) + (f_{bx}/F_{bx})^2 + (f_{by}/F_{by})^2 + (f_s/F_s)^2 \leq 1.0$$

which then becomes:

$$(4.1 \text{ ksi}/15.1 \text{ ksi}) + (0.91 \text{ ksi}/18.8 \text{ ksi})^2 + (0.13 \text{ ksi}/28 \text{ ksi})^2$$
$$+ (0.33 \text{ ksi}/12 \text{ ksi})^2 = 0.27 < 1.0$$

Member no. 34 satisfies all of the combined stress checks, with each of the

TABLE 10.8 Summary of Stresses and Allowable Stresses for Member No. 34

Stress State	Allowable Stress (ksi)	Stress (ksi)
Overall column buckling	15.1	
Column element buckling	17.6	
Column buckling	$F_a = 15.1$	$f_a = 4.1$
Bending tension (major axis)	$F_{btx} = 19$	$f_{btx} = 1.15$
Lateral buckling	18.8	
Local buckling	22.7	
Bending compression (major axis)	$F_{bx} = 18.8$	$f_{bx} = 0.91$
Bending tension (minor axis)	$F_{bty} = 28$	$f_{bty} = 0.13$
Bending compression (minor axis)	$F_{by} = 28$	$f_{by} = 0.13$
Shear	$F_s = 12$	$f_s = 0.33$

interaction ratios being far less than 1.0. While this may seem to suggest that the trial section is much too heavy, remember that it was chosen to satisfy the general buckling criteria of Equation 10.7.

Check Member No. 84 Next, we will check member no. 84, which is located in the bottom bay of diagonal members (see Figure 10.20). The forces and moments in this member are summarized in Table 10.9.

Member No. 84—Axial Compression Since we are using the same trial member size as for member no. 34, the only parameter that will change in determining the overall column slenderness ratio is the member length (L). The overall column slenderness ratio, is then:

$$kL/r = 1.0 (126 \text{ in.})/3.37 \text{ in.} = 37.4$$

The slenderness ratios of the cross-sectional elements, as before, are available from Appendix D.

The overall slenderness ratio of 37.4 is between S_1 and S_2, so from Table 7.4 we determine the overall column allowable stress as:

$$20.2 - 0.126 \, kL/r = 20.2 - 0.126 (37.4) = 15.5 \text{ ksi}$$

We compare this value to the allowable stress for the elements from Appendix D:

Overall column allowable stress =	15.5 ksi
Element allowable stress =	17.6 ksi
Column allowable axial stress (F_a) =	15.5 ksi

The axial compressive stress (f_a) is calculated from the maximum axial force of 23.0 k (Table 10.9) and the cross-sectional area of 5.26 in.² (Appendix B). The axial compressive stress is:

TABLE 10.9 Forces and Moments in Member No. 84

Member No. 84 $L = 126$ in.	Compressive Axial Force (k)	Major Axis Moment (in.-k)	Minor Axis Moment (in.-k)	Shear (k)
Top end	22.5	−48.2	−1.58	0.73
Mid span	22.7	−81.2	1.99	0.32
Bottom end	23.0	−88.6	−3.96	−0.09

$$f_a = 23.0 \text{ k}/5.26 \text{ in.}^2 = 4.4 \text{ ksi}$$

Member No. 84—Major Axis Bending   Next, check bending beginning with the tension flange, for which the allowable tensile bending stress (F_{btx}) is 19 ksi (from Section 7.3.1).

Recalling the net section modulus of 11.8 in.^{3} that we calculated for member no. 34 and applying the maximum moment from Table 10.9, the tensile bending stress (f_{btx}) is:

$$f_{btx} = 88.6 \text{ in.-k}/11.8 \text{ in.}^3 = 7.51 \text{ ksi}$$

This is greater than the axial compressive stress (f_a), so the resultant state of stress in this flange will be tension.

Once again, we must determine the unbraced length of the compression flange (L_b) to calculate the lateral buckling allowable stress. The moment diagram for member no. 84 (see Figure 10.22) shows that the top flange is in compression for the entire length of the member. Because this is the flange braced by the panels, the unbraced length (L_b) is equal to 0. From Table 7.9 we see that the allowable stress for lateral buckling is then 21 ksi. We compare this to the allowable stress for elements of beams from Appendix D:

Lateral buckling allowable stress =	21 ksi
Local buckling allowable stress =	22.0 ksi
Allowable compressive bending stress =	21 ksi = F_{bx}

You can determine compressive bending stress (f_{bx}) by dividing the major

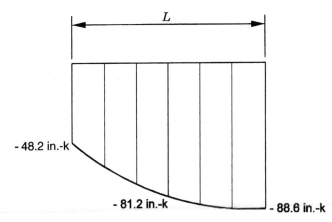

Figure 10.22   Moment diagram for member no. 84.

axis moment from Table 10.9 by the gross section modulus from Appendix B:

$$f_{bx} = 88.6 \text{ in.-k}/14.9 \text{ in.}^3 = 5.95 \text{ ksi}$$

Member No. 84—Minor Axis Bending We will now turn to minor axis bending; we can obtain the allowable tensile bending stress (F_{bty}) for an I-beam bent about its weak axis from Figure 5.43 as 28 ksi.

Dividing the minor axis bending moment (from Table 10.9) by the minor axis net section modulus that we calculated previously, the minor axis tensile bending stress (f_{bty}) is:

$$f_{bty} = 3.96 \text{ in.-k}/2.8 \text{ in.}^3 = 1.41 \text{ ksi}$$

Checking minor axis bending compression, we remember that a beam bent about its weak axis is not subject to lateral buckling. Recalling that the b/t ratio for the flanges, given in Appendix D, is 5.96, and that this is less than S_1, the allowable minor axis local buckling stress (F_{by}) is 28 ksi (from Figure 5.56, as before). The minor axis bending stress (f_{by}) is obtained from the minor axis moment (Table 10.9) and the minor axis gross section modulus (from Appendix B):

$$f_{by} = 3.96 \text{ in.-k}/2.92 \text{ in.}^3 = 1.36 \text{ ksi}$$

Member No. 84—Shear We are left with only shear stress to evaluate. The maximum axial force and moments for member no. 84 have all occurred at the lower end of the member. The shear force at this end is 0.09 k. Recalling the approximate formula for shear stress:

$$f_s = \text{shear}/(\text{area of the web}) = 0.09 \text{ k}/[(7.3 \text{ in.})(0.23 \text{ in.})] = 0.05 \text{ ksi}$$

Member No. 84—Combined Stresses As we did with member no. 34, let's summarize the calculated and allowable stresses (see Table 10.10).

Now we're ready to check combined stresses. Again using Equation 7.9:

$$\frac{f_a}{F_a} + \frac{C_{mx}f_{bx}}{F_{bx}(1 - f_a/F_{ex})} + \frac{C_{my}f_{by}}{F_{by}(1 - f_a/F_{ey})} \leq 1.0$$

As with member no. 34, the values for the stresses are in the summary table (Table 10.10 for member no. 84), and we'll use C_m equal to 0.85. F_{ey} is again equal to (51,100/0), so we need only calculate F_{ex}. Having already determined that (kL/r) equals 37.4:

TABLE 10.10   Summary of Stresses and Allowable Stresses for Member No. 84

Stress State	Allowable Stress (ksi)	Stress (ksi)
Overall column buckling	15.5	
Column element buckling	17.6	
Column buckling	$F_a = 15.5$	$f_a = 4.4$
Bending tension (major axis)	$F_{btx} = 19$	$f_{btx} = 7.51$
Lateral buckling	21	
Local buckling	22.0	
Bending compression (major axis)	$F_{bx} = 21$	$f_{bx} = 5.95$
Bending tension (minor axis)	$F_{bty} = 28$	$f_{bty} = 1.41$
Bending compression (minor axis)	$F_{by} = 28$	$f_{by} = 1.36$
Shear	$F_s = 12$	$f_s = 0.05$

$$F_{ex} = [\pi^2 (10,100 \text{ ksi})]/[1.95 (37.4)^2] = 36.5 \text{ ksi}$$

We can now solve Equation 7.9:

$$\frac{4.4}{15.5} + \frac{0.85(5.95)}{21(1 - 4.4/36.5)} + \frac{0.85(1.36)}{28(1 - 4.4(0/51,100))}$$

$$\frac{4.4}{15.5} + \frac{0.85(5.95)}{21(0.88)} + \frac{0.85(1.36)}{28(1)}$$

$$= 0.284 + 0.274 + 0.041 = 0.60 < 1.0$$

Now let's perform the second check (Equation 7.10):

$$(f_a/F_{ao}) + (f_{bx}/F_{bx}) + (f_{by}/F_{by}) \leq 1.0$$

Remember that F_{ao} is the allowable compressive stress of the member considered as if it has an overall slenderness ratio less than S_1. We are using the same trial section as for member no. 34, so the allowable buckling stress once again 17.6 ksi (Table 10.10). Substituting the stresses for member no. 84:

$$(4.4 \text{ ksi}/17.6 \text{ ksi}) + (5.95 \text{ ksi}/21 \text{ ksi}) + (1.36 \text{ ksi}/28 \text{ ksi})$$

$$= 0.58 < 1.0$$

We had also checked the combined shear, compression, and bending for member no. 34, but with a shear stress in member no. 84 of 0.05 ksi, this check would be trivial. We will, however, check combined tension and bending in accordance with Equation 7.12:

$$(f_t/F_t) + (f_{btx}/F_{btx}) + (f_{bty}/F_{bty}) \leq 1.0$$

We will take the ratio (f_t/F_t) as 0 since the axial load is compression. This is quite conservative because the axial compression would act to reduce the tensile stresses. Our tensile bending stresses are not large, however, so we can stand a conservative simplification. Applying the values from Table 10.10, we have:

$$(7.51 \text{ ksi}/19 \text{ ksi}) + (1.41 \text{ ksi}/28 \text{ ksi}) = 0.45 < 1.0$$

Although member no. 84 satisfies all of the combined stress checks, the interaction ratios are about twice those for member no. 34. The larger stresses in member no. 84 are due to the moment induced by the outrigger detail at the support. Alternate support details could minimize this boundary condition.

Note also that even the higher interaction ratios of member no. 84 are still well below 1.0. The general buckling criteria of Equation 10.7 govern the sizing of these two members for this load case. These members were selected as representing the more highly stressed members in the dome, so it appears that general buckling is a more critical limit state than the calculated stresses, for all members other than those in the tension ring.

While we selected a standard I-beam for our design checks, recall that the product of its area (A) and radius of gyration (r) was 17.7 in.[3] and the requirement for resisting general buckling was 14 in.[3]. The standard I-beam, therefore, has more than 25% excess material, and it does not have the special ribs on the top flange to engage the panel and batten bar. Real-world design engineers would modify the section to make it more efficient, add the ribs to the top flange, and order it as a custom extrusion.

This example is intended to demonstrate the application of the design requirements of the Aluminum *Specification*—it should not be considered a comprehensive review of the design requirements of aluminum domes. Assumptions made for the sake of the illustration do not apply to every member in the dome nor to every load case that may be applied to it.

10.3 ALUMINUM COMPOSITE MEMBERS

Aluminum is often used in combination with other materials, for example, an aluminum I-beam might be reinforced with a steel cover plate, or an aluminum bar might be used to reinforce a wood member. You have to address potential corrosion issues that dissimilar material contact can cause (see Section 10.3.3), but once that's done you have to do what you're really paid for—the structural design. This section provides some examples. Once you have determined the forces and moments in the various materials, use the appropriate design standard for each material to determine if it's adequately sized.

10.3.1 Composite Beams

You can predict the behavior of composite beams by using the equivalent-width method to account for the different properties of the composite parts. By this method, the width of each part is factored by the modular ratio n, the ratio of the part's modulus of elasticity to the modulus chosen as the base material. It doesn't matter which material you chose to be the base. For example, in an aluminum-steel composite beam, the steel parts' widths can be factored up by E_s/E_a and the aluminum widths left alone, or the aluminum parts' widths can be factored down by E_a/E_s and the steel parts left alone. Next, determine the location of the neutral axis and moment of inertia of the composite beam using the equivalent widths. Finally, multiply the calculated stresses in the various parts by the ratio used to calculate their widths. The only assumptions are that the members don't slip in relation to each other and that stresses remain in the elastic range.

Example 10.2: What are the maximum stresses in the aluminum and the steel in the beam in Figure 10.23, assuming they are joined so as to act compositely? The applied bending moment is 100 in.-k.

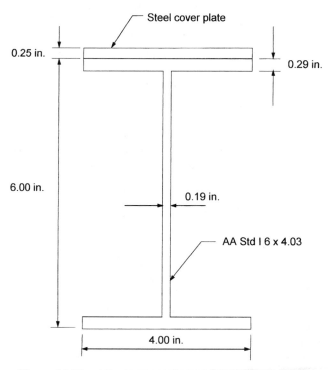

Figure 10.23 Aluminum-steel composite beam example.

The modular ratio of steel to aluminum is $n = E_s/E_a = 29,000/10,000 = 2.9$, so the aluminum-equivalent width of the steel cover plate is (4 in.) (2.9) = 11.6 in. The aluminum-equivalent area of the steel is (11.6 in.)(0.25 in.) = 2.9 in.^{2} and the aluminum-equivalent moment of inertia of the steel is $(1/12)(11.9$ in.)(0.25 in.)$^3 = 0.015$ in.^{4}. Next, determine the location of the neutral axis and the moment of inertia of the equivalent beam.

	A	y	Ay	I	Ad^2
Aluminum	3.43	3.0	10.29	22.0	$3.43(4.43 - 3.0)^2 = 7.01$
Steel	2.90	6.125	17.76	0.015	$2.90(6.125 - 4.43)^2 = 8.33$
	6.33		28.05	22.0	15.34

The neutral axis is $\Sigma Ay/\Sigma A = 28.05/6.33 = 4.43$ in. above the bottom of the section. The moment of inertia is $I + Ad^2 = 22.0 + 15.34 = 37.3$ in.^{4}.

The section modulus at the bottom flange is $S = I/c_b = (37.3$ in.$^4)/(4.43$ in.$) = 8.42$ in.^{3}, so the stress there is $f_b = M/S = (100$ in.-k$)/(8.42$ in.$^3) = 11.9$ ksi.

The section modulus at the top flange of the aluminum is $S = I/c_t = (37.3$ in.$^4)/(6 - 4.43)$ in. $= 23.8$ in.^{3}, so the stress there is $f_b = M/S = (100$ in.-k$)/(23.8$ in.$^3) = 4.2$ ksi.

The section modulus at the top of the steel is $S = I/c_t = (37.3$ in.$^4)/(6.25 - 4.43)$ in. $= 20.5$ in.^{3}, so the stress there is $f_b = M/S = (100$ in.-k$)/(20.5$ in.$^3)2.9 = 14.1$ ksi.

The equivalent-width method also works for the simpler case of columns.

10.3.2 Thermal Stresses

Thermal stresses arise when the temperature changes after materials with different coefficients of thermal expansion are joined. Since aluminum's thermal coefficient (a) is about twice that of steel or concrete (see Section 4.8), people tend to get concerned when aluminum is made composite with these other materials.

Figure 10.24 shows an example of an aluminum composite member. The aluminum jackets a steel core, and the aluminum and steel are joined at their ends so that both must expand or contract the same distance when the temperature changes. If the temperature goes up, the aluminum wants to expand twice the distance the steel does, so the aluminum is compressed by the steel and the steel is placed in tension by the aluminum. But how much expansion occurs, and what are the stresses in the materials?

We know two facts: the two materials expand the same amount, and the compressive force in the aluminum must equal the tension in the steel since no external force is applied. Using these, we can prove (mercifully, elsewhere) that the expansion is:

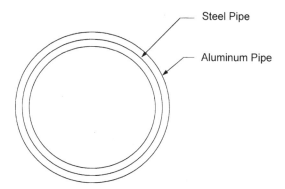

Figure 10.24   Aluminum-steel composite member.

$$\delta = (\Delta T)L\left[\frac{\alpha_a A_a E_a + \alpha_s A_s E_s}{A_a E_a + A_s E_s}\right]$$

The stress in the aluminum is:

Figure 10.25   Corrosion caused by contact with wood.

$$f_a = \frac{(\Delta T)(\alpha_a - \alpha_s)E_a A_s E_s}{E_a A_a + E_s A_s}$$

The stress in the steel is:

$$f_s = \frac{(\Delta T)(\alpha_s - \alpha_a)E_s A_a E_a}{E_a A_a + E_s A_s}$$

where:

   δ = change in length
   ΔT = temperature change
   α_a = coefficient of thermal expansion for aluminum
   α_s = coefficient of thermal expansion for steel
   L = length of the member before the temperature changed
   A_a = cross-sectional area of the aluminum
   A_s = cross-sectional area of the steel
   E_a = modulus of elasticity of aluminum
   E_s = modulus of elasticity of steel

The expansion equation passes the reality check since the expansion approaches the amount that the aluminum would expand if it was not connected to the steel $(\Delta T \alpha_a L)$ when the axial stiffness of the aluminum $(A_a E_a)$ is much larger than that of the steel $(A_s E_s)$. Conversely, the expansion approaches the amount that the steel would expand if it was not connected to the aluminum $(\Delta T \alpha_s L)$ when the axial stiffness of the steel is much larger than that of the aluminum. What may be slightly harder to swallow is that the thermal stresses are independent of the length of the member.

One thing is clear: Thermal stresses are directly proportional to the temperature change, so establishing the proper ΔT is important. ΔT is a function of not just the maximum and minimum temperatures you expect the structure to be subjected to, but also the maximum and minimum installation temperatures. The maximum temperature increase $(+\Delta T)$ is the difference between the structure's maximum temperature and the minimum installation temperature. The maximum temperature decrease $(-\Delta T)$ is the difference between the maximum installation temperature and the structure's minimum temperature.

When the neutral axes of the aluminum and the other material don't coincide, bending stresses are also induced by the temperature change. Bending stresses can also be induced when a temperature gradient acts on the composite cross section, heating some components more than others.

Thermal stresses can also be induced when aluminum is used alone but constrained from thermal movement by support conditions. Interestingly, although aluminum has twice the expansion coefficient steel does, aluminum stresses will be two-thirds that of steel in the same restraint since aluminum's modulus of elasticity is one-third that of steel.

Thermal stresses are self-limiting because if they exceed the yield strength, the force is relieved by the resulting deformation. The Aluminum *Specification* doesn't address this by providing different allowable stresses for thermal loads, but ASCE 7's load and resistance factor design (LRFD) combinations prescribe a low load factor (1.2) on self-straining forces (T), which include thermal loads.

10.3.3 Dissimilar Material Contact

Dissimilar material contact is one of the most talked about and least understood aspects of using aluminum. We mentioned it in Section 8.1.3 on fasteners, but the situation arises in a different manner in composite members and wherever aluminum is placed immediately adjacent to other materials. In these situations, contact isn't limited to a small area as for fasteners.

Dissimilar materials can cause corrosion of the aluminum in several ways. If the material is cathodic to aluminum and moisture is present, galvanic corrosion of the aluminum will occur. Other materials, such as wood or concrete, can cause corrosion of aluminum by holding moisture against the aluminum. Galvanic corrosion occurs when materials of different electrical potential are in contact through an electrolyte, which can be provided by moisture. Most metals other than zinc, magnesium, and cadmium are cathodic to aluminum, although austenitic stainless steels tend to have little galvanic effect on aluminum despite their position on the galvanic series. Contact with brass, bronze, and copper should be especially avoided as they cause vigorous galvanic attack of aluminum. Some practical guidelines for contact with various materials are listed below. Aluminum *Specification* Sections 6.6 and 6.7 also address this issue. In all cases, you can improve the corrosion resistance by sealing the contact area from moisture. Silicone, polysulfide, or butyl sealants can be used with aluminum; the choice depends on the intended service.

Carbon Steel   Prevent contact with carbon steel by coating the steel with a zinc-based primer. Zinc provides cathodic protection for both the aluminum and the steel. Other paints suitable for steel in the intended service are also used.

Concrete and Masonry   Where aluminum is in contact with concrete, masonry, plaster, or other alkaline material and the assembly is exposed to moisture, the aluminum should be coated. You can use zinc-based paint, clear lacquer, or bituminous paint. Bituminous paint is relatively inexpensive, so a thick coating can be economically applied. Some concrete additives, such as chlorides, are corrosive to aluminum and should be avoided. Products, such as magnesium phosphate grout, that are designed to remain at a relatively low pH during and after curing are compatible with uncoated aluminum.

Wood   Aluminum can be corroded by the moisture in wood or chemical preservatives in treated wood and can be protected by painting the aluminum with any of the paints used for aluminum or a bituminous paint.

Soil   In many cases, aluminum may be used in direct contact with soil without protection. While the 2xxx series alloys are not suitable for this, such alloys as Alclad 3004, 5052, 6061 and 6063 are used in the ground without coatings and generally suffer only slight pitting corrosion whose rate decreases over time. Aluminum light poles have been driven directly into the ground, buried aluminum pipe has been used in the oil industry, and corrugated aluminum pipe and culvert are common (see Section 3.1.5). The Aluminum Association *Drainage Products Manual*(s) recommends the following conditions be met to minimize corrosion in soils:

1) Soil and water pH should be between 4 and 9.
2) Soil and water resistivities should be 500 ohm-cm or greater.
3) Aluminum should not be in electrical contact with dissimilar metals.
4) Aluminum should not be used in clay-muck soils containing organic materials.
5) Aluminum should be isolated from stray electrical currents.

Runoff from Heavy Metals   Heavy metals, such as copper and lead, are corrosive to aluminum, so aluminum should not be exposed to rain water runoff from copper or terne-coated steel, which has a lead and tin coating.

10.4   ALUMINUM PRESSURE PIPING

Aluminum pressure piping used in: petroleum refineries; chemical, pharmaceutical, textile, paper, semiconductor, and cryogenic plants; and related processing plants and terminals is addressed by ASME B31.3, *Process Piping*, the ASME code for piping with internal gage pressures of 15 psi [105 kPa] or greater (84). This code uses the allowable stress method for structural design and includes certain aluminum alloys and tempers in certain product forms.

For straight pipe under internal pressure and with a design wall thickness less than one-sixth the outside diameter ($t < D/6$), B31.3 establishes the design wall thickness as the greater of:

$$t = \frac{PD}{2(SE + PY)}$$

and

$$t = \frac{P(d + 2c)}{2[SE - P(1 - Y)]}$$

where:

t = design wall thickness
P = design internal gage pressure
D = outside diameter of the pipe
S = allowable stress (provided in B31.3 and discussed further below)
E = quality factor (provided in B31.3 and summarized below)
Y = ductility coefficient = 0.4 for aluminum alloys covered by B31.3
d = maximum inside diameter of pipe considering tolerances
c = the sum of mechanical allowances (thread or groove depth) plus corrosion and erosion allowances

So, for aluminum pipe, the B31.3 design wall thickness is the greater of:

$$t = \frac{PD}{2SE + 0.8P}$$

and

$$t = \frac{P(d + 2c)}{2SE - 1.2P}$$

The first of these two expressions is very similar to the one given in *Aluminum Design Manual*, Part III, Section 3.12, which gives the pipe burst pressure as:

$$P = \frac{2tF_{tu}K}{D - 0.8t}$$

and which can be rewritten for the design wall thickness as:

$$t = \frac{PD}{2SK + 0.8P}$$

The B31.3 and *Aluminum Design Manual* expressions differ in that the K factor in the *Aluminum Design Manual* is a ductility factor equal to 1.06 or less ($K = 0.73 + 0.33F_{ty}/F_{tu}$), whereas the E factor in the ASME piping code is a quality factor equal to 1 or less (see Table 10.12).

The aluminum alloys, tempers, and product forms covered by B31.3 are listed in Table 10.11.

The allowable stresses for aluminum pipe material are given for all alloys at temperatures of up to 100°F [38°C] and 150°F; and for all alloys other than

TABLE 10.11   Aluminum Products and Tempers in ASME B31.3

ASTM Specification	Subject	Alloy tempers
B26	Aluminum-Alloy Sand Castings	443.0-F, 356.0-T6, 356.0-T71
B209	Aluminum and Aluminum-Alloy Sheet and Plate	1060-O, H112, H12, H14
		1100-O, H112, H12, H14
		3003-O, H112, H12, H14
		Alclad 3003-O, H112, H12, H14
		3004-O, H112, H32, H34
		Alclad 3004-O, H112, H32, H34
		5050-O, H112, H32, H34
		5052-O, H112, H32, H34
		5083-O, H321
		5086-O, H112, H32, H34
		5154-O, H112, H32, H34
		5254-O, H112, H32, H34
		5454-O, H112, H32, H34
		5456-O, H321
		5652-O, H112, H32, H34
		6061-T4, T6, T651, welded
		Alclad 6061-T4, T451, T6, T651, welded
B210	Aluminum-Alloy Drawn Seamless Tubes	1060-O, H112, H113, H14
		1100-H113, H14
		3003-O, H112, H14, H18
		Alclad 3003-O, H112, H14, H18
		5052-O, H32, H34
		5083-O, H112
		5086-O, H112, H32, H34
		5154-O, H34
		5456-O, H112
		6061-T4, T6, welded
		6063-T4, T6, welded
B221	Aluminum-Alloy Extruded Bars, Rods, Wire, Shapes, and Tubes	1060-O, H112
		1100-O, H112
		3003-O, H112
		Alclad 3003-O, H112
		5052-O
		5083-O
		5086-O
		5154-O
		5454-O
		5456-O
		6061-T4, T6, welded
		6063-T4, T5, T6, welded

TABLE 10.11   Aluminum Products and Tempers in ASME B31.3 (*continued*)

ASTM Specification	Subject	Alloy tempers
B241	Aluminum-Alloy Seamless Pipe and Seamless Extruded Tube	1060-O, H112, H113 1100-O, H112 3003-O, H112, H18 Alclad 3003-O, H112 5052-O 5083-O, H112 5086-O, H112 5454-O, H112 5456-O, H112 5652-O, H112 6061-T4, T6, welded 6063-T4, T5, T6, welded
B247	Aluminum-Alloy Die, Hand, and Rolled Ring Forgings	3003-H112, welded 5083-O, H112, welded 6061-T6, welded
B345	Aluminum-Alloy Seamless Extruded Tube and Seamless Pipe for Gas and Oil Transmission and Distribution Piping Systems	1060-O, H112, H113 3003-O, H112, H18 Alclad 3003-O, H112 5083-O, H112 5086-O, H112 6061-T4, T6, welded 6063-T4, T5, T6, welded
B361	Factory-Made Wrought Aluminum and Aluminum-Alloy Welded Fittings	1060-O, H112 1100-O, H112 3003-O, H112 Alclad 3003-O, H112 5083-O, H112 5154-O, H112 6061-T4, T6, welded 6063-T4, T6, welded
B491	Aluminum and Aluminum-Alloy Extruded Round Tubes for General-Purpose Applications	3003-O, H112

5083, 5086, 5154, 5254, and 5456 at temperatures of: 200°F, 250°F, 300°F, 350°F, and 400°F [204°C]. (The alloys excluded from use above 150°F are those with more than 3% magnesium and, therefore, are subject to exfoliation corrosion when held at elevated temperatures.) The minimum temperature for all aluminum alloys is −452°F [−269°C]—in other words, colder than you'll ever need to go. The allowable stress is the lesser of $F_{tu}/3$ and $2F_{ty}/3$ for temperatures up to 100°F; above that, additional criteria with safety factors on creep and rupture strengths are also applied:

1) 100% of the average stress for a creep rate of 0.01% per 1,000 hr
2) 67% of the average stress for rupture at the end of 100,000 hr
3) 80% of the minimum stress for rupture at the end of 100,000 hr.

The quality factor E from B31.3 for aluminum piping components is given in Table 10.12.

10.5   ALUMINUM PLATE STRUCTURES

Aluminum plate structures, such as storage tanks, pressure vessels, ship hulls, and liquid transportation tanks, are usually welded for leak-tightness. We'll address a few of the special considerations for these structures here.

10.5.1   Stiffeners

Plate structures are frequently reinforced with stiffeners. Knowing how much of the plate can be considered to act with the stiffener is useful when you're evaluating loads on the stiffener. Unlike the AISC Steel *Specification* in its Table B5.1, the Aluminum *Specification* doesn't provide limiting width-to-thickness ratios for compression elements. The Steel *Specification* establishes the maximum b/t ratio for plates projecting from compact compression members as:

$$\frac{b}{t} \le \frac{95}{\sqrt{F_y}}$$

Embedded in the 95 are material properties appropriate to steel. What's the proper expression for aluminum?

TABLE 10.12   Quality Factor E for Aluminum Pipe in B31.3

Material	Description	E
B26 temper F	Castings	1.00
B26 tempers T6, T71	Castings	0.80
B210	Seamless tube	1.00
B241	Seamless pipe and tube	1.00
B247	Forgings and fittings	1.00
B345	Seamless pipe and tube	1.00
B361	Seamless fittings	1.00
	Welded fittings, 100% radiograph	1.00
	Welded fittings, double butt	0.85
	Welded fittings, single butt	0.80

The buckling strength of a plate is:

$$F_{cr} = \frac{k\pi^2 E}{12(1 - v^2)(b/t)^2}$$

Setting the buckling strength of the plate equal to the yield strength will prevent buckling of the plate, which means the member is compact. Doing this, setting

$$b/t = \frac{95}{\sqrt{F_y}},$$

and using the appropriate material properties for steel ($E = 29,000$ ksi and $v = 0.3$) gives:

$$F_y = \frac{k\pi^2(29,000)}{12(1 - 0.3^2)95^2/F_y}$$

Solving this expression for k:

$$k = \frac{12(1 - 0.3^2)95^2}{\pi^2(29,000)} = 0.344$$

Next, evaluating the buckling strength equation for aluminum with $k = 0.344$ and setting the buckling strength equal to the yield strength:

$$F_y = \frac{k\pi^2 E}{12(1 - v^2)(b/t)^2} = \frac{0.344\pi^2(10,000)}{12(1 - 0.3^2)(b/t)^2}$$

gives the expression for b/t:

$$\frac{b}{t} = \pi \sqrt{\frac{0.344(10,000)}{12(1 - 0.3^2)F_y}} = \frac{56}{\sqrt{F_y}} \text{ for aluminum.}$$

For 6061-T6 with a yield strength $F_y = 35$ ksi, the length of the plate on either side of the stiffener attachment point that can be considered to act with the stiffener is:

$$b = \frac{56t}{\sqrt{F_y}} = \frac{56t}{\sqrt{35}} = 9.5t$$

compared to:

$$b = \frac{95t}{\sqrt{F_y}} = \frac{95t}{\sqrt{36}} = 16t \quad \text{for A36 steel.}$$

So, even though 6061-T6 aluminum and A36 steel have very similar yield strengths (35 and 36 ksi, respectively), the length of plate acting with the stiffener is considerably less for aluminum than for steel. From the above, you can see that this is due almost entirely to aluminum's lower modulus of elasticity. However, an interesting aspect of the expression for the length of plate acting with the stiffener is that it's inversely related to the yield strength of the plate—in other words, the stronger the plate, the shorter the length that acts with the stiffener—and this is true for both steel and aluminum. Since aluminum yield strengths tend to be lower than steel yield strengths, this somewhat offsets aluminum's lower coefficient (56 versus 95) in the length expression. An example is 3003-O aluminum plate; with its yield strength of 5 ksi, the length of plate that act with a stiffener for it is:

$$b = \frac{56}{\sqrt{F_y}} = \frac{56}{\sqrt{5}} = 25t, \text{ compared to } 16t \text{ for A36 steel.}$$

10.5.2   Compressive Strengths

Stored liquid and internal pressure usually put plate structures in tension. In cylindrical vessels with longitudinal seams, the circumferential stresses are limited to the strength of the seams. If the seams are welded, the strength is limited to the tensile welded strength. So there's no sense in using cold-worked or precipitation-heat-treated tempers, such as 5454-H34 or 6061-T6 (see Section 9.1.1). At least the strength is easy to determine.

Compressive loads are caused by wind pressure, vacuum inside the vessel, and lateral loads from such dynamic effects as earthquakes. Unfortunately, compressive strengths are not as easy to determine as tensile strengths because the provisions of the Aluminum *Specification* don't apply to thin-walled

TABLE 10.13   Minimum Diameter Without Shaping (in.-lb units)

Nominal Thickness t (in.)	Steel Tank Nominal Diameter (ft)	Aluminum Tank Nominal Diameter (ft)
$\frac{3}{16} \le t < \frac{3}{8}$	40	30
$\frac{3}{8} \le t < \frac{1}{2}$	60	50
$\frac{1}{2} \le t < \frac{5}{8}$	120	90
$t \ge \frac{5}{8}$	All	All

TABLE 10.14   Minimum Diameter Without Shaping (SI units)

Nominal Thickness t (mm)	Steel Tank Nominal Diameter (m)	Aluminum Tank Nominal Diameter (m)
$5 \leq t < 10$	12	9
$10 \leq t < 13$	18	15
$13 \leq t < 16$	36	27
$t \geq 16$	All	All

$(R_b/t > 20)$ round shapes with circumferential welds (Sections 3.4.10, 3.4.12, and 3.4.16.1). Common pipe sections are thick-walled enough to be covered by the *Specification*, but most welded plate structures aren't. (For example, an 8 in. schedule 40 pipe has $R_b/t = (8.63 + 7.98)/2/2/0.322 = 12.9$, while a 30 ft diameter $\frac{1}{2}$ in. thick aluminum storage tank has $R_b/t = (30)(12)/2/(0.5) = 360$). One reason for the *Specification* limitation on applicability to thin-walled, round elements is that their compressive strengths are very dependent on geometric imperfections. Therefore, establishing their strengths needs to go hand in glove with establishing construction tolerances on the structure. Information on this topic is available in the SSRC *Guide to Stability Design Criteria for Metal Structures* (107).

10.5.3   Fabrication

Because aluminum has a lower modulus of elasticity than steel, it's more flexible. This means that it can be made to fit a tighter radius of curvature in the elastic range than steel and needs less fabrication. This is illustrated in Tables 10.13 and 10.14, which show the minimum diameters for cylinders of aluminum and steel that can be made without shaping (roll-forming) the material as a function of plate thickness. The values in the tables are from ASME B96.1, *Welded Aluminum-Alloy Storage Tanks* (85), and API 650, *Welded Steel Tanks for Oil Storage* (45).

PART V
Load and Resistance Factor Design

Truss-supported aluminum cover. (Courtesy of Conservatek Industries, Inc.)

11 Load and Resistance Factor Design

11.1 NEW TRICKS FOR OLD DOGS

By now you've undoubtedly heard of *load and resistance factor design* (LRFD), and you're probably either already using it with steel or else you're hoping that if you ignore it, it will go away. (After all, that seems to have worked so far for converting to metric.) The whole subject was largely a moot point for aluminum designers, however, since no aluminum LRFD code was available in the U.S. up until 1994. Now that there's an aluminum LRFD specification, those of you who have been able to avoid it may be eyeing the gathering cloud uneasily.

Actually, it isn't all that bad. You really don't need to relearn structural engineering to use LRFD. If you're still skeptical, give us a chance to prove this to you. You may, in fact, have already learned the fundamentals of LRFD under its alias when used for concrete: *strength design*. It may be comforting to know that load and resistance factor design is simply the metal equivalent to strength design, taught in most engineering colleges since around 1970 for concrete.

In the beginning, designers in all materials used *allowable stress design* (ASD), a method of proportioning structural components in order to limit stresses under *nominal loads* to the yield or ultimate strength divided by a *safety factor*. Strength design, or LRFD, started in the 1960s in concrete where dead load is a significant part of the total load. Concrete designers seeking more efficient structures reasoned that there should to be a way to justify lower safety factors for well-known loads, such as that big dead load, versus the margins required for the more variable live load. They also found it inconvenient to calculate stresses in concrete, with its nonlinear stress-strain curve and low tensile strength. Instead, they preferred to determine the capacity of concrete members in terms of the ultimate force or moment (strength) that the members were able to sustain.

While strength design replaced the working stress method for concrete design, steel designers resisted the new philosophy. The benefits were not as potentially great for them due to the lower dead loads of steel structures. Slowly, hot-rolled steel came around, however, and the first LRFD version of the AISC Steel *Specification* was published in 1986. The lightest materials, such as cold-formed steel, wood, and aluminum, followed suit last. Their reluctance was based in part on the fear that higher safety factors on live load than dead load might actually erode their competitiveness. We'll examine this issue and compare ASD and LRFD outcomes in aluminum.

11.2   LRFD—THE CONCEPT

LRFD has been a bit like the Wizard of Oz: a fearsome persona hidden behind a veil of mystery. What's hidden behind the curtain, however, is much more familiar-looking than the projected image would suggest. The science of statistics that LRFD employs has not altered the laws of physics. The mission of engineers remains to design the structure to be stronger than the effect of the loads it supports. This may be expressed as:

$$R/Q \geq 1.0 \tag{11.1}$$

where:

R = strength of the structure
Q = effect of the applied loads.

This fundamental relationship forms the basis for both ASD and LRFD. The difference is in how this relationship is expressed. These forms of expression are compared as follows:

$$R \div Q \geq 1.0$$

$$\text{ASD:} \quad (R/\text{SF}) \div Q \geq 1.0 \tag{11.2}$$

$$\text{LRFD:} \quad \phi R \quad \div \gamma Q \geq 1.0 \tag{11.3}$$

where:

SF = safety factor
ϕ = resistance factor
γ = load factor.

Both methods apply factors as an allowance for uncertainty. Whereas ASD applies factors only to the strength of the structure, LRFD also factors the loads. This may be expressed for an individual member of a structure as:

$$\text{ASD:} \quad F_{tu}/n_u; \, F_{ty}/n_y; \, F_{cy}/n_y, \text{ etc.} \div Q \geq 1.0 \tag{11.4}$$

$$\text{LRFD:} \quad \phi_u F_{tu}; \, \phi_y F_{ty}; \, \phi_y F_{cy}, \text{ etc.} \div \Sigma(\gamma_D D + \gamma_L L, \text{ etc.}) \geq 1.0 \tag{11.5}$$

where:

F_{tu} = tensile ultimate strength
F_{ty} = tensile yield strength
F_{cy} = compressive yield strength
n_u = safety factor on ultimate strength
n_y = safety factor on yield strength
ϕ_u = resistance factor on ultimate strength
ϕ_y = resistance factor on yield strength
γ_D = load factor for dead load
D = the effect of dead load
γ_L = load factor for live load
L = effect of live load.

The 'new' concept in LRFD is that it applies statistics and probability to refine the factors on strength, as well as to assign factors to the load effects.

11.3   WHAT'S NEW: LOAD FACTORS

Equation 11.3 may be expressed as:

$$\phi R \geq \Sigma \gamma Q \qquad (11.6)$$

LRFD requires that the *nominal strength* of a member, multiplied by a *resistance factor* ϕ less than one, be greater than or equal to the forces in the members produced by the sum of the *nominal loads*, each multiplied by a *load factor* γ.

The left-hand side of Equation 11.6 (nominal strength times the resistance factor) is called the *design strength* (or, by the more wordy, *factored limit state strength*). The load combinations for the right-hand side are specified in the LRFD Aluminum *Specification* commentary (*Aluminum Design Manual*, Part IIB, Section 2.3) as load combinations numbered:

1) $\Sigma \gamma Q = 1.4\,(D + F)$

2) $\Sigma \gamma Q = 1.2\,(D + F + T) + 1.6(L + H) + 0.5(L_r \text{ or } S \text{ or } R)$

3) $\Sigma \gamma Q = 1.2D + 1.6(L_r \text{ or } S \text{ or } R) + (0.5L \text{ or } 0.8W \text{ or } 1.0F_a)$

4) $\Sigma \gamma Q = 1.2D + 1.6W + 0.5L + 0.5(L_r \text{ or } S \text{ or } R)$

5) $\Sigma \gamma Q = 1.2D + 1.0E + 0.5L + 0.2S$

6) $\Sigma \gamma Q = 0.9D + 1.6W + 1.6H + 1.0F_a$

7) $\Sigma \gamma Q = 0.9D + 1.0E + 1.6H$

where:

D = dead load
E = earthquake load
F = loads due to fluids with well-defined pressures and maximum heights
F_a = flood load
H = load due to weight of lateral pressure of soil and water in soil
L = live load due to occupancy and use
L_r = roof live load
R = roof rain load
S = snow load
T = self-straining force
W = wind load.

These loads are the nominal loads prescribed for the structure by, for example, the owner's specifications, building codes, or ASCE Standard 7-98, *Minimum Design Loads for Buildings and Other Structures* (83).

Comparing equations 11.4 and 11.5, you see that ASD distinguishes between failure modes (called *limit states*), by assigning different safety factors to yield (n_y) versus ultimate strength (n_u). LRFD makes a similar distinction by applying different resistance factors (ϕ) to different limit states, but also differentiates between loads producing force in the member by using different load factors (γ) for different loads.

The load factors are related to the likelihood that a particular load combination will actually occur. The less predictable loads have a higher probability of exceeding their nominal value and, thus, are assigned higher load factors. Conversely, reduced load factors are applied when combining loads that have a relatively low probability of acting at the same time.

The Aluminum *Specification* refers to ASCE Standard 7-98, *Minimum Design Loads for Buildings and Other Structures* (83) for load factors, given in ASCE 7, Section 2.3.2.

Sometimes the limit state in question is due to serviceability considerations, such as when the span of roofing or checkered plate is limited by deflection rather than strength. When this is the case, the load factors to be applied to the nominal loads are all to be taken as 1.0 (in accordance with Aluminum *Specification* Section 2.2). In other words, it is not necessary to limit deflections to those produced by factored loads but rather only those due to nominal loads. This is consistent with ASCE 7-98, which notes that the load factors provided there are not intended for serviceability requirements. Recent work (106) suggests using $D + L$ and $D + 0.5S$ as load combinations for serviceability, but these have not yet been adopted in ASCE 7.

11.4 WHAT'S THE SAME

It's worthwhile to note that the expressions for strengths (properly called *nominal strengths*) of aluminum members are the same in the LRFD *Speci-*

fication as in the ASD *Specification*. For example, in the ASD *Specification* Section 3.4.11 for certain beams with a slenderness greater than S_2, the allowable stress of the beam is:

$$\frac{\pi^2 E}{n_y \left(\dfrac{L_b}{1.2r_y}\right)^2}$$

The corresponding expression for design strength in the LRFD *Specifications* Section 3.4.11 is:

$$\frac{\phi_b \pi^2 E}{\left(\dfrac{L_b}{1.2r_y}\right)^2}$$

In each case, the strength term is the same:

$$\frac{\pi^2 E}{\left(\dfrac{L_b}{1.2r_y}\right)^2}$$

Not only are the expressions for strengths the same in ASD and LRFD, but the values employed for mechanical properties are also the same. Aluminum *Specification* Tables 3.3-1 and 3.3-2 are identical in the respective specification for each method.

Similarly, the loads (properly called *nominal loads*) used for LRFD are the same as those for ASD. It's just that in LRFD, these loads are multiplied by load factors, and in ASD, they're not. In both cases, stresses are calculated by conventional *elastic analysis*. LRFD is *not* an attempt to introduce plastic design under another name.

That's all that you need to know to understand LRFD. From this discussion, you can see that the only additional effort LRFD requires over ASD is that you apply different load factors to the various loads during analysis to determine the force in a member for design check purposes. If you'd like to gain additional insights and pick up some tools to help you use LRFD, read on.

11.5   WHEN DO I USE LRFD?

As far as the Aluminum Association is concerned, designers may use either LRFD or ASD. However, it is not permissible to mix and match ASD and LRFD; whichever method is chosen must then be used for all of the members in a structure. You can't use LRFD for the tension chord of a truss, for example, when using ASD for the compression elements of the same structure. (In Canada, where they call it *strength design* (96), *only* LRFD is allowed.) The day may be coming when U.S. codes require LRFD, especially for certain loads, such as earthquakes, but it isn't here yet.

While the aluminum ASD specifications may be used for either *building-* or *bridge-type structures* (with higher safety factors for bridges), the aluminum LRFD specifications are intended only to be used for building-type structures. If you want to design an aluminum highway bridge using LRFD, use Section 7 (aluminum structures) of the AASHTO *LRFD Bridge Design Specifications* (35).

Provisions for fatigue are identical in the LRFD and ASD specifications. Both provide an allowable stress range for fatigue design. Fatigue stresses are calculated for unfactored loads for both ASD and LRFD.

11.6   WHICH WAY LETS ME USE LESS METAL?

The obvious question is: Which method of design requires less aluminum? As you may suspect, the answer is: it depends. That is, in an individual instance of stress and load, it depends (100), but overall, if enough instances are considered, it has been predetermined that the two methods require about the same amount of metal.

One conclusion that can be stated is that as the dead load decreases in relation to the live load, the chances improve that LRFD will require more metal than ASD. Because aluminum has a high strength-to-weight ratio, dead-to-live-load ratios for aluminum structures tend to be lower than for steel structures. An example is an aluminum space frame designed for a roof live load of 25 psf with a 2.5 psf dead load, giving a dead-to-live-load ratio of 0.1. Aluminum roofing and siding typically weigh less than 5% of their design live loads. It is very rare for an aluminum structure to have a dead-to-live-load ratio greater than 0.3 (111). By contrast, the development of resistance factors for hot-rolled steel used a dead-to-live-load ratio of 0.33 (38), which is about three times that for aluminum. This, of course, is related to the fact that structural grades of steel have similar strengths compared to structural aluminum, but weigh nearly three times as much per unit volume (0.283 lb/in.^{3} versus 0.098 lb/in.^{3}). The dead-to-live-load ratio that was chosen as the basis for aluminum resistance factors was 0.2.

Other factors that will affect the comparison between ASD and LRFD include the type of stress the member is subjected to and the type of load being considered. For example, for axial tension, LRFD will require less metal than ASD for live loads, such as snow load and minimum roof live load, but more metal for wind load. However, the differences between LRFD and ASD will generally be less than about 20%.

Let's examine the premise that when wind loads are considered, LRFD will typically require more metal than would be required by ASD. The Aluminum *Specification* for ASD allows a one-third increase in allowable stresses when wind or seismic loads are considered, either when they act alone or in combination with other live loads (see *Specification* Section 2.3). This provision is similar to provisions in the AISC and AISI specifications and is so ancient that its origin is unknown, and factors usually offered to justify it are already included in the determination of the magnitude of wind loads. (For some interesting reading on this, see Ellifritt's "The Mysterious 1/3 Stress Increase" [101]). The Aluminum ASD *Specification* bans the one-third increase, however, when the code used to determine the wind or seismic load prohibits it. If the wind load is determined in accordance with ASCE 7-98, which does, in fact, prohibit the one-third allowable stress increase, you can't take it. ASCE 7 throws you a bone, though, by reducing the wind loads from what they were back in the days when the one-third increase was allowed, by introducing a directionality factor (less than 1) on wind. This factor (K_d) accounts for the fact that the direction of the design wind may not coincide with the most vulnerable direction of the structure and ranges from 0.85 for buildings to 0.95 for chimneys. While the 15% load reduction on buildings helps to soften the blow, it doesn't completely offset loss of the 33% allowable-stress increase.

Additionally, though, the Aluminum LRFD *Specification* allows a 10% increase in resistance factors for secondary members subjected to wind or seismic loads, with the limitation that the resulting resistance factor may not exceed one. Secondary members are those members that serve to carry load from its point of application on the structure to the main force-resisting frame, and include purlins and building siding. This provision is based on the same reasoning as the 0.9 factor on wind load allowed in the AISI LRFD *Specification for Cold-Formed Steel Members* (40) for these same secondary members.

Example 11.1: Compare ASD and LRFD for an extruded 6061-T6 aluminum bar in tension. The bar is to carry a 10 kip axial force, of which 1 kip is due to dead load and the rest is from snow load. Assume that the bar's net area and gross area are equal.

The limit states for the bar are failure by fracture, which occurs at 38 ksi, and yielding, which occurs at 35 ksi. For allowable stress design, the safety factor applied to fracture is 1.95, and the safety factor applied to yielding is 1.65; the allowable stress is the lesser of

$$(38 \text{ ksi}/1.95 = 19.5 \text{ ksi})$$

and

$$(35 \text{ ksi}/1.65 = 21.2 \text{ ksi}),$$

or 19.5 ksi. The fracture, or ultimate strength, limit state governs (since net area and gross area are the same) with an allowable stress of 19.5 ksi. The net area required, then, is (10 k/19.5 ksi), or 0.513 in.^{2}.

Now let's apply the LRFD approach to the same bar. The limit states are the same as before since the bar doesn't know if we're using ASD or LRFD. For the limit state of fracture, the strength is 38 ksi, for which the resistance factor is 0.85, giving a design strength of 38 ksi $\times$ 0.85 = 32.3 ksi. For the limit state of yielding, the strength is 35 ksi and the resistance factor is 0.95, giving a design strength of 35 ksi $\times$ 0.95 = 33.2 ksi. The design strength is the lesser of these two, or 32.3 ksi.

The LRFD design strength is higher than the ASD allowable force in the bar, but whereas the allowable force is compared to the force in the bar produced by the nominal load, the design strength is compared to the nominal loads times load factors, which are typically greater than one.

The factored load is:

$$1.2D + 1.6S = 1.2(1 \text{ k}) + 1.6(9 \text{ k}) = 15.6 \text{ kips}$$

The required net area for LRFD is, then, (15.6 k/32.3 ksi) = 0.483 in.^{2}, or 6% less than the 0.513 in.^{2} required by ASD.

If the live load on the bar had instead been a wind load, ASD would have required (10 k/[(19.5 ksi)(4/3)]) = 0.385 in.^{2}. The factored load for LRFD would have been:

$$1.2D + 1.6W = 1.2(1 \text{ k}) + 1.6(9 \text{ k}) = 15.6 \text{ kips}$$

The required net area for LRFD is, then, (15.6 k/32.3 ksi) = 0.483 in.^{2}, or 25% more than the 0.385 in.^{2} required by ASD. For an even comparison, however, you should recognize that wind loads to be used with LRFD have been reduced by about 15% by the directionality factor below wind loads used for ASD. Taking this into account, the required area is 0.411 in.^{2}, a net increase of 7%.

If, however, the bar were a secondary member, the LRFD resistance factors could be increased by 10%, as long as they do not exceed 1.0. For the limit state of fracture, $1.1\phi = 1.1 (0.85) = 0.93$, which does not exceed 1.0, so the design strength for fracture is 0.93 (38 ksi) = 35.5 ksi. For the limit state of yield, $1.1\phi = 1.1 (0.95) = 1.04$, which exceeds 1.0, so the design strength is 1.0 (35 ksi) = 35 ksi. Using the lesser of 35.5 and 35 ksi, the required net area is (15.6 k) /(35 ksi) = 0.446 in.^{2}. When the wind directionality factor is accounted for, this is actually 0.379 in.^{2}, or 2% less than required by ASD. The area that ASD and LRFD require for each case is compared in Table 11.1.

So, you see that the answer to which method requires less material truly is: it depends.

TABLE 11.1   Comparison of Material Required by LRFD Versus ASD in Example 11.1

Live Load Case	ASD	LRFD	Percent Difference to Use LRFD
Snow	0.513 in.^{2}	0.483 in.^{2}	−6%
Wind, primary member	0.385 in.^{2}	0.411 in.^{2}	7%
Wind, secondary member	0.385 in.^{2}	0.379 in.^{2}	−2%

11.7   THE GENERAL EXPRESSION FOR COMPARING LRFD TO ASD

Once you have determined the governing limit state for a particular application in accordance with one method, you may want to check whether the other method would require more or less material. You can do this by determining the *nominal resistance* (R) required by each method for the load combination under consideration. You can then compare the resistance required by LRFD to that required by ASD for the given limit state.

This may be illustrated by considering the basic case of dead load plus live load, which may be obtained for LRFD from load combination 2 above by ignoring the $0.5(L_r$ or S or $R)$ term (Table 11.2).

The ratio of the resistance required by LRFD to that required by ASD is, then:

$$\frac{(1.2D/L + 1.6)}{[\phi(SF)(D/L + 1)]} \tag{11.7}$$

You can determine required resistance ratio (LRFD to ASD) for a given dead-to-live-load ratio by inserting the values of ϕ and SF for the governing limit state. You can follow this procedure for any load combination by substituting the appropriate load factors. Consider, for example, the combination of dead load and wind. The LRFD load combination, again ignoring the $0.5(L_r$ or

TABLE 11.2   Comparison of LRFD and ASD Resistance

	LRFD	ASD
Resistance Equation	$\phi R = 1.2D + 1.6L$	$(R/SF) = D + L$
Solve for R	$R = (1.2D + 1.6L)/\phi$	$R = (D + L)\,SF$
R in terms of D/L	$R = (1.2D/L + 1.6)(L/\phi)$	$R = (D/L + 1)\,L\,(SF)$

S or *R*) term, would be (1.2*D* + 1.6*W*), and the ASD form would be simply (*D* + *W*). This would yield the following form of the LRFD/ASD ratio:

$$\frac{(1.2D/W + 1.6))}{[\phi(SF)(D/W + 1)]} \qquad (11.8)$$

Example 11.2: Use the general expression to perform the comparisons of Example 11.1 for a 6061-T6 aluminum bar in tension. The dead load resulted in a 1 kip axial force, and the live load added 9 kips, for a total force of 10 kips. The *D/L* ratio, then, is 1/9, or 0.11. We have determined by ASD that the governing limit state is fracture, for which SF = 1.95. Next, we will compare this result to LRFD, where ϕ = 0.85. Substituting these values of *D/L*, SF, and ϕ into Equation 11.7:

$$\frac{1.2(0.11) + 1.6}{0.85(1.95)(0.11 + 1)} = 0.94$$

The resistance the LRFD method requires is shown to be 6% less than that for the ASD method, so we can see that in this case LRFD would require less material.

In the case of the live load being replaced with a wind load, you can divide the ASD safety factor by four-thirds to account for the one-third increase in allowable stresses. We have already accounted for the wind in LRFD by adjusting the load factors in Equation 11.8. The LRFD resistance factor, which is a function of the limit state, remains 0.85. Equation 11.8, then, becomes:

$$\frac{1.2(0.11) + 1.6}{0.85 \left(\dfrac{1.95}{4/3} \right)(0.11 + 1)} = 1.25$$

At first glance, it appears that the resistance LRFD requires for this wind load case is 25% greater than for ASD, indicating that LRFD would require considerably more material. This is offset, however, because wind loads have been decreased by about 15% through the load directionality factor (see Section 11.6 above).

Looking back to the values in the summary table in Example 11.1, we see that we have determined the same results with these general expressions as we did with the longhand calculations. This approach is sometimes helpful in evaluating the impact on a given structure of changing the design method from ASD to LRFD.

11.8   **HOW THEY CAME UP WITH THE LRFD** *SPECIFICATION*

You don't need to know this to use LRFD. Chances are, you won't be alone if you skip this section. If you want to reach a level of understanding beyond the ability to plug through the formulas, however, this section will help. We will introduce the statistical term *standard deviation* (σ) as a variable, but will not get into how to calculate it. If you're intimidated by statistics, you'll be glad that we don't get into this in more detail. On the other hand, if you are already familiar with basic statistical terms, you don't need us to explain them. For those remaining few who have a yen to understand statistics, take a look at Appendix M.

The basic concept to keep in mind is that ASD and LRFD are two methods of accounting for uncertainty. Testing samples of a particular material will indicate its typical strength, but we don't know what the actual strength of the next beam made from that material will be. In a similar manner, we may have collected many years of snowfall data for a particular location, but we still don't know how much snow will fall there next year.

Based on the data that we do have, we can predict both the strength of the structure and the magnitude of the loads, but we recognize that we aren't certain that our predictions will be accurate. A safety factor may be thought of as a margin for error in our predictions. This margin has been adjusted over the years in response to the observed service history of structures. The present values for ASD factors of safety are generally accepted as a sound basis for designing safe structures, but their development may be viewed as a trial-and-error process. The LRFD method also recognizes that we don't know the actual values for either the strength of the structure or the loads that it will be subjected to, but LRFD applies statistics to account for this uncertainty.

Regardless of the method, the goal is to produce a structure whose resistance (R) is greater than the effect of the loads (Q). While neither R nor Q is precisely known, the LRFD statistician represents each by a frequency distribution, as shown in Figure 11.1. This plot shows the resistance curve to the right of (greater than) the load-effect curve. While the actual values for R and Q could fall anywhere along their respective curves, the height of the curve at a given point along the x-axis indicates the likelihood of that point being the true value. The true value will typically be close to where the curve peaks, then, and will seldom occur toward either of the tailing-off extremes. When the resistance of a structure is greater than (to the right of) the load effect, the structure is safe. The resistance can only lie to the left of the load effect when both R and Q fall in the area where the two curves overlap. When the curves are spaced sufficiently far apart, they overlap only in their areas of low frequency. In that the chances of either R or Q falling in this area are very small, it is not likely that both would fall in the overlap region at the same time. When this does happen, however, R can be less than Q. If the

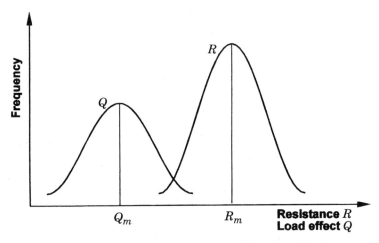

Figure 11.1   Frequency distribution for load effect (Q) and resistance (R).

resistance of a structure is less than the effect of the loads that it experiences, the structure fails.

This situation may be expressed in the form of a single curve if we rewrite $R > Q$ as $R/Q > 1$ and take the logarithm of both sides of the equation. The resulting curve shown in Figure 11.2 includes the uncertainties in both R and Q. The probability that the structure will collapse is, then, the probability that $\ln(R/Q) < 0$ and is represented by the shaded area of the graph.

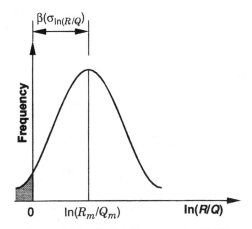

Figure 11.2   Reliability curve ln (R/Q).

The reliability of the structure is expressed as the number of standard deviations ($\sigma_{\ln(R/Q)}$) between the mean value and zero on the $\ln(R/Q)$ distribution (38). This multiple is termed β.

$$\beta(\sigma_{\ln(R/Q)}) = \ln(R_m/Q_m) \tag{11.9}$$

where:

β = reliability index
$\sigma_{\ln(R/Q)}$ = standard deviation of the $\ln(R/Q)$ distribution
R_m = mean resistance
Q_m = mean load effect.

The standard deviation of a distribution is a statistician's tool for measuring how peaked the curve is, and β, a dimensionless quantity, is called the *reliability index*. The value of β increases either as the mean of the $\ln(R/Q)$ distribution increases or as the scatter of the $\ln(R/Q)$ distribution is reduced. (in other words, if the curve shown in Figure 11.2 is either moved to the right or is made more spike-like, as opposed to flattened).

Another statistical term that we will use is the *coefficient of variation* (V), which is simply the standard deviation divided by the mean. We approximate the standard deviation of the $\ln(R/Q)$ distribution of Figure 11.2 from the coefficients of variation of the individual curves for R and Q in Figure 11.1:

$$\sigma_{\ln(R/Q)} = \sqrt{V_R^2 + V_Q^2} \tag{11.10}$$

$$V_R = \frac{\text{(standard deviation of resistance)}}{\text{(mean resistance)}}$$

$$V_Q = \frac{\text{(standard deviation of load effect)}}{\text{(mean load effect)}}$$

where:

V_R = coefficient of variation of resistance
V_Q = coefficient of variation of load effect.

Substituting for $\sigma_{\ln(R/Q)}$ in Equation 11.9:

$$\beta = \frac{\ln(R_m/Q_m)}{\sqrt{V_R^2 + V_Q^2}} \tag{11.11}$$

The three factors that are considered in evaluating variations in resistance are:

1) The material strength factor (M), equal to the ratio of actual material strengths to the nominal material strengths (which are called *minimum mechanical properties* in aluminum)
2) The fabrication factor (F), equal to the ratio of fabricated member dimensions to nominal dimensions
3) The professional factor (P), equal to the ratio of strengths measured in experiments to strengths predicted by the design specifications.

The terms for the mean and coefficient of variation for each of these factors are summarized in Table 11.3. From statistics:

$$V_R = \sqrt{V_M^2 + V_F^2 + V_P^2} \qquad (11.12)$$

and

$$R_m/R = M_m F_m P_m \qquad (11.13)$$

where R is the nominal resistance.

Having established these statistical relationships for LRFD, we need to determine values for each of the variables. Since the ASD method has a demonstrated history of resulting in safe structures, LRFD factors have been developed from an analysis of the ASD method. The nominal resistance for ASD may be expressed as:

$$R = (D + L)(\text{SF}) \qquad (11.14)$$

Combining Equations 11.13 and 11.14:

$$R_m = (M_m F_m P_m)(D + L)(\text{SF}) \qquad (11.15)$$

TABLE 11.3 Terms Used for the Mean and the Coefficient of Variation for Variable Factors in Determining Resistance

Factor	Mean	Coefficient of Variation
Material Strength Factor (M)	M_m	V_M
Fabrication Factor (F)	F_m	V_F
Professional Factor (P)	P_m	V_P

or

$$R_m = (M_m F_m P_m)L(\alpha + 1)(SF) \tag{11.16}$$

where α equals the dead-to-live-load ratio. Next, we must gain a measure of how these factors vary. This information, developed by Galambos (108), is summarized in Table 11.4.

For the basic dead plus live load combination, the mean load effect is:

$$Q_m = D_m + L_m \tag{11.17}$$

where:

Q_m = total mean load effect
D_m = mean dead load effect
L_m = mean live load effect.

Using the information from Table 11.4 for D_m and L_m, Equation 11.17 can be written as:

$$Q_m = L(1.05\alpha + 1) \tag{11.18}$$

Again from statistics, the coefficient of variation of the load effect is:

$$V_Q = \frac{\sqrt{(D_m V_D)^2 + (L_m V_L)^2}}{Q_m} \tag{11.19}$$

Using the information in Table 11.4 for D_m, V_D, L_m, and V_L, Equation 11.19 can be rewritten as:

TABLE 11.4   Values Used for the Variables in the Resistance Factor Equation

Factor	Expression	Value
Material	M_m/M	1.10 for yield and fracture; 1.0 for buckling
	V_m	0.06
Fabrication	F_m/F	1.0
	V_f	0.05
Professional	P_m/P	Varies from 0.93 to 1.24
	V_p	Varies from 0.0 for yield and fracture to 0.27 for some conditions of elastic buckling
Dead Load	D_m/D	1.05
	V_D	0.10
Live Load	L_m/L	1.0
	V_L	0.25

$$V_Q = \frac{\sqrt{(1.05\alpha(0.1))^2 + 0.25^2}}{(1.05\alpha + 1)} \tag{11.20}$$

Using Equations 11.12, 16, 18, and 20 and the values given in Table 11.4, the equation for β (11.11) can be evaluated for any given dead-to-live-load ratio α. With these values, we can calculate the reliability index (β) for any limit state (for example, tensile fracture) in the current ASD specification.

Example 11.3: Calculate the reliability index β for the ASD Aluminum *Specification* for axial tensile fracture under dead load plus live load, assuming a dead-to-live-load ratio of 0.2.

Substituting values from Table 11.4 into Equation 11.12:

$$V_R = \sqrt{V_M^2 + V_F^2 + V_P^2} = \sqrt{0.06^2 + 0.05^2 + 0^2} = 0.08 \tag{11.21}$$

Substituting 0.2 for the dead-to-live-load ratio α in Equation 11.20:

$$V_Q = \frac{\sqrt{[(1.05)(0.2)(0.1)]^2 + 0.25^2}}{(1.05)(0.2) + 1} = 0.207 \tag{11.22}$$

From Equations 11.16 and 11.18:

$$\frac{R_m}{Q_m} = \frac{(M_m F_m P_m)L(\alpha + 1)(\mathrm{SF})}{L(1.05\alpha + 1)} \tag{11.23}$$

The safety factor (SF) for axial tension fracture is given in the Aluminum *Specification* Section 3.4.1 as $n_u = 1.95$:

$$\frac{R_m}{Q_m} = \frac{(1.1)(1.0)(1.0)L(0.2 + 1)(1.95)}{L[1.05(0.2) + 1]} = 2.127 \tag{11.24}$$

So, by Equation 11.11:

$$\beta = \frac{\ln(2.127)}{\sqrt{0.08^2 + 0.207^2}} = 3.40 \tag{11.25}$$

This calculation was performed for each of the limit states in the 5th edition of the Aluminum *Specifications*. The limit states were grouped into 11 basic categories, and the reliability index was calculated for each category. The reliability indices inherent in the provisions of the 5th edition of the Aluminum *Specifications* were thus determined for dead-to-live-load ratios of 0.1 and 0.2, based on the dead and live load combination (1.2D + 1.6L). These calculations thereby took into account the ASD factors of safety and the material, fabrication, professional, and load effect factors described above. The resulting reliability indices ranged from 1.65 to 3.16 for a dead-to-live-load ratio of 0.1, and 1.75 to 3.40 for a D/L ratio of 0.2.

Based on these results, target reliability indices of 2.5 for yielding and buckling limit states and 3.0 for fracture limit states were judged to be representative of current aluminum practice and selected to give similar (but more uniform) reliabilities for the LRFD *Specification*. Equation 11.9 was reformulated with β_T set equal to the appropriate target (2.5 or 3.0, depending on the limit state in question) and the nominal resistance expressed as:

$$R = (1.2D + 1.6L)/\phi \qquad (11.26)$$

The only unknown variable in the equation was ϕ, which is solved for as:

$$\phi = \frac{(R_m/R)(1.2\alpha + 1.6)}{(1.05\alpha + 1)\ln^{-1}(\beta_T\sqrt{V_R^2 + V_Q^2})}$$

where β_T = the target reliability index.

The resistance factors (ϕ) were thus calculated for each of the 11 categories of limit states, rounded to the nearest 0.05, and used for the LRFD *Specification*. This process of developing values for LRFD from current ASD practice is referred to as calibrating the resistance factors.

Reliability indices of 2.0 and 2.5 were chosen for secondary members subjected to wind or seismic loads, which is achieved by multiplying resistance factors by 1.1 for these cases, as noted previously in this chapter.

If you followed this somewhat tortuous process, you'll see that the basic premise used in setting up LRFD is that the overall reliability of structures designed to ASD specifications has been acceptable. So why tinker with ASD? Because the reliability in ASD specifications is not uniform for the various failure modes, and the LRFD approach can be more so. Some people care about that.

11.9   HOW DO I ACTUALLY START USING LRFD?

It's evident that quite a lot of effort has gone into the development of LRFD. While the development of the LRFD method may be complicated, the use of it in practice need not be any more difficult than that of ASD. The design steps may be summarized as:

1) Analyze the structure in the same manner as with ASD, but factor the applied loads.
2) Perform all the design checks that ASD requires, but use the LRFD design stresses rather than the ASD allowable stresses.

The *Aluminum Design Manual*, however, while including handy allowable stress tables for ASD, offers no corresponding design stress tables for LRFD. To remedy this, Appendix I contains design stress tables for some common aluminum alloys, similar to the allowable stress tables in *Aluminum Design Manual*, Part VII, Design Aids, Tables 2-2 through 2-23. We have also provided the LRFD design stresses for local buckling of Aluminum Association standard channels and I-beams in Appendix E. With these tools, LRFD can be undertaken just as readily as ASD.

11.10   THE FUTURE OF THE ASD AND LRFD ALUMINUM SPECIFICATIONS

The AISC stopped updating the ASD Steel *Specification* after the 9th edition was issued in 1989 while continuing to update the LRFD specification. It now appears the AISC is moving toward a unified specification allowing both ASD and LRFD, to be published around 2005. The Aluminum Association, however, has chosen a different approach, and plans to continue to present both the ASD and LRFD specifications in the *Aluminum Design Manual*.

While this might seem to represent a fundamental difference in philosophy, it isn't. The AISC *Specification for Structural Steel Buildings* is, as its title says, exclusively for designing buildings, for which load factors have been developed and which are designed by structural engineers almost exclusively. The ASD Aluminum *Specification*, on the other hand, is used to design all sorts of structural components in addition to building structures, such as piping, crane hoists, automotive parts, and concrete forms, for which load factors haven't been developed. These components are often designed by mechanical engineers who never heard of LRFD and were taught by professors who probably never will. This isn't too surprising, considering that dead loads are usually negligible for such components and a single overriding service load is often the sole criterion for their design. For these uses, an ASD Aluminum *Specification* will be both necessary and useful for the foreseeable future. Although it may have that effect, maintenance of the ASD Aluminum *Specification* is not intended to discourage transition to LRFD for building structures.

Maintenance of the parallel ASD and LRFD specifications, however, does result in one anomaly. In order to keep the specifications as parallel as possible, the LRFD *Specification* has been written so that the strength of members is expressed in terms of stress (the "factored limit state stress"), rather than in terms of force and moment (the "design strength"), the way the steel and Canadian and European aluminum design specifications are expressed.

APPENDIX A
Pre-1954 Wrought Alloy Designations

The current Aluminum Association alloy and temper designation systems were adopted on July 1, 1954. Believe it or not, some references still use the previous designation system. If you encounter an unfamiliar designation, try looking it up here.

TABLE A.1   Wrought Alloy Designations New (Aluminum Association) and Old (Pre-1954)

New Designation	Old Designation	New Designation	Old Designation
1100	2S	5056	56S
1350	EC	5086	K186
2014	14S	5154	A54S
2017	17S	6051	51S
2024	24S	6053	53S
2025	25S	6061	61S
2027	27S	6063	63S
2117	A17S	6066	66S
3003	3S	6101	No. 2 EC
3004	4S	6951	J51S
4043	43S	7072	72S
5050	50S	7075	75S
5052	52S	7076	76S

TABLE A.2 Wrought Alloy Designations Old (Pre-1954) and New (Aluminum Association)

Old Designation	New Designation	Old Designation	New Designation
2S	1100	52S	5052
3S	3003	53S	6053
4S	3004	A54S	5154
14S	2014	56S	5056
17S	2017	6061	61S
A17S	2117	6063	63S
24S	2024	6066	66S
25S	2025	72S	7072
27S	2027	75S	7075
43S	4043	76S	7076
50S	5050	K186	5086
51S	6051	EC	1350
J51S	6951	No. 2 EC	6101

APPENDIX B
Section Properties of Common
Aluminum Shapes

Aluminum Association Standard I-Beams

Designation (Depth × Weight) (in. × lb/ft)	Depth d (in.)	Width b_f (in.)	Flange Thickness t_f (in.)	Web Thickness t_w (in.)	Fillet Radius R (in.)	Area A (in.²)	Axis X-X			Axis Y-Y			Warping Constant C_w (in.⁶)	Torsion Constant J (in.⁴)
							I_x (in.⁴)	S_x (in.³)	r_x (in.)	I_y (in.⁴)	S_y (in.³)	r_y (in.)		
I 3 × 1.64	3	2.50	0.20	0.13	0.25	1.39	2.24	1.49	1.27	0.522	0.418	0.613	1.02	0.019
I 3 × 2.03	3	2.50	0.26	0.15	0.25	1.73	2.71	1.81	1.25	0.679	0.543	0.627	1.27	0.037
I 4 × 2.31	4	3.00	0.23	0.15	0.25	1.96	5.62	2.81	1.69	1.04	0.691	0.727	3.68	0.033
I 4 × 2.79	4	3.00	0.29	0.17	0.25	2.38	6.71	3.36	1.68	1.31	0.872	0.742	4.50	0.061
I 5 × 3.70	5	3.50	0.32	0.19	0.30	3.15	13.9	5.58	2.11	2.29	1.31	0.853	12.5	0.098
I 6 × 4.03	6	4.00	0.29	0.19	0.30	3.43	22.0	7.33	2.53	3.10	1.55	0.951	25.3	0.089
I 6 × 4.69	6	4.00	0.35	0.21	0.30	3.99	25.5	8.50	2.53	3.74	1.87	0.968	29.8	0.145
I 7 × 5.80	7	4.50	0.38	0.23	0.30	4.93	42.9	12.3	2.95	5.78	2.57	1.08	63.3	0.206
I 8 × 6.18	8	5.00	0.35	0.23	0.30	5.26	59.7	14.9	3.37	7.30	2.92	1.18	107	0.188
I 8 × 7.02	8	5.00	0.41	0.25	0.30	5.97	67.8	16.9	3.37	8.55	3.42	1.20	123	0.286
I 9 × 8.36	9	5.50	0.44	0.27	0.30	7.11	102	22.7	3.79	12.2	4.44	1.31	224	0.386
I 10 × 8.65	10	6.00	0.41	0.25	0.40	7.35	132	26.4	4.24	14.8	4.93	1.42	340	0.360
I 10 × 10.3	10	6.00	0.50	0.29	0.40	8.75	156	31.2	4.22	18.0	6.01	1.44	407	0.620
I 12 × 11.7	12	7.00	0.47	0.29	0.40	9.92	256	42.6	5.07	26.9	7.69	1.65	894	0.621
I 12 × 14.3	12	7.00	0.62	0.31	0.40	12.2	317	52.9	5.11	35.5	10.1	1.71	1149	1.26

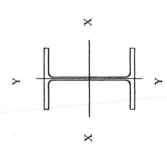

Aluminum Association Standard Channels

Designation (Depth × Weight) (in. × lb/ft)	Depth d (in.)	Width b_f (in.)	Flange Thickness t_f (in.)	Web Thickness t_w (in.)	Fillet Radius R (in.)	Area A (in.²)	Axis X-X			Axis Y-Y				Warping Constant C_w (in.⁶)	Torsion Constant J (in.⁴)
							I_x (in.⁴)	S_x (in.³)	r_x (in.)	I_y (in.⁴)	S_y (in.³)	r_y (in.)	x (in.)		
CS 2 × 0.577	2	1.00	0.13	0.13	0.10	0.490	0.288	0.288	0.766	0.0450	0.0639	0.303	0.296	0.0324	0.003
CS 2 × 1.07	2	1.25	0.26	0.17	0.15	0.911	0.546	0.546	0.774	0.139	0.178	0.390	0.471	0.0894	0.017
CS 3 × 1.14	3	1.50	0.20	0.13	0.25	0.965	1.41	0.940	1.21	0.217	0.215	0.474	0.494	0.332	0.010
CS 3 × 1.60	3	1.75	0.26	0.17	0.25	1.36	1.97	1.31	1.20	0.417	0.368	0.554	0.617	0.626	0.025
CS 4 × 1.74	4	2.00	0.23	0.15	0.25	1.48	3.91	1.95	1.63	0.601	0.446	0.638	0.653	1.65	0.020
CS 4 × 2.33	4	2.25	0.29	0.19	0.25	1.98	5.21	2.60	1.62	1.02	0.692	0.717	0.775	2.76	0.044
CS 5 × 2.21	5	2.25	0.26	0.15	0.30	1.88	7.88	3.15	2.05	0.975	0.642	0.720	0.731	4.17	0.031
CS 5 × 3.09	5	2.75	0.32	0.19	0.30	2.63	11.1	4.45	2.06	2.05	1.14	0.884	0.955	8.70	0.070
CS 6 × 2.83	6	2.50	0.29	0.17	0.30	2.41	14.4	4.78	2.44	1.53	0.896	0.798	0.788	9.52	0.050
CS 6 × 4.03	6	3.25	0.35	0.21	0.30	3.43	21.0	7.01	2.48	3.76	1.76	1.05	1.12	23.1	0.109
CS 7 × 3.21	7	2.75	0.29	0.17	0.30	2.73	22.1	6.31	2.85	2.10	1.10	0.878	0.842	17.8	0.055
CS 7 × 4.72	7	3.50	0.38	0.21	0.30	4.01	33.8	9.65	2.90	5.13	2.23	1.13	1.20	43.0	0.147
CS 8 × 4.15	8	3.00	0.35	0.19	0.30	3.53	37.4	9.35	3.26	3.25	1.57	0.959	0.934	36.0	0.102
CS 8 × 5.79	8	3.75	0.41	0.25	0.35	4.92	52.7	13.2	3.27	7.12	2.82	1.20	1.22	78.5	0.210
CS 9 × 4.98	9	3.25	0.35	0.23	0.35	4.24	54.4	12.1	3.58	4.40	1.89	1.02	0.928	62.8	0.127
CS 9 × 6.97	9	4.00	0.44	0.29	0.35	5.93	78.3	17.4	3.63	9.60	3.49	1.27	1.25	135	0.293
CS 10 × 6.14	10	3.50	0.41	0.25	0.35	5.22	83.2	16.6	3.99	6.33	2.55	1.10	1.02	111	0.209
CS 10 × 8.36	10	4.25	0.50	0.31	0.40	7.11	116	23.2	4.04	13.0	4.46	1.35	1.34	226	0.444
CS 12 × 8.27	12	4.00	0.47	0.29	0.40	7.04	160	26.6	4.77	11.0	3.85	1.25	1.14	281	0.367
CS 12 × 11.8	12	5.00	0.62	0.35	0.45	10.1	240	39.9	4.88	25.7	7.59	1.60	1.61	639	0.948

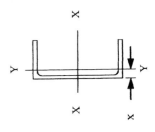

APPENDIX C
Minimum Mechanical Properties of Aluminum Alloys

Minimum Mechanical Properties of Aluminum Alloys (Unwelded) (U.S. Units)

Alloy	Temper	Product	Thickness Range from (in.)	to (in.)	Tension F_u (ksi)	F_{ty} (ksi)	Compression F_{cy} (ksi)	Shear F_{su} (ksi)	F_{sy} (ksi)	Compressive Modulus of Elasticity E (ksi)
1100	H12	Sheet and Plate	All		14	11	10	9	6.5	10,100
1100	H14	Drawn Tube	All		16	14	13	10	8	10,100
2014	T6	Sheet	0.040	0.249	66	58	59	40	33	10,900
2014	T651	Plate	0.250	2.000	67	59	58	40	34	10,900
2014	T6, T6510, T6511	Extrusions	All		60	53	52	35	31	10,900
2014	T6, T651	Cold-Finished Rod and Bar, Drawn Tube	All		65	55	53	38	32	10,900
Alclad 2014	T6	Sheet	0.025	0.039	63	55	56	38	32	10,800
Alclad 2014	T6	Sheet	0.040	0.249	64	57	58	39	33	10,800
Alclad 2014	T651	Plate	0.250	0.499	64	57	56	39	33	10,800
3003	H12	Sheet and Plate	0.017	2.000	17	12	10	11	7	10,100
3003	H14	Sheet and Plate	0.009	1.000	20	17	14	12	10	10,100
3003	H16	Sheet	0.006	0.162	24	21	18	14	12	10,100
3003	H18	Sheet	0.006	0.128	27	24	20	15	14	10,100
3003	H12	Drawn Tube	All		17	12	11	11	7	10,100
3003	H14	Drawn Tube	All		20	17	16	12	10	10,100
3003	H16	Drawn Tube	All		24	21	19	14	12	10,100
3003	H18	Drawn Tube	All		27	24	21	15	14	10,100
Alclad 3003	H12	Sheet and Plate	0.017	2.000	16	11	9	10	6.5	10,100
Alclad 3003	H14	Sheet and Plate	0.009	1.000	19	16	13	12	9	10,100
Alclad 3003	H16	Sheet	0.006	0.162	23	20	17	14	12	10,100
Alclad 3003	H18	Sheet	0.006	0.128	26	23	19	15	13	10,100
Alclad 3003	H14	Drawn Tube	0.025	0.259	19	16	15	12	9	10,100
Alclad 3003	H18	Drawn Tube	0.010	0.500	26	23	20	15	13	10,100
3004	H32	Sheet and Plate	0.017	2.000	28	21	18	17	12	10,100

Alloy	Temper		Product Form								Modulus
3004	H34		Sheet and Plate	0.009	1.000	32	25	22	19	14	10,100
3004	H36		Sheet	0.006	0.162	35	28	25	20	16	10,100
3004	H38		Sheet	0.006	0.128	38	31	29	21	18	10,100
3004	H34		Drawn Tube	0.018	0.450	32	25	24	19	14	10,100
3004	H36		Drawn Tube	0.018	0.450	35	28	27	20	16	10,100
Alclad 3004	H32		Sheet	0.017	0.249	27	20	17	16	12	10,100
Alclad 3004	H34		Sheet	0.009	0.249	31	24	21	18	14	10,100
Alclad 3004	H36		Sheet	0.006	0.162	34	27	24	19	16	10,100
Alclad 3004	H38		Sheet	0.006	0.128	37	30	28	21	17	10,100
Alclad 3004	H131, H241, H341		Sheet	0.024	0.050	31	26	22	18	15	10,100
Alclad 3004	H151, H261, H361		Sheet	0.024	0.050	34	30	28	19	17	10,100
3005	H25		Sheet	0.013	0.050	26	22	20	15	13	10,100
3005	H28		Sheet	0.006	0.080	31	27	25	17	16	10,100
3105	H25		Sheet	0.013	0.080	23	19	17	14	11	10,100
5005	H12		Sheet and Plate	0.017	2.000	18	14	13	11	8	10,100
5005	H14		Sheet and Plate	0.009	1.000	21	17	15	12	10	10,100
5005	H16		Sheet	0.006	0.162	24	20	18	14	12	10,100
5005	H32		Sheet and Plate	0.017	2.000	17	12	11	11	7	10,100
5005	H34		Sheet and Plate	0.009	1.000	20	15	14	12	8.5	10,100
5005	H36		Sheet	0.006	0.162	23	18	16	13	10	10,100
5050	H32		Sheet	0.017	0.249	22	16	14	14	9	10,100
5050	H34		Sheet	0.009	0.249	25	20	18	15	12	10,100
5050	H32		Drawn Tube	All		22	16	15	13	9	10,100
5050	H34		Drawn Tube	All		25	20	19	15	12	10,100
5052	O		Sheet and Plate	0.006	3.000	25	9.5	9.5	16	5.5	10,200
5052	H32		Sheet and Plate	All		31	23	21	19	13	10,200
5052	H34		Drawn Tube	All		34	26	24	20	15	10,200

Minimum Mechanical Properties of Aluminum Alloys (Unwelded) (U.S. Units) (continued)

Alloy	Temper	Product	Thickness Range from (in.)	to (in.)	Tension F_{tu} (ksi)	F_{ty} (ksi)	Compression F_{cy} (ksi)	Shear F_{su} (ksi)	F_{sy} (ksi)	Compressive Modulus of Elasticity E (ksi)
5052	H36	Sheet	0.006	0.162	37	29	26	22	17	10,200
5083	O	Extrusions		5.000	39	16	16	24	9	10,400
5083	H111	Extrusions		0.500	40	24	21	24	14	10,400
5083	H111	Extrusions	0.501	5.000	40	24	21	23	14	10,400
5083	O	Sheet and Plate	0.051	1.500	40	18	18	25	10	10,400
5083	H116	Sheet and Plate	0.188	1.500	44	31	26	26	18	10,400
5083	H321	Sheet and Plate	0.188	1.500	44	31	26	26	18	10,400
5083	H116	Plate	1.501	3.000	41	29	24	24	17	10,400
5083	H321	Plate	1.501	3.000	41	29	24	24	17	10,400
5086	O	Extrusions		5.000	35	14	14	21	8	10,400
5086	H111	Extrusions		0.500	36	21	18	21	12	10,400
5086	H111	Extrusions	0.501	5.000	36	21	18	21	12	10,400
5086	O	Sheet and Plate	0.020	2.000	35	14	14	21	8	10,400
5086	H112	Plate	0.250	0.499	36	18	17	22	10	10,400
5086	H112	Plate	0.500	1.000	35	16	16	21	9	10,400
5086	H112	Plate	1.001	2.000	35	14	15	21	8	10,400
5086	H112	Plate	2.001	3.000	34	14	15	21	8	10,400
5086	H116	Sheet and Plate	All		40	28	26	24	16	10,400
5086	H32	Sheet and Plate	All		40	28	26	24	16	10,400
5086	H34	Drawn Tube Sheet and Plate	All		44	34	32	26	20	10,400
5154	H38	Drawn Tube Sheet	0.006	0.128	45	35	33	24	20	10,300
5454	O	Extrusions		5.000	31	12	12	19	7	10,400
5454	H111	Extrusions		0.500	33	19	16	20	11	10,400

Alloy	Temper	Product	Thickness (from)	Thickness (to)						Modulus
5454	H111	Extrusions	0.501	5.000	33	19	16	19	11	10,400
5454	H112	Extrusions		5.000	31	12	13	19	7	10,400
5454	O	Sheet and Plate	0.020	3.000	31	12	12	19	7	10,400
5454	H32	Sheet and Plate	0.020	2.000	36	26	24	21	15	10,400
5454	H34	Sheet and Plate	0.020	1.000	39	29	27	23	17	10,400
5456	O	Sheet and Plate	0.051	1.500	42	19	19	26	11	10,400
5456	H116	Sheet and Plate	0.188	1.250	46	33	27	27	19	10,400
5456	H321	Sheet and Plate	0.188	1.250	46	33	27	27	19	10,400
5456	H116	Plate	1.251	1.500	44	31	25	25	18	10,400
5456	H321	Plate	1.251	1.500	44	31	25	25	18	10,400
5456	H116	Plate	1.501	3.000	41	29	25	25	17	10,400
5456	H321	Plate	1.501	3.000	41	29	25	25	17	10,400
6005	T5	Extrusions		1.000	38	35	35	24	20	10,100
6105	T5	Extrusions		0.500	38	35	35	24	20	10,100
6061	T6, T651	Sheet and Plate	0.010	4.000	42	35	35	27	20	10,100
6061	T6, T6510, T6511	Extrusions	All		38	35	35	24	20	10,100
6061	T6, T651	Cold-Finished Rod and Bar		8.000	42	35	35	25	20	10,100
6061	T6	Drawn Tube	0.025	0.500	42	35	35	27	20	10,100
6061	T6	Pipe	All		38	35	35	24	20	10,100
6063	T5	Extrusions		0.500	22	16	16	13	9	10,100
6063	T5	Extrusions	0.500	1.000	21	15	15	12	8.5	10,100
6063	T52	Extrusions		1.000	22	16	16	13		10,100
6063	T6	Extrusions and Pipe	All		30	25	25	19	14	10,100
6066	T6, T6510, T6511	Extrusions	All		50	45	45	27	26	10,100
6070	T6, T62	Extrusions		2.999	48	45	45	29	26	10,100
6351	T5	Extrusions		1.000	38	35	35	24	20	10,100
6351	T6	Extrusions		0.750	42	37	37	27		10,100
6463	T6	Extrusions		0.500	30	25	25	19	14	10,100
7005	T53	Extrusions		0.750	50	44	43	28		10,500

Minimum Mechanical Properties of Aluminum Alloys (Welded) (U.S. Units)

Alloy	Temper	Product	Thickness Range from (in.)	to (in.)	Tension F_{tu} (ksi)	F_{ty} (ksi)	Compression F_{cy} (ksi)	Shear F_{su} (ksi)
1100	H12, H14	All			11	3.5	3.5	8
3003	H12, H14, H16, H18	All			14	5	5	10
Alclad 3003	H12, H14, H16, H18	All			13	4.5	4.5	10
3004	H32, H34, H36, H38	All			22	8.5	8.5	14
Alclad 3004	H32, H34, H36, H38	All			21	8	8	13
3005	H25	Sheet			17	6.5	6.5	12
5005	H12, H14, H32, H34	All			15	5	5	9
5050	H32, H34	All			18	6	6	12
5052	H32, H34, O	All			25	9.5	9.5	16
5083	H111, O	Extrusions			39	16	15	23
5083	H116, H321, O	Sheet and Plate	0.188	1.500	40	18	18	24
5083	H116, H321, O	Plate	1.501	3.000	39	17	17	24
5086	H111, O	Extrusions			35	14	13	21
5086	H112, H32, H34, H116, O	Sheet and Plate			35	14	14	21
5154	H38	Sheet			30	11	11	19
5454	H111, O	Extrusions			31	12	11	19
5454	H112	Extrusions			31	12	12	19
5454	H32, H34, O	Sheet and Plate			31	12	12	19
5456	H116, H321, O	Sheet and Plate	0.188	1.500	42	19	18	25
5456	H116, H321, O	Plate	1.501	3.000	41	18	17	25
6005	T5	Extrusions		0.250	24	13	13	15
6061	T6, T651, T6510, T6511(2)	All	0.375		24	15	15	15
6061	T6, T651, T6510, T6511(3)	All			24	11	11	15
6063	T5, T52, T6	All			17	8	8	11
6351	T5, T6(2)	Extrusions	0.375		24	15	15	15
6351	T5, T6(3)	Extrusions			24	11	11	15
6463	T6	Extrusions	0.125	0.500	17	8	8	11
7005	T53	Extrusions		0.750	40	24	24	22

Notes

(1) Yield strengths ared based on a 2 in. gauge length.

(2) Values when welded with 5183, 5356, or 5556 filler, regardless of thickness, and to thicknesses ≤ 0.375 in. when welded with 4043, 5554, or 5654 filler.

(3) Values when welded with 4043, 5554, or 5654 filler.

Minimum Mechanical Properties of Aluminum Alloys (Unwelded) (SI Units)

Alloy	Temper	Product	Thickness Range from (mm)	to (mm)	Tension F_{tu} (MPa)	F_{ty} (MPa)	Compression F_{cy} (MPa)	Shear F_{su} (MPa)	F_{sy} (MPa)	Compressive Modulus of Elasticity E (MPa)
1100	H12	Sheet and Plate	All		95	75	70	62	45	69,600
1100	H14	Drawn Tube	All		110	95	90	70	55	69,600
2014	T6	Sheet	1.00	6.30	455	400	405	275	230	75,200
2014	T651	Plate	6.30	50.00	460	405	400	275	235	75,200
2014	T6, T6510, T6511	Extrusions	All		415	365	360	240	215	75,200
2014	T6, T651	Cold-Finished Rod and Bar, Drawn Tube	All		450	380	365	260	220	75,200
Alclad 2014	T6	Sheet	0.63	1.00	435	380	385	260	220	74,500
Alclad 2014	T6	Sheet	1.00	6.30	440	395	400	270	230	74,500
Alclad 2014	T651	Plate	6.30	12.50	440	395	385	270	230	74,500
3003	H12	Sheet and Plate	0.40	50.00	120	85	70	75	48	69,600
3003	H14	Sheet and Plate	0.20	25.00	140	115	95	85	70	69,600
3003	H16	Sheet	0.15	4.00	165	145	125	95	85	69,600
3003	H18	Sheet	0.15	3.20	185	165	140	105	95	69,600
3003	H12	Drawn Tube	All		120	85	75	75	48	69,600
3003	H14	Drawn Tube	All		140	115	110	85	70	69,600
3003	H16	Drawn Tube	All		165	145	130	95	85	69,600
3003	H18	Drawn Tube	All		185	165	145	105	95	69,600
Alclad 3003	H12	Sheet and Plate	0.40	50.00	115	80	62	70	45	69,600
Alclad 3003	H14	Sheet and Plate	0.20	25.00	135	110	90	85	62	69,600
Alclad 3003	H16	Sheet	0.15	4.00	160	140	115	95	85	69,600
Alclad 3003	H18	Sheet	0.15	3.20	180	160	130	105	90	69,600
Alclad 3003	H14	Drawn Tube	0.63	6.30	135	110	105	85	62	69,600
Alclad 3003	H18	Drawn Tube	0.25	12.50	180	160	140	105	90	69,600
3004	H32	Sheet and Plate	0.40	50.00	190	145	125	115	85	69,600

Minimum Mechanical Properties of Aluminum Alloys (Unwelded) (SI Units) (continued)

Alloy	Temper	Product	Thickness Range from (mm)	to (mm)	Tension F_{tu} (MPa)	F_{ty} (MPa)	Compression F_{cy} (MPa)	Shear F_{su} (MPa)	F_{sy} (MPa)	Compressive Modulus of Elasticity E (MPa)
3004	H34	Sheet and Plate	0.20	25.00	220	170	150	130	95	69,600
3004	H36	Sheet	0.15	4.00	240	190	170	140	110	69,600
3004	H38	Sheet	0.15	3.20	260	215	200	145	125	69,600
3004	H34	Drawn Tube	0.45	11.50	220	170	165	130	95	69,600
3004	H36	Drawn Tube	0.45	11.50	240	190	185	140	110	69,600
Alclad 3004	H32	Sheet	0.40	6.30	185	140	115	110	85	69,600
Alclad 3004	H34	Sheet	0.20	6.30	215	165	145	125	95	69,600
Alclad 3004	H36	Sheet	0.15	4.00	235	185	165	130	110	69,600
Alclad 3004	H38	Sheet	0.15	3.20	255	205	195	145	115	69,600
Alclad 3004	H131, H241, H341	Sheet	0.60	1.20	215	180	150	125	105	69,600
Alclad 3004	H151, H261, H361	Sheet	0.60	1.20	235	205	195	130	115	69,600
3005	H25	Sheet	0.32	1.20	180	150	140	105	90	69,600
3005	H28	Sheet	0.15	2.00	215	185	170	115	110	69,600
3105	H25	Sheet	0.32	2.00	160	130	115	95	75	69,600
5005	H12	Sheet and Plate	0.40	50.00	125	95	90	75	55	69,600
5005	H14	Sheet and Plate	0.20	25.00	145	115	105	85	70	69,600
5005	H16	Sheet	0.15	4.00	165	135	125	95	85	69,600
5005	H32	Sheet and Plate	0.40	50.00	120	85	75	75	48	69,600
5005	H34	Sheet and Plate	0.20	25.00	140	105	95	85	59	69,600
5005	H36	Sheet	0.15	4.00	160	125	110	90	75	69,600
5050	H32	Sheet	0.40	6.30	150	110	95	95	62	69,600
5050	H34	Sheet	0.20	6.30	170	140	125	105	85	69,600
5050	H32	Drawn Tube	All		150	110	105	90	62	69,600

Alloy	Temper	Product	Thickness (mm)						E (MPa)
5050	H34	Drawn Tube	All	170	140	130	105	85	69,600
5052	O	Sheet and Plate	0.15–80.00	170	65	66	110	38	70,300
5052	H32	Sheet and Plate	All	215	160	145	130	90	70,300
5052	H34	Drawn Tube	All	235	180	165	140	105	70,300
5052	H36	Sheet	0.15–4.00	255	200	180	150	115	70,300
5083	O	Extrusions	130.00	270	110	110	165	62	71,700
5083	H111	Extrusions	12.70	275	165	145	165	95	71,700
5083	H111	Extrusions	130.00	275	165	145	160	95	71,700
5083	O	Sheet and Plate	1.20–6.30	275	125	125	170	70	71,700
5083	H116	Sheet and Plate	4.00–40.00	305	215	180	180	125	71,700
5083	H321	Sheet and Plate	4.00–40.00	305	215	180	180	125	71,700
5083	H116	Plate	40.00–80.00	285	200	165	165	115	71,700
5083	H321	Plate	40.00–80.00	285	200	165	165	115	71,700
5086	O	Extrusions	130.00	240	95	95	145	55	71,700
5086	H111	Extrusions	130.00	250	145	125	145	85	71,700
5086	O	Sheet and Plate	0.50–50.00	240	95	95	145	55	71,700
5086	H112	Sheet and Plate	4.00–12.50	250	125	115	150	70	71,700
5086	H112	Plate	12.50–40.00	240	105	110	145	62	71,700
5086	H112	Plate	40.00–80.00	235	95	105	145	55	71,700
5086	H116	Sheet and Plate	1.60–50.00	275	195	180	165	110	71,700
5086	H32	Sheet and Plate	1.60–50.00	275	195	180	165	110	71,700
5086		Drawn Tube	All						71,700
5086	H34	Sheet and Plate	All	300	235	220	180	140	71,700
		Drawn Tube							71,700
5154	H38	Sheet	0.15–3.20	310	240	230	165	140	71,000
5454	O	Extrusions	130.00	215	85	85	130	48	71,700
5454	H111	Extrusions	12.70	230	130	110	140	75	71,700

Minimum Mechanical Properties of Aluminum Alloys (Unwelded) (SI Units) (continued)

Alloy	Temper	Product	Thickness Range from (mm)	to (mm)	Tension F_{tu} (MPa)	F_{ty} (MPa)	Compression F_{cy} (MPa)	Shear F_{su} (MPa)	F_{sy} (MPa)	Compressive Modulus of Elasticity E (MPa)
5454	H111	Extrusions	12.70	130.00	230	130	110	130	75	71,700
5454	H112	Extrusions	12.70	130.00	215	85	90	130	48	71,700
5454	O	Sheet and Plate	0.50	80.00	215	85	85	130	48	71,700
5454	H32	Sheet and Plate	0.50	50.00	250	180	165	145	105	71,700
5454	H34	Sheet and Plate	0.50	25.00	270	200	185	160	115	71,700
5456	O	Sheet and Plate	1.20	6.30	290	130	130	180	75	71,700
5456	H116	Sheet and Plate	4.00	12.50	315	230	185	185	130	71,700
5456	H321	Sheet and Plate	4.00	12.50	315	230	185	185	130	71,700
5456	H116, H321	Plate	12.50	40.00	305	215	170	170	125	71,700
5456	H116, H321	Plate	40.00	80.00	285	200	170	170	115	71,700
6005	T5	Extrusions	0.25	25.00	260	240	240	165	140	69,600
6105	T5	Extrusions		12.50	260	240	240	165	140	69,600
6061	T6, T651	Sheet and Plate	0.25	100.00	290	240	240	185	140	69,600
6061	T6, T6510, T6511	Extrusions	All		260	240	240	165	140	69,600
6061	T6, T651	Cold-Finished Rod and Bar		200	290	240	240	170	140	69,600
6061	T6	Drawn Tube	0.63	12.50	290	240	240	185	140	69,600
6061	T6	Pipe	All		260	240	240	165	140	69,600
6063	T5	Extrusions		12.50	150	110	110	90	62	69,600
6063	T5	Extrusions	12.50	25.00	145	105	105	85	59	69,600
6063	T52	Extrusions		25.00	150	110	110	90		69,600
6063	T6	Extrusions and Pipe	All		205	170	170	130	95	69,600
6066	T6, T6510, T6511	Extrusions	All		345	310	310	185	180	69,600
6070	T6, T62	Extrusions		80.00	330	310	310	200	180	69,600
6351	T5	Extrusions		25.00	260	240	240	165	140	69,600
6351	T6	Extrusions		20.00	290	255	255	185		69,600
6463	T6	Extrusions		12.50	205	170	170	130	95	69,600
7005	T53	Extrusions		20.00	345	305	295	195		72,400

Minimum Mechanical Properties of Aluminum Alloys (Welded) (SI Units)

Alloy	Temper	Product	Thickness Range from (mm)	Thickness Range to (mm)	Tension F_{tu} (MPa)	Tension F_{ty} (MPa)	Compression F_{cy} (MPa)	Shear F_{su} (MPa)
1100	H12, H14	All			75	25	25	55
3003	H12, H14, H16, H18	All			95	35	35	70
Alclad 3003	H12, H14, H16, H18	All			90	30	30	70
3004	H32, H34, H36, H38	All			150	60	60	95
Alclad 3004	H32, H34, H36, H38	All			145	55	55	90
3005	H25	Sheet			115	45	45	85
5005	H12, H14, H32, H34	All			105	35	35	62
5050	H32, H34	All			125	40	40	85
5052	H32, H34, O	All			170	65	65	110
5083	H111, O	Extrusions			270	110	110	160
5083	H116, H321, O	Sheet and Plate	6.30	80.00	270	115	115	165
5086	H111, O	Extrusions			240	95	85	145
5086	H112, H32, H34, H116, O	Sheet and Plate	6.30	50.00	240	95	95	145
5154	H38	Sheet			205	75	75	130
5454	H111, O	Extrusions			215	85	80	130
5454	H112	Extrusions			215	85	85	130
5454	H32, H34, O	Sheet and Plate			215	85	85	130
5456	H116, H321, O	Sheet and Plate	6.30	38.00	285	125	115	170
5456	H116, H321, O	Plate	38.00	80.00	285	125	120	170
6005	T5	Extrusions		12.50	165	90	90	105
6061	T6, T651, T6510, T6511(2)	All	9.50		165	105	105	105
6061	T6, T651, T6510, T6511(3)	All			165	80	80	105
6063	T5, T52, T6	All			115	55	55	75
6351	T5, T6(2)	Extrusions	9.50		165	105	105	105
6351	T5, T6(3)	Extrusions			165	80	80	105
6463	T6	Extrusions	3.20	12.50	115	55	55	75
7005	T53	Extrusions		20.00	275	165	165	155

Notes

(1) Yield strengths are based on a 50 mm gauge length.

(2) Values when welded with 5183, 5356, or 5556 filler, regardless of thickness, and to thicknesses $\leq$ 9.5 mm when welded with 4043, 5554, or 5654 filler.

(3) Values when welded with 4043, 5554, or 5654 filler.

APPENDIX D
Allowable Stresses for Elements of Common Aluminum Shapes

ALUMINUM ASSOCIATION STANDARD I-BEAMS (6061-T6)

Allowable Stresses for Elements of Columns

Designation (Depth × Weight) (in. × lb/ft)	Web (3.4.9)			Flange (3.4.8)			Weighted Average F_{ca} (4.7.2a) (ksi)
	b/t	$>S_2 = 33$?	F_{cw} (ksi)	b/t	$>S_2 = 10$?	F_{cf} (ksi)	
I 3 × 1.64	16.2	No	19.1	4.68	No	19.4	19.3
I 3 × 2.03	13.2	No	19.8	3.56	No	20.3	20.2
I 4 × 2.31	20.3	No	18.0	5.11	No	19.1	18.8
I 4 × 2.79	17.2	No	18.8	4.02	No	19.9	19.6
I 5 × 3.70	19.8	No	18.2	4.23	No	19.8	19.3
I 6 × 4.03	25.4	No	16.8	5.53	No	18.7	18.1
I 6 × 4.69	22.4	No	17.5	4.56	No	19.5	18.9
I 7 × 5.80	24.5	No	17.0	4.83	No	19.3	18.6
I 8 × 6.18	29.1	No	15.8	5.96	No	18.4	17.6
I 8 × 7.02	26.3	No	16.5	5.06	No	19.1	18.3
I 9 × 8.36	27.9	No	16.1	5.26	No	18.9	18.1
I 10 × 8.65	33.5	Yes	14.6	6.04	No	18.3	17.2
I 10 × 10.3	28.3	No	16.0	4.91	No	19.2	18.3
I 12 × 11.7	35.4	Yes	13.8	6.29	No	18.1	16.7
I 12 × 14.3	32.1	No	15.1	4.75	No	19.3	18.2

Allowable Stresses for Elements of Beams (Major Axis Bending)

Designation (Depth × Weight) (in. × lb/ft)	Web (3.4.18)			Flange (3.4.15)			Weighted Average F_{bxa} (4.7.2b) (ksi)
	b/t	$>S_2 = 75$?	F_{bxw} (ksi)	b/t	$>S_2 = 10$?	F_{bxf} (ksi)	
I 3 × 1.64	16.2	No	28.0	4.68	No	21.0	21.4
I 3 × 2.03	13.2	No	28.0	3.56	No	21.0	21.3
I 4 × 2.31	20.3	No	28.0	5.11	No	21.0	21.4
I 4 × 2.79	17.2	No	28.0	4.02	No	21.0	21.4
I 5 × 3.70	19.8	No	28.0	4.23	No	21.0	21.4
I 6 × 4.03	25.4	No	28.0	5.53	No	21.0	21.5
I 6 × 4.69	22.4	No	28.0	4.56	No	21.0	21.4
I 7 × 5.80	24.5	No	28.0	4.83	No	21.0	21.5
I 8 × 6.18	29.1	No	28.0	5.96	No	21.0	21.5
I 8 × 7.02	26.3	No	28.0	5.06	No	21.0	21.5
I 9 × 8.36	27.9	No	28.0	5.26	No	21.0	21.5
I 10 × 8.65	33.5	No	28.0	6.04	No	21.0	21.5
I 10 × 10.3	28.3	No	28.0	4.91	No	21.0	21.5
I 12 × 11.7	35.4	No	28.0	6.29	No	21.0	21.5
I 12 × 14.3	32.1	No	28.0	4.75	No	21.0	21.4

Note
  Specification for Aluminum Structures section numbers shown in parentheses.

ALUMINUM ASSOCIATION STANDARD CHANNELS (6061-T6)

Allowable Stresses for Elements of Columns

Designation (Depth × Weight) (in. × lb/ft)	Web (3.4.9)			Flange (3.4.8)			Weighted Average F_{ca} (4.7.2a) (ksi)
	b/t	$>S_2 = 33?$	F_{cw} (ksi)	b/t	$>S_2 = 10?$	F_{cf} (ksi)	
CS 2 × 0.577	11.8	No	20.1	5.92	No	18.4	19.2
CS 2 × 1.07	6.94	No	21.0	3.58	No	20.3	20.5
CS 3 × 1.14	16.2	No	19.1	5.60	No	18.7	18.8
CS 3 × 1.60	11.6	No	20.2	5.12	No	19.1	19.4
CS 4 × 1.74	20.3	No	18.0	6.96	No	17.6	17.8
CS 4 × 2.33	15.4	No	19.3	6.24	No	18.2	18.5
CS 5 × 2.21	25.9	No	16.6	6.92	No	17.6	17.3
CS 5 × 3.09	19.8	No	18.2	7.06	No	17.5	17.7
CS 6 × 2.83	28.4	No	16.0	7.00	No	17.6	17.0
CS 6 × 4.03	22.4	No	17.5	7.83	No	16.9	17.1
CS 7 × 3.21	34.2	Yes	14.3	7.86	No	16.9	15.8
CS 7 × 4.72	26.9	No	16.4	7.87	No	16.9	16.7
CS 8 × 4.15	35.3	Yes	13.9	7.17	No	17.4	16.0
CS 8 × 5.79	25.9	No	16.6	7.68	No	17.0	16.9
CS 9 × 4.98	33.0	Yes	14.8	7.63	No	17.1	16.0
CS 9 × 6.97	25.6	No	16.7	7.64	No	17.1	16.9
CS 10 × 6.14	33.9	Yes	14.4	7.07	No	17.5	16.1
CS 10 × 8.36	26.5	No	16.5	7.08	No	17.5	17.1
CS 12 × 8.27	35.4	Yes	13.8	7.04	No	17.5	15.8
CS 12 × 11.8	28.2	No	16.1	6.77	No	17.7	17.1

Note

   Allowable stresses for elements of Aluminum Association Standard Channels are the same whether the buckling axis is the major or minor axis.

Allowable Stresses for Elements of Beams (Major Axis Bending)

Designation (Depth × Weight) (in. × lb/ft)	Web (3.4.18)			Flange (3.4.15)			Weighted Average F_{bxa} (4.7.2b) (ksi)
	b/t	$>S_2 = 75$?	F_{bxw} (ksi)	b/t	$>S_2 = 10$?	F_{bxf} (ksi)	
CS 2 × 0.577	11.8	No	28.0	5.92	No	21.0	21.9
CS 2 × 1.07	6.94	No	28.0	3.58	No	21.0	21.4
CS 3 × 1.14	16.2	No	28.0	5.60	No	21.0	21.6
CS 3 × 1.60	11.6	No	28.0	5.12	No	21.0	21.5
CS 4 × 1.74	20.3	No	28.0	6.96	No	20.8	21.5
CS 4 × 2.33	15.4	No	28.0	6.24	No	21.0	21.5
CS 5 × 2.21	25.9	No	28.0	6.92	No	20.9	21.5
CS 5 × 3.09	19.8	No	28.0	7.06	No	20.7	21.3
CS 6 × 2.83	28.4	No	28.0	7.00	No	20.8	21.5
CS 6 × 4.03	22.4	No	28.0	7.83	No	20.0	20.6
CS 7 × 3.21	34.2	No	28.0	7.86	No	20.0	20.8
CS 7 × 4.72	26.9	No	28.0	7.87	No	20.0	20.6
CS 8 × 4.15	35.3	No	28.0	7.17	No	20.6	21.4
CS 8 × 5.79	25.9	No	28.0	7.68	No	20.2	20.9
CS 9 × 4.98	33.0	No	28.0	7.63	No	20.2	21.2
CS 9 × 6.97	25.6	No	28.0	7.64	No	20.2	21.0
CS 10 × 6.14	33.9	No	28.0	7.07	No	20.7	21.6
CS 10 × 8.36	26.5	No	28.0	7.08	No	20.7	21.4
CS 12 × 8.27	35.4	No	28.0	7.04	No	20.8	21.7
CS 12 × 11.8	28.2	No	28.0	6.77	No	21.0	21.6

Note
Specification for Aluminum Structures section numbers shown in parentheses.

APPENDIX E
LRFD Design Stresses for Elements of Common Aluminum Shapes

ALUMINUM ASSOCIATION STANDARD I-BEAMS (6061-T6)

LRFD Design Stresses for Elements of Columns

Designation (Depth × Weight) (in. × lb/ft)	Web (3.4.9)			Flange (3.4.8)			Weighted Average F_{ca} (4.7.2a) (ksi)
	b/t	$>S_2 = 33?$	F_{cw} (ksi)	b/t	$>S_2 = 10?$	F_{cf} (ksi)	
I 3 × 1.64	16.2	No	31.6	4.68	No	32.2	32.0
I 3 × 2.03	13.2	No	32.9	3.56	No	33.3	33.2
I 4 × 2.31	20.3	No	30.0	5.11	No	31.6	31.2
I 4 × 2.79	17.2	No	31.2	4.02	No	33.0	32.6
I 5 × 3.70	19.8	No	30.2	4.23	No	32.7	32.0
I 6 × 4.03	25.4	No	27.9	5.53	No	31.1	30.1
I 6 × 4.69	22.4	No	29.1	4.56	No	32.3	31.4
I 7 × 5.80	24.5	No	28.2	4.83	No	32.0	30.9
I 8 × 6.18	29.1	No	26.3	5.96	No	30.5	29.2
I 8 × 7.02	26.3	No	27.5	5.06	No	31.7	30.4
I 9 × 8.36	27.9	No	26.9	5.26	No	31.4	30.0
I 10 × 8.65	33.5	Yes	24.2	6.04	No	30.4	28.4
I 10 × 10.3	28.3	No	26.7	4.91	No	31.9	30.3
I 12 × 11.7	35.4	Yes	22.9	6.29	No	30.1	27.7
I 12 × 14.3	32.1	No	25.1	4.75	No	32.1	30.1

LRFD Design Stresses for Elements of Beams (Major Axis Bending)

Designation (Depth × Weight) (in. × lb/ft)	Web (3.4.18)			Flange (3.4.15)			Weighted Average F_{bxa} (4.7.2b) (ksi)
	b/t	$>S_2 = 75?$	F_{bxw} (ksi)	b/t	$>S_2 = 10?$	F_{bxf} (ksi)	
I 3 × 1.64	16.2	No	43.2	4.68	No	32.2	32.8
I 3 × 2.03	13.2	No	43.2	3.56	No	33.3	33.7
I 4 × 2.31	20.3	No	43.2	5.11	No	31.6	32.3
I 4 × 2.79	17.2	No	43.2	4.02	No	33.0	33.6
I 5 × 3.70	19.8	No	43.2	4.23	No	32.7	33.4
I 6 × 4.03	25.4	No	43.2	5.53	No	31.1	31.9
I 6 × 4.69	22.4	No	43.2	4.56	No	32.3	33.0
I 7 × 5.80	24.5	No	43.2	4.83	No	32.0	32.7
I 8 × 6.18	29.1	No	43.2	5.96	No	30.5	31.4
I 8 × 7.02	26.3	No	43.2	5.06	No	31.7	32.5
I 9 × 8.36	27.9	No	43.2	5.26	No	31.4	32.2
I 10 × 8.65	33.5	No	43.2	6.04	No	30.4	31.3
I 10 × 10.3	28.3	No	43.2	4.91	No	31.9	32.6
I 12 × 11.7	35.4	No	43.2	6.29	No	30.1	31.1
I 12 × 14.3	32.1	No	43.2	4.75	No	32.1	32.7

Note

Specification for Aluminum Structures section numbers shown in parentheses.

ALUMINUM ASSOCIATION STANDARD CHANNELS (6061-T6)

LRFD Design Stresses for Elements of Columns

Designation (Depth × Weight) (in. × lb/ft)	Web (3.4.9)			Flange (3.4.8)			Weighted Average F_{ca} (4.7.2a) (ksi)
	b/t	$>S_2 = 33?$	F_{cw} (ksi)	b/t	$>S_2 = 10?$	F_{cf} (ksi)	
CS 2 × 0.577	11.8	No	33.3	5.92	No	30.6	31.8
CS 2 × 1.07	6.94	No	33.3	3.58	No	33.3	33.3
CS 3 × 1.14	16.2	No	31.6	5.60	No	31.0	31.2
CS 3 × 1.60	11.6	No	33.3	5.12	No	31.6	32.1
CS 4 × 1.74	20.3	No	30.0	6.96	No	29.2	29.5
CS 4 × 2.33	15.4	No	32.0	6.24	No	30.1	30.7
CS 5 × 2.21	25.9	No	27.7	6.92	No	29.3	28.7
CS 5 × 3.09	19.8	No	30.2	7.06	No	29.1	29.4
CS 6 × 2.83	28.4	No	26.7	7.00	No	29.2	28.2
CS 6 × 4.03	22.4	No	29.1	7.83	No	28.1	28.4
CS 7 × 3.21	34.2	Yes	23.7	7.86	No	28.0	26.3
CS 7 × 4.72	26.9	No	27.3	7.87	No	28.0	27.8
CS 8 × 4.15	35.3	Yes	23.0	7.17	No	28.9	26.6
CS 8 × 5.79	25.9	No	27.6	7.68	No	28.3	28.0
CS 9 × 4.98	33.0	Yes	24.5	7.63	No	28.3	26.6
CS 9 × 6.97	25.6	No	27.8	7.64	No	28.3	28.1
CS 10 × 6.14	33.9	Yes	23.9	7.07	No	29.1	26.8
CS 10 × 8.36	26.5	No	27.4	7.08	No	29.0	28.4
CS 12 × 8.27	35.4	Yes	22.9	7.04	No	29.1	26.2
CS 12 × 11.8	28.2	No	26.7	6.77	No	29.4	28.4

Note

LRFD design stresses for elements of Aluminum Association Standard Channels are the same whether the buckling axis is the major or minor axis.

LRFD Design Stresses for Elements of Beams (Major Axis Bending)

Designation (Depth × Weight) (in. × lb/ft)	Web (3.4.18)			Flange (3.4.15)			Weighted Average F_{bxa} (4.7.2b) (ksi)
	b/t	$>S_2 = 75?$	F_{bxw} (ksi)	b/t	$>S_2 = 10?$	F_{bxf} (ksi)	
CS 2 × 0.577	11.8	No	43.2	5.92	No	30.6	32.2
CS 2 × 1.07	6.94	No	43.2	3.58	No	33.3	33.9
CS 3 × 1.14	16.2	No	43.2	5.60	No	31.0	32.0
CS 3 × 1.60	11.6	No	43.2	5.12	No	31.6	32.4
CS 4 × 1.74	20.3	No	43.2	6.96	No	29.2	30.4
CS 4 × 2.33	15.4	No	43.2	6.24	No	30.1	31.1
CS 5 × 2.21	25.9	No	43.2	6.92	No	29.3	30.5
CS 5 × 3.09	19.8	No	43.2	7.06	No	29.1	30.1
CS 6 × 2.83	28.4	No	43.2	7.00	No	29.2	30.5
CS 6 × 4.03	22.4	No	43.2	7.83	No	28.1	29.2
CS 7 × 3.21	34.2	No	43.2	7.86	No	28.0	29.6
CS 7 × 4.72	26.9	No	43.2	7.87	No	28.0	29.2
CS 8 × 4.15	35.3	No	43.2	7.17	No	28.9	30.3
CS 8 × 5.79	25.9	No	43.2	7.68	No	28.3	29.6
CS 9 × 4.98	33.0	No	43.2	7.63	No	28.3	30.2
CS 9 × 6.97	25.6	No	43.2	7.64	No	28.3	29.8
CS 10 × 6.14	33.9	No	43.2	7.07	No	29.1	30.7
CS 10 × 8.36	26.5	No	43.2	7.08	No	29.0	30.4
CS 12 × 8.27	35.4	No	43.2	7.04	No	29.1	30.9
CS 12 × 11.8	28.2	No	43.2	6.77	No	29.4	30.7

Note

Specification for Aluminum Structures section numbers shown in parentheses.

APPENDIX F
Column Buckling Allowable Stress

For F_{cy} = 35 ksi [240MPa], E = 10,100 ksi [69,600 MPa] (6005-T5, 6061-T6, 6105-T5, 6351-T5)

kL/r	F_c (ksi)	F_c (MPa)	kL/r	F_c (ksi)	F_c (MPa)	kL/r	F_c (ksi)	F_c (MPa)	kL/r	F_c (ksi)	F_c (MPa)
1	20.1	138	41	15.0	104	81	7.8	54	121	3.5	24
2	19.9	138	42	14.9	103	82	7.6	52	122	3.4	24
3	19.8	137	43	14.8	102	83	7.4	51	123	3.4	23
4	19.7	136	44	14.7	101	84	7.2	50	124	3.3	23
5	19.6	135	45	14.5	100	85	7.1	49	125	3.3	23
6	19.4	134	46	14.4	99	86	6.9	48	126	3.2	22
7	19.3	133	47	14.3	98	87	6.8	47	127	3.2	22
8	19.2	132	48	14.2	98	88	6.6	45	128	3.1	22
9	19.1	131	49	14.0	97	89	6.5	44	129	3.1	21
10	18.9	131	50	13.9	96	90	6.3	43	130	3.0	21
11	18.8	130	51	13.8	95	91	6.2	43	131	3.0	21
12	18.7	129	52	13.6	94	92	6.0	42	132	2.9	20
13	18.6	128	53	13.5	93	93	5.9	41	133	2.9	20
14	18.4	127	54	13.4	92	94	5.8	40	134	2.8	20
15	18.3	126	55	13.3	91	95	5.7	39	135	2.8	19
16	18.2	125	56	13.1	91	96	5.5	38	136	2.8	19
17	18.1	125	57	13.0	90	97	5.4	37	137	2.7	19
18	17.9	124	58	12.9	89	98	5.3	37	138	2.7	19
19	17.8	123	59	12.8	88	99	5.2	36	139	2.6	18
20	17.7	122	60	12.6	87	100	5.1	35	140	2.6	18
21	17.6	121	61	12.5	86	101	5.0	35	141	2.6	18
22	17.4	120	62	12.4	85	102	4.9	34	142	2.5	17
23	17.3	119	63	12.3	85	103	4.8	33	143	2.5	17
24	17.2	118	64	12.1	84	104	4.7	33	144	2.5	17
25	17.0	118	65	12.0	83	105	4.6	32	145	2.4	17
26	16.9	117	66	11.7	81	106	4.5	31	146	2.4	17
27	16.8	116	67	11.4	78	107	4.5	31	147	2.4	16
28	16.7	115	68	11.1	76	108	4.4	30	148	2.3	16
29	16.5	114	69	10.7	74	109	4.3	30	149	2.3	16
30	16.4	113	70	10.4	72	110	4.2	29	150	2.3	16
31	16.3	112	71	10.1	70	111	4.1	29	151	2.2	15
32	16.2	111	72	9.9	68	112	4.1	28	152	2.2	15
33	16.0	111	73	9.6	66	113	4.0	28	153	2.2	15
34	15.9	110	74	9.3	64	114	3.9	27	154	2.2	15
35	15.8	109	75	9.1	63	115	3.9	27	155	2.1	15
36	15.7	108	76	8.8	61	116	3.8	26	156	2.1	14
37	15.5	107	77	8.6	59	117	3.7	26	157	2.1	14
38	15.4	106	78	8.4	58	118	3.7	25	158	2.0	14
39	15.3	105	79	8.2	56	119	3.6	25	159	2.0	14
40	15.2	105	80	8.0	55	120	3.5	24	160	2.0	14

Note
Local buckling must also be checked.

APPENDIX G
Summary of the Aluminum *Specification* Design Provisions for Columns and Beams

Columns

Cross Sections	Overall Buckling	Local Buckling		Weighted Average	Local/Overall Interaction
I-beams	3.4.7	Web 3.4.9	Flange 3.4.8	4.7.2a	4.7.4
Channels	3.4.7	Web 3.4.9	Flange 3.4.8 or 3.4.8.1	4.7.2a	4.7.4
Round or oval tubes	3.4.7	Wall 3.4.10			
Rectangular tubes	3.4.7	Walls 3.4.9		4.7.2a	4.7.4
Angles	3.4.7	Legs 3.4.8.1			

Beams

Cross Section	Overall Buckling	Local Buckling		Weighted Average	Local/ Overall Average	Deflection Calculation
Single web beams bent about strong axis	3.4.11	Web 3.4.18, 3.4.20	Flange 3.4.15	4.7.2b	4.7.5	4.7.6
Round or oval tubes	3.4.12	Wall 3.4.16.1, 4.2				
Rectangular bars	3.4.13	Web 3.4.20				
Rectangular tubes	3.4.14	Web 3.4.18, 3.4.20	Flange 3.4.16	4.7.2b	*	4.7.6

*The Aluminum *Specification* does not require this check, which would only govern for very thin walled tubes.

APPENDIX H
Cross Reference to the Aluminum *Specification*

Chapter 7 of this book tells you which sections of the *Specification for Aluminum Structures* to use to design a particular structural member (for example, a doubly symmetric I-beam). You may instead find yourself looking at a specific section in the Aluminum *Specification* and wondering what it's about. Below is a cross reference that locates discussion in this book with respect to sections in the Aluminum *Specification*.

Aluminum *Specification* Section Number	Title	Section Number in This Book
1.2	Materials	4.10
1.3	Safety Factors	6.3
2.1	Properties of Sections	5.4
2.3	Loads	11.3
3.2	Nomenclature	Glossary, 5.2.1, 5.2.2
3.3	Tables Relating to Mechanical Properties and Buckling Constants	4.2, 5.2.1, App C, App K
Table 3.4-2	Coefficient k_t	4.4, 5.5.1
Table 3.4-4	Allowable Stresses for Bridge Type Structures Casting Alloys	3.1.4
3.4.1	Tension, Axial	5.1, 8.1.13
3.4.2	Tension in Extreme Fibers of Beams— Structural Shapes Bent About Strong Axis, Rectangular Tubes	5.3.1
3.4.3	Tension in Extreme Fibers of Beams— Round or Oval Tubes	5.3.1
3.4.4	Tension in Extreme Fibers of Beams— Shapes Bent About Weak Axis, Rectangular Bars, Solid Round Bars and Plates	5.3.1
3.4.5	Bearing on Rivets and Bolts	8.1.12
3.4.6	Bearing on Flat Surfaces and Pins and on Bolts in Slotted Holes	8.1.12
3.4.7	Compression in Columns, Axial	5.2.1

438

APPENDIX I
LRFD Design Stresses for Various Alloys

ALLOY 1100-H14

F_{tu} = 16 ksi	F_{tuw} = 11 ksi
F_{ty} = 14 ksi	F_{tyw} = 3.5 ksi
F_{cy} = 13 ksi	F_{cyw} = 3.5 ksi
F_{su} = 10 ksi	F_{suw} = 8 ksi
F_{sy} = 8 ksi	k_t = 1.00
E = 10,100 ksi	

Specification	Design Stress (ksi)		
3.4.1	13.5	Gross Area	
	13.5	Net Area	
3.4.2	13.5	Tubes	13.5
3.4.3	15.5		
3.4.4	17.5		
3.4.5	27		
3.4.6	18		

	Design Stress (ksi), Slenderness $<S_1$	S_1	Design Stress (ksi), $S_1 <$ Slenderness $< S_2$	S_2	Design Stress (ksi), Slenderness $> S_2$
3.4.8	12.5	5.4	$14.5 - 0.370\,(b/t)$	20	$141/(b/t)$
3.4.8.1	12.5	5.4	$14.5 - 0.370\,(b/t)$	26	$3{,}300/(b/t)^2$
3.4.9	12.5	17	$14.5 - 0.116\,(b/t)$	62	$450/(b/t)$
3.4.9.2	12.5	0	$12.3 - 0.057\,\lambda_s$	144	$85{,}000/\lambda_s^2$
3.4.10	12.5	14.3	$14.2 - 0.450\,\sqrt{R/t}$	(1)	$5{,}000\left/\left[(R/t)\left(1 + \dfrac{\sqrt{R/t}}{35}\right)^2\right]\right.$
3.4.11	12.5	0	$12.3 - 0.048\,(L_b/r_y)$	173	$122{,}000/(L_b/r_y)^2$
3.4.12	14.5	40	$21.3 - 1.07\,\sqrt{R/t}$	134	See 3.4.10
3.4.13	16	12	$19.2 - 0.260\left(\dfrac{d}{t}\sqrt{\dfrac{L_b}{d}}\right)$	50	$16{,}000/[(d/t)^2(L_b/d)]$
3.4.14	12.5	0	$12.3 - 0.092\sqrt{\dfrac{L_b S_c}{0.5\sqrt{I_y J}}}$	8,100	$33{,}000/(L_b S_c/0.5\sqrt{I_y J})$
3.4.15	12.5	5.4	$14.5 - 0.370\,(b/t)$	20	$141/(b/t)$
3.4.16	12.5	17	$14.5 - 0.116\,(b/t)$	62	$450/(b/t)$
3.4.16.1	14.5	0	$14.2 - 0.450\,\sqrt{R/t}$	(1)	$5{,}300\left/\left[(R/t)\left(1 + \dfrac{\sqrt{R/t}}{35}\right)^2\right]\right.$
3.4.16.3	12.5	0	$12.3 - 0.057\,\lambda_s$	144	$85{,}000/\lambda_s^2$
3.4.17	16	8.2	$19.2 - 0.390\,(b/t)$	33	$6{,}900/(b/t)^2$
3.4.18	16	43	$19.2 - 0.075\,(h/t)$	129	$1{,}240/(h/t)$
3.4.19	16	100	$19.2 - 0.032\,(h/t)$	300	$2{,}900/(h/t)$
3.4.20	7.5	43	$9.5 - 0.047\,(h/t)$	(1)	$51{,}000/(h/t)^2$
3.4.21	7.5	86	$13.1 - 0.065(a_e/t)$	(1)	$70{,}000/(a_e/t)^2$

Note

(1) Use the lesser of the design stresses for (a) slenderness between S_1 and S_2, and (b) slenderness greater than S_2.

ALLOY 2014-T6

F_{tu} =	60 ksi
F_{ty} =	53 ksi
F_{cy} =	52 ksi
F_{su} =	35 ksi
F_{sy} =	31 ksi
E =	10,900 ksi
k_t =	1.25

Specification	Design Stress (ksi)		
3.4.1	50	Gross Area	
	41	Net Area	
3.4.2	41	Tubes	43
3.4.3	50.5		
3.4.4	58		
3.4.5	102		
3.4.6	68		

	Design Stress (ksi), Slenderness $<S_1$	S_1	Design Stress (ksi), $S_1 <$ Slenderness $< S_2$	S_2	Design Stress (ksi), Slenderness $> S_2$
3.4.8	49.5	3.9	$58.7 - 2.38\,(b/t)$	8.6	$330/(b/t)$
3.4.8.1	49.5	3.9	$58.7 - 2.38\,(b/t)$	10	$3,500/(b/t)^2$
3.4.9	49.5	12	$58.7 - 0.750\,(b/t)$	27	$1,050/(b/t)$
3.4.9.2	49.5	3.7	$50.9 - 0.380\,\lambda_s$	55	$91,000/\lambda_s^2$
3.4.10	49.5	6.9	$55.4 - 2.24\,\sqrt{R/t}$	(1)	$5,400\Big/\left[(R/t)\left(1 + \dfrac{\sqrt{R/t}}{35}\right)^2\right]$
3.4.11	49.5	4.5	$50.9 - 0.310\,(L_b/r_y)$	66	$132,000/(L_b/r_y)^2$
3.4.12	58.0	15	$83.1 - 6.40\,\sqrt{R/t}$	44	See 3.4.10
3.4.13	64.0	10	$88.1 - 2.42\left(\dfrac{d}{t}\sqrt{\dfrac{L_b}{d}}\right)$	24	$17,300/[(d/t)^2(L_b/d)]$
3.4.14	49.5	5.4	$50.9 - 0.600\sqrt{\dfrac{L_b S_c}{0.5\sqrt{I_y J}}}$	1,180	$36,000/(L_b S_c/0.5\sqrt{I_y J})$
3.4.15	49.5	3.9	$58.7 - 2.38\,(b/t)$	8.6	$330/(b/t)$
3.4.16	49.5	12	$58.7 - 0.750\,(b/t)$	27	$1,050/(b/t)$
3.4.16.1	58.0	0	$55.4 - 2.24\,\sqrt{R/t}$	(1)	$5,700\Big/\left[(R/t)\left(1 + \dfrac{\sqrt{R/t}}{35}\right)^2\right]$
3.4.16.3	49.5	3.7	$50.9 - 0.380\,\lambda_s$	55	$91,000/\lambda_s^2$
3.4.17	64.0	6.5	$88.1 - 3.68\,(b/t)$	16	$7,500/(b/t)^2$
3.4.18	64.0	34	$88.1 - 0.700\,(h/t)$	63	$2,800/(h/t)$
3.4.19	64.0	80	$88.1 - 0.300\,(h/t)$	144	$6,400/(h/t)$
3.4.20	29.5	27	$37.4 - 0.290\,(h/t)$	(1)	$55,000/(h/t)^2$
3.4.21	29.5	55	$51.4 - 0.400\,(a_e/t)$	(1)	$76,000/(a_e/t)^2$

Notes

(1) Use the lesser of the design stresses for (a) slenderness between S_1 and S_2, and (b) slenderness greater than S_2.

(2) Alloy 2014 is generally not welded.

ALLOY 3003-H14

			F_{tu} =	20 ksi	F_{tuw} = 14 ksi
			F_{ty} =	17 ksi	F_{tyw} = 5 ksi
			F_{cy} =	14 ksi	F_{cyw} = 5 ksi
			F_{su} =	12 ksi	F_{suw} = 10 ksi
			F_{sy} =	10 ksi	k_t = 1.00
			E =	10,100 ksi	

Specification	Design Stress (ksi)		
3.4.1	16	Gross Area	
	17	Net Area	
3.4.2	16.0	Tubes	16.0
3.4.3	19.0		
3.4.4	21.0		
3.4.5	34		
3.4.6	23		

	Design Stress (ksi), Slenderness $<S_1$	S_1	Design Stress (ksi), $S_1 <$ Slenderness $< S_2$	S_2	Design Stress (ksi), Slenderness $> S_2$
3.4.8	13.5	4.0	$15.6 - 0.420 \, (b/t)$	19	$147/(b/t)$
3.4.8.1	13.5	4.0	$15.6 - 0.420 \, (b/t)$	25	$3,300/(b/t)^2$
3.4.9	13.5	16	$15.6 - 0.131 \, (b/t)$	60	$470/(b/t)$
3.4.9.2	13.5	0	$13.3 - 0.064 \, \lambda_x$	138	$85,000/\lambda_x^2$
3.4.10	13.5	14	$15.4 - 0.510 \, \sqrt{R/t}$	(1)	$5,000 \left/ \left[(R/t)\left(1 + \dfrac{\sqrt{R/t}}{35}\right)^2 \right] \right.$
3.4.11	13.5	0	$13.3 - 0.054 \, (L_b/r_y)$	166	$122,000/(L_b/r_y)^2$
3.4.12	15.5	40	$23 - 1.19 \, \sqrt{R/t}$	126	See 3.4.10
3.4.13	17.5	11	$20.8 - 0.290 \left(\dfrac{d}{t} \sqrt{\dfrac{L_b}{d}} \right)$	48	$16,000/[(d/t)^2(L_b/d)]$
3.4.14	13.5	0	$13.3 - 0.103 \sqrt{\dfrac{L_b S_c}{0.5\sqrt{I_y J}}}$	7,440	$33,000/(L_b S_c/0.5\sqrt{I_y J})$
3.4.15	13.5	5.0	$15.6 - 0.420 \, (b/t)$	19	$147/(b/t)$
3.4.16	13.5	16	$15.6 - 0.131 \, (b/t)$	60	$470/(b/t)$
3.4.16.1	15.5	0	$15.4 - 0.510 \, \sqrt{R/t}$	(1)	$5,300 \left/ \left[(R/t)\left(1 + \dfrac{\sqrt{R/t}}{35}\right)^2 \right] \right.$
3.4.16.3	13.5	0	$13.3 - 0.064 \, \lambda_x$	138	$85,000/\lambda_x^2$
3.4.17	17.5	7.5	$20.8 - 0.440 \, (b/t)$	32	$6,900/(b/t)^2$
3.4.18	17.5	39	$20.8 - 0.084 \, (h/t)$	124	$1,290/(h/t)$
3.4.19	17.5	92	$20.8 - 0.036 \, (h/t)$	290	$3,000/(h/t)$
3.4.20	9.5	40	$12.2 - 0.068 \, (h/t)$	(1)	$51,000/(h/t)^2$
3.4.21	9.5	78	$16.8 - 0.094 \, (a_c/t)$	(1)	$70,000/(a_c/t)^2$

Note

(1) Use the lesser of the design stresses for (a) slenderness between S_1 and S_2, and (b) slenderness greater than S_2.

ALLOY 3003-H16

F_{tu} =	24 ksi	F_{tuw} = 14 ksi	
F_{ty} =	21 ksi	F_{tyw} = 5 ksi	
F_{cy} =	18 ksi	F_{cyw} = 5 ksi	
F_{su} =	14 ksi	F_{suw} = 10 ksi	
F_{sy} =	12 ksi	k_t = 1.00	
E =	10,100 ksi		

Specification	Design Stress (ksi)		
3.4.1	23	Gross Area	
	18	Net Area	
3.4.2	20.0	Tubes	20.0
3.4.3	23.5		
3.4.4	26.0		
3.4.5	36		
3.4.6	24		

	Design Stress (ksi), Slenderness $<S_1$	S_1	Design Stress (ksi), S_1 < Slenderness < S_2	S_2	Design Stress (ksi), Slenderness $> S_2$
3.4.8	17	5.7	$20.6 - 0.630\,(b/t)$	16	$168/(b/t)$
3.4.8.1	17	5.7	$20.6 - 0.630\,(b/t)$	22	$3,300/(b/t)^2$
3.4.9	17	18	$20.6 - 0.197\,(b/t)$	52	$540/(b/t)$
3.4.9.2	17	3.2	$17.3 - 0.095\,\lambda_x$	121	$85,000/\lambda_x^2$
3.4.10	17	17	$20 - 0.720\,\sqrt{R/t}$	(1)	$5,000 \Big/ \left[(R/t)\left(1 + \dfrac{\sqrt{R/t}}{35}\right)^2\right]$
3.4.11	17.0	3.8	$17.3 - 0.080\,(L_b/r_y)$	145	$122,000/(L_b/r_y)^2$
3.4.12	20.0	35	$30 - 1.69\,\sqrt{R/t}$	107	See 3.4.10
3.4.13	22.0	12	$27.4 - 0.440\left(\dfrac{d}{t}\sqrt{\dfrac{L_b}{d}}\right)$	42	$16,000/[(d/t)^2(L_b/d)]$
3.4.14	17.0	3.8	$17.3 - 0.153\sqrt{\dfrac{L_b S_c}{0.5\sqrt{I_y J}}}$	5,720	$33,000/(L_b S_c/0.5\sqrt{I_y J})$
3.4.15	17.0	5.7	$20.6 - 0.630\,(b/t)$	16	$168/(b/t)$
3.4.16	17.0	18	$20.6 - 0.197\,(b/t)$	52	$540/(b/t)$
3.4.16.1	20.0	0	$20 - 0.720\,\sqrt{R/t}$	(1)	$5,300 \Big/ \left[(R/t)\left(1 + \dfrac{\sqrt{R/t}}{35}\right)^2\right]$
3.4.16.3	17.0	3.2	$17.3 - 0.095\,\lambda_x$	121	$85,000/\lambda_x^2$
3.4.17	22.0	8.2	$27.4 - 0.660\,(b/t)$	27	$6,900/(b/t)^2$
3.4.18	22.0	43	$27.4 - 0.127\,(h/t)$	108	$1,480/(h/t)$
3.4.19	22.0	98	$27.4 - 0.055\,(h/t)$	249	$3,400/(h/t)$
3.4.20	11.5	36	$14.8 - 0.091\,(h/t)$	(1)	$51,000/(h/t)^2$
3.4.21	11.5	70	$20.3 - 0.125\,(a_e/t)$	(1)	$70,000/(a_e/t)^2$

Note

(1) Use the lesser of the design stresses for (a) slenderness between S_1 and S_2, and (b) slenderness greater than S_2.

ALLOY 3004-H16 Alclad

F_{tu} = 35 ksi	F_{tuw} = 21 ksi
F_{ty} = 30 ksi	F_{tyw} = 8 ksi
F_{cy} = 28 ksi	F_{cyw} = 8 ksi
F_{su} = 20 ksi	F_{suw} = 13 ksi
F_{sy} = 17 ksi	k_t = 1.00
E = 10,100 ksi	

Specification	Design Stress (ksi)		
3.4.1	29	Gross Area	
	30	Net Area	
3.4.2	28.5	Tubes	28.5
3.4.3	33.5		
3.4.4	37.0		
3.4.5	60		
3.4.6	40		

	Design Stress (ksi), Slenderness $<S_1$	S_1	Design Stress (ksi), $S_1 <$ Slenderness $< S_2$	S_2	Design Stress (ksi), Slenderness $> S_2$
3.4.8	26.5	5.2	$33.3 - 1.300\,(b/t)$	13	$210/(b/t)$
3.4.8.1	26.5	5.2	$33.3 - 1.30\,(b/t)$	17	$3,300/(b/t)^2$
3.4.9	26.5	17	$33.3 - 0.410\,(b/t)$	41	$680/(b/t)$
3.4.9.2	26.5	6.2	$27.8 - 0.194\,\lambda_x$	96	$85,000/\lambda_x^2$
3.4.10	26.5	16	$31.8 - 1.33\,\sqrt{R/t}$	(1)	$5,000 \left/ \left[(R/t)\left(1 + \dfrac{\sqrt{R/t}}{35}\right)^2 \right]\right.$
3.4.11	26.5	8.1	$27.8 - 0.161\,(L_b/r_y)$	115	$122,000/(L_b/r_y)^2$
3.4.12	31.0	28	$47.7 - 3.13\,\sqrt{R/t}$	78	See 3.4.10
3.4.13	34.5	11	$44.4 - 0.900\left(\dfrac{d}{t}\sqrt{\dfrac{L_b}{d}}\right)$	33	$16,000/[(d/t)^2(L_b/d)]$
3.4.14	26.5	18	$27.8 - 0.310\sqrt{\dfrac{L_b S_c}{0.5\sqrt{I_y J}}}$	3,600	$33,000/(L_b S_c/0.5\sqrt{I_y J})$
3.4.15	26.5	5.2	$33.3 - 1.30\,(b/t)$	13	$210/(b/t)$
3.4.16	26.5	17	$33.3 - 0.410\,(b/t)$	41	$680/(b/t)$
3.4.16.1	31.0	0.4	$31.8 - 1.33\,\sqrt{R/t}$	(1)	$5,300 \left/ \left[(R/t)\left(1 + \dfrac{\sqrt{R/t}}{35}\right)^2 \right]\right.$
3.4.16.3	26.5	6.7	$27.8 - 0.194\,\lambda_x$	96	$85,000/\lambda_x^2$
3.4.17	34.5	7.2	$44.4 - 1.37\,(b/t)$	22	$6,900/(b/t)^2$
3.4.18	34.5	38	$44.4 - 0.260\,(h/t)$	85	$1,880/(h/t)$
3.4.19	34.5	88	$44.4 - 0.113\,(h/t)$	196	$4,300/(h/t)$
3.4.20	16.0	35	$21.7 - 0.162\,(h/t)$	(1)	$51,000/(h/t)^2$
3.4.21	16.0	63	$29.8 - 0.220\,(a_e/t)$	(1)	$70,000/(a_e/t)^2$

Note

(1) Use the lesser of the design stresses for (a) slenderness between S_1 and S_2 and (b) slenderness greater than S_2.

ALLOY 3004-H34 Alclad

F_{tu} =	31 ksi		F_{tuw} =	21 ksi
F_{ty} =	24 ksi		F_{tyw} =	8 ksi
F_{cy} =	21 ksi		F_{cyw} =	8 ksi
F_{su} =	18 ksi		F_{suw} =	13 ksi
F_{sy} =	14 ksi		k_t =	1.00
E =	10,100 ksi			

Specification	Design Stress (ksi)		
3.4.1	23	Gross Area	
	26	Net Area	
3.4.2	23.0	Tubes	23.0
3.4.3	26.5		
3.4.4	29.5		
3.4.5	53		
3.4.6	35		

	Design Stress (ksi), Slenderness $<S_1$	S_1	Design Stress (ksi), $S_1 <$ Slenderness $< S_2$	S_2	Design Stress (ksi), Slenderness $> S_2$
3.4.8	20	5.3	$24.3 - 0.810\,(b/t)$	15	$183/(b/t)$
3.4.8.1	20	5.3	$24.3 - 0.810\,(b/t)$	20	$3,300/(b/t)^2$
3.4.9	20	17	$24.3 - 0.250\,(b/t)$	48	$580/(b/t)$
3.4.9.2	20	3.3	$20.4 - 0.122\,\lambda_x$	112	$85,000/\lambda_x^2$
3.4.10	20	15	$23.5 - 0.890\,\sqrt{R/t}$	(1)	$5,000 \Big/ \left[(R/t)\left(1 + \dfrac{\sqrt{R/t}}{35}\right)^2 \right]$
3.4.11	20.0	4	$20.4 - 0.101\,(L_b/r_y)$	134	$122,000/(L_b/r_y)^2$
3.4.12	23.5	32	$35.3 - 2.09\,\sqrt{R/t}$	95	See 3.4.10
3.4.13	26.0	11	$32.4 - 0.560\left(\dfrac{d}{t}\sqrt{\dfrac{L_b}{d}}\right)$	39	$16,000/[(d/t)^2(L_b/d)]$
3.4.14	20.0	4.2	$20.4 - 0.195\sqrt{\dfrac{L_b S_c}{0.5\sqrt{I_y J}}}$	4,900	$33,000/(L_b S_c/0.5\sqrt{I_y J})$
3.4.15	20.0	5.3	$24.3 - 0.810\,(b/t)$	15	$183/(b/t)$
3.4.16	20.0	17	$24.3 - 0.250\,(b/t)$	48	$580/(b/t)$
3.4.16.1	23.5	0	$23.5 - 0.890\,\sqrt{R/t}$	(1)	$5,300 \Big/ \left[(R/t)\left(1 + \dfrac{\sqrt{R/t}}{35}\right)^2 \right]$
3.4.16.3	20.0	3.3	$20.4 - 0.122\,\lambda_x$	112	$85,000/\lambda_x^2$
3.4.17	26.0	7.5	$32.4 - 0.850\,(b/t)$	25	$6,900/(b/t)^2$
3.4.18	26.0	39	$32.4 - 0.163\,(h/t)$	99	$1,610/(h/t)$
3.4.19	26.0	90	$32.4 - 0.071\,(h/t)$	229	$3,700/(h/t)$
3.4.20	13.5	34	$17.5 - 0.117\,(h/t)$	(1)	$51,000/(h/t)^2$
3.4.21	13.5	66	$24.1 - 0.161(a_e/t)$	(1)	$70,000/(a_e/t)^2$

Note
(1) Use the lesser of the design stresses for (a) slenderness between S_1 and S_2, and (b) slenderness greater than S_2.

ALLOY 5005-H14

				F_{tu} = 21 ksi		F_{tuw} = 15 ksi
				F_{ty} = 17 ksi		F_{tyw} = 5 ksi
				F_{cy} = 15 ksi		F_{cyw} = 5 ksi
				F_{su} = 12 ksi		F_{suw} = 9 ksi
Specification	Design Stress (ksi)			F_{sy} = 10 ksi E = 10,100 ksi		k_t = 1.00
3.4.1	16	Gross Area				
	18	Net Area				
3.4.2	16.0	Tubes	16.0			
3.4.3	19.0					
3.4.4	21.0					
3.4.5	36					
3.4.6	24					

	Design Stress (ksi), Slenderness $<S_1$	S_1	Design Stress (ksi), $S_1 <$ Slenderness $< S_2$	S_2	Design Stress (ksi), Slenderness $> S_2$
3.4.8	14	6.2	$16.9 - 0.470\,(b/t)$	18	$152/(b/t)$
3.4.8.1	14	6.2	$16.9 - 0.470\,(b/t)$	24	$3,300/(b/t)^2$
3.4.9	14	20	$16.9 - 0.147\,(b/t)$	57	$490/(b/t)$
3.4.9.2	14	4.2	$14.3 - 0.071\,\lambda_s$	133	$85,000/\lambda_s^2$
3.4.10	14	21	$16.5 - 0.550\,\sqrt{R/t}$	(1)	$5,000 \left/ \left[(R/t)\left(1 + \dfrac{\sqrt{R/t}}{35}\right)^2 \right] \right.$
3.4.11	14.0	5.1	$14.3 - 0.059\,(L_b/r_y)$	160	$122,000/(L_b/r_y)^2$
3.4.12	16.5	40	$24.8 - 1.31\,\sqrt{R/t}$	122	See 3.4.10
3.4.13	18.5	12	$22.4 - 0.320\left(\dfrac{d}{t}\sqrt{\dfrac{L_b}{d}}\right)$	46	$16,000/[(d/t)^2(L_b/d)]$
3.4.14	14.0	6.9	$14.3 - 0.114\sqrt{\dfrac{L_b S_c}{0.5\sqrt{I_y J}}}$	6,910	$33,000/(L_b S_c/0.5\sqrt{I_y J})$
3.4.15	14.0	6.2	$16.9 - 0.470\,(b/t)$	18	$152/(b/t)$
3.4.16	14.0	20	$16.9 - 0.147\,(b/t)$	57	$490/(b/t)$
3.4.16.1	16.5	0	$16.5 - 0.550\,\sqrt{R/t}$	(1)	$5,300 \left/ \left[(R/t)\left(1 + \dfrac{\sqrt{R/t}}{35}\right)^2 \right] \right.$
3.4.16.3	14.0	4.2	$14.3 - 0.071\,\lambda_s$	133	$85,000/\lambda_s^2$
3.4.17	18.5	8.0	$22.4 - 0.490\,(b/t)$	30	$6,900/(b/t)^2$
3.4.18	18.5	41	$22.4 - 0.094\,(h/t)$	119	$1,340/(h/t)$
3.4.19	18.5	95	$22.4 - 0.041\,(h/t)$	280	$3,100/(h/t)$
3.4.20	9.5	40	$12.2 - 0.068\,(h/t)$	(1)	$51,000/(h/t)^2$
3.4.21	9.5	78	$16.8 - 0.094\,(a_e/t)$	(1)	$70,000/(a_e/t)^2$

Note

(1) Use the lesser of the design stresses for (a) slenderness between S_1 and S_2, and (b) slenderness greater than S_2.

ALLOY 5050-H34

F_{tu} =	25 ksi	F_{tuw} = 18 ksi	
F_{ty} =	20 ksi	F_{tyw} = 6 ksi	
F_{cy} =	18 ksi	F_{cyw} = 6 ksi	
F_{su} =	15 ksi	F_{suw} = 12 ksi	
F_{sy} =	12 ksi	k_t = 1.00	
E =	10,100 ksi		

Specification	Design Stress (ksi)		
3.4.1	19	Gross Area	
	21	Net Area	
3.4.2	19.0	Tubes	19.0
3.4.3	22.0		
3.4.4	24.5		
3.4.5	43		
3.4.6	28		

	Design Stress (ksi), Slenderness $<S_1$	S_1	Design Stress (ksi), $S_1 <$ Slenderness $< S_2$	S_2	Design Stress (ksi), Slenderness $> S_2$
3.4.8	17	5.7	$20.6 - 0.630\,(b/t)$	16	$168/(b/t)$
3.4.8.1	17	5.7	$20.6 - 0.630\,(b/t)$	22	$3,300/(b/t)^2$
3.4.9	17	18	$20.6 - 0.197\,(b/t)$	52	$540/(b/t)$
3.4.9.2	17	3.2	$17.3 - 0.095\,\lambda_s$	121	$85,000/\lambda_s^2$
3.4.10	17	17	$20 - 0.720\,\sqrt{R/t}$	(1)	$5,000 \Big/ \left[(R/t)\left(1 + \dfrac{\sqrt{R/t}}{35}\right)^2\right]$
3.4.11	17.0	3.8	$17.3 - 0.080\,(L_b/r_y)$	145	$122,000/(L_b/r_y)^2$
3.4.12	20.0	35	$30 - 1.69\,\sqrt{R/t}$	107	See 3.4.10
3.4.13	22.0	12	$27.4 - 0.440 \left(\dfrac{d}{t}\sqrt{\dfrac{L_b}{d}}\right)$	42	$16,000/[(d/t)^2(L_b/d)]$
3.4.14	17.0	3.8	$17.3 - 0.153\,\sqrt{\dfrac{L_b S_c}{0.5\sqrt{I_y J}}}$	5,720	$33,000/(L_b S_c/0.5\sqrt{I_y J})$
3.4.15	17.0	5.7	$20.6 - 0.630\,(b/t)$	16	$168/(b/t)$
3.4.16	17.0	18	$20.6 - 0.197\,(b/t)$	52	$540/(b/t)$
3.4.16.1	20.0	0	$20 - 0.720\,\sqrt{R/t}$	(1)	$5,300 \Big/ \left[(R/t)\left(1 + \dfrac{\sqrt{R/t}}{35}\right)^2\right]$
3.4.16.3	17.0	3.2	$17.3 - 0.095\,\lambda_s$	121	$85,000/\lambda_s^2$
3.4.17	22.0	8.2	$27.4 - 0.660\,(b/t)$	27	$6.900/(b/t)^2$
3.4.18	22.0	43	$27.4 - 0.127\,(h/t)$	108	$1,480/(h/t)$
3.4.19	22.0	98	$27.4 - 0.055\,(h/t)$	249	$3,400/(h/t)$
3.4.20	11.5	36	$14.8 - 0.091\,(h/t)$	(1)	$51,000/(h/t)^2$
3.4.21	11.5	70	$20.3 - 0.125\,(a_e/t)$	(1)	$70,000/(a_e/t)^2$

Note

(1) Use the lesser of the design stresses for (a) slenderness between S_1 and S_2, and (b) slenderness greater than S_2.

ALLOY 5052-H34

F_{tu} =	34 ksi		F_{tuw} = 25 ksi		
F_{ty} =	26 ksi		F_{tyw} = 9.5 ksi		
F_{cy} =	24 ksi		F_{cyw} = 9.5 ksi		
F_{su} =	20 ksi		F_{suw} = 16 ksi		
F_{sy} =	15 ksi		k_t = 1.00		
E =	10,200 ksi				

Specification	Design Stress (ksi)				
3.4.1	25	Gross Area			
	29	Net Area			
3.4.2	24.5	Tubes	24.5		
3.4.3	29.0				
3.4.4	32.0				
3.4.5	58				
3.4.6	39				

	Design Stress (ksi), Slenderness $<S_1$	S_1	Design Stress (ksi), S_1 < Slenderness < S_2	S_2	Design Stress (ksi), Slenderness > S_2
3.4.8	23	5.1	$28.1 - 1.00\ (b/t)$	14	$198/(b/t)$
3.4.8.1	23	5.1	$28.1 - 1.00\ (b/t)$	19	$3,300/(b/t)^2$
3.4.9	23	16	$28.1 - 0.310\ (b/t)$	45	$630/(b/t)$
3.4.9.2	23	3.3	$23.5 - 0.150\ \lambda_s$	104	$86,000/\lambda_s^2$
3.4.10	23	14	$27 - 1.07\ \sqrt{R/t}$	(1)	$5,000\left/\left[(R/t)\left(1 + \dfrac{\sqrt{R/t}}{35}\right)^2\right]\right.$
3.4.11	23.0	4.0	$23.5 - 0.125\ (L_b/r_y)$	125	$123,000/(L_b/r_y)^2$
3.4.12	26.5	31	$40.5 - 2.51\ \sqrt{R/t}$	88	See 3.4.10
3.4.13	29.5	12	$37.5 - 0.690\left(\dfrac{d}{t}\sqrt{\dfrac{L_b}{d}}\right)$	36	$16,200/[(d/t)^2(L_b/d)]$
3.4.14	23.0	4.3	$23.5 - 0.240\sqrt{\dfrac{L_b S_c}{0.5\sqrt{I_y J}}}$	4,230	$33,000/(L_b S_c/0.5\sqrt{I_y J})$
3.4.15	23.0	5.1	$28.1 - 1.00\ (b/t)$	14	$198/(b/t)$
3.4.16	23.0	16	$28.1 - 0.310\ (b/t)$	45	$630/(b/t)$
3.4.16.1	26.5	0.2	$27 - 1.07\ \sqrt{R/t}$	(1)	$5,300\left/\left[(R/t)\left(1 + \dfrac{\sqrt{R/t}}{35}\right)^2\right]\right.$
3.4.16.3	23.0	3.3	$23.5 - 0.150\ \lambda_s$	104	$86,000/\lambda_s^2$
3.4.17	29.5	7.5	$37.5 - 1.06\ (b/t)$	24	$7,000/(b/t)^2$
3.4.18	29.5	40	$37.5 - 0.200\ (h/t)$	93	$1,740/(h/t)$
3.4.19	29.5	91	$37.5 - 0.088\ (h/t)$	214	$4,000/(h/t)$
3.4.20	14.0	37	$18.9 - 0.131\ (h/t)$	(1)	$52,000/(h/t)^2$
3.4.21	14.0	66	$26.0 - 0.181\ (a_e/t)$	(1)	$71,000/(a_e/t)^2$

Note

(1) Use the lesser of the design stresses for (a) slenderness between S_1 and S_2, and (b) slenderness greater than S_2.

ALLOY 5083-H111

F_{tu} = 40 ksi		F_{tuw} = 39 ksi	
F_{ty} = 24 ksi		F_{tyw} = 16 ksi	
F_{cy} = 21 ksi		F_{cyw} = 15 ksi	
F_{su} = 23 ksi		F_{suw} = 23 ksi	
F_{sy} = 14 ksi		k_t = 1.00	
E = 10,400 ksi			

Specification	Design Stress (ksi)		
3.4.1	23	Gross Area	
	34	Net Area	
3.4.2	23.0	Tubes	23.0
3.4.3	26.5		
3.4.4	29.5		
3.4.5	68		
3.4.6	45		

	Design Stress (ksi), Slenderness $<S_1$	S_1	Design Stress (ksi), S_1 < Slenderness < S_2	S_2	Design Stress (ksi), Slenderness $> S_2$
3.4.8	20	5.4	$24.3 - 0.800\ (b/t)$	15	$185/(b/t)$
3.4.8.1	20	5.4	$24.3 - 0.800\ (b/t)$	20	$3,400/(b/t)^2$
3.4.9	20	17	$24.3 - 0.250\ (b/t)$	49	$590/(b/t)$
3.4.9.2	20	3.3	$20.4 - 0.120\ \lambda_x$	113	$87,000/\lambda_s^2$
3.4.10	20	16	$23.5 - 0.880\ \sqrt{R/t}$	(1)	$5,100 \bigg/ \left[(R/t)\left(1 + \dfrac{\sqrt{R/t}}{35}\right)^2\right]$
3.4.11	20.0	4.0	$20.4 - 0.100\ (L_b/r_y)$	136	$126,000/(L_b/r_y)^2$
3.4.12	23.5	32	$35.3 - 2.07\ \sqrt{R/t}$	97	See 3.4.10
3.4.13	26.0	12	$32.4 - 0.550\left(\dfrac{d}{t}\sqrt{\dfrac{L_b}{d}}\right)$	39	$16,500/[(d/t)^2(L_b/d)]$
3.4.14	20.0	4.3	$20.4 - 0.192\sqrt{\dfrac{L_b S_c}{0.5\sqrt{I_y J}}}$	4,990	$34,000/(L_b S_c/0.5\sqrt{I_y J})$
3.4.15	20.0	5.4	$24.3 - 0.800\ (b/t)$	15	$185/(b/t)$
3.4.16	20.0	17	$24.3 - 0.250\ (b/t)$	49	$590/(b/t)$
3.4.16.1	23.5	0	$23.5 - 0.880\ \sqrt{R/t}$	(1)	$5,500 \bigg/ \left[(R/t)\left(1 + \dfrac{\sqrt{R/t}}{35}\right)^2\right]$
3.4.16.3	20.0	3.3	$20.4 - 0.120\ \lambda_x$	113	$87,000/\lambda_s^2$
3.4.17	26.0	7.6	$32.4 - 0.840\ (b/t)$	26	$7,100/(b/t)^2$
3.4.18	26.0	40	$32.4 - 0.161\ (h/t)$	101	$1,630/(h/t)$
3.4.19	26.0	91	$32.4 - 0.070\ (h/t)$	233	$3,800/(h/t)$
3.4.20	13.5	35	$17.5 - 0.115\ (h/t)$	(1)	$53,000/(h/t)^2$
3.4.21	13.5	67	$24.1 - 0.159\ (a_e/t)$	(1)	$72,000/(a_e/t)^2$

Note

(1) Use the lesser of the design stresses for (a) slenderness between S_1 and S_2, and (b) slenderness greater than S_2.

ALLOY 5083-H116

F_{tu} =	44 ksi		F_{tuw} = 40 ksi	
F_{ty} =	31 ksi		F_{tyw} = 14 ksi	
F_{cy} =	26 ksi		F_{cyw} = 14 ksi	
F_{su} =	26 ksi		F_{suw} = 21 ksi	
F_{sy} =	18 ksi		k_t = 1.00	
E =	10,400 ksi			

Specification	Design Stress (ksi)		
3.4.1	29	Gross Area	
	37	Net Area	
3.4.2	29.5	Tubes	29.5
3.4.3	34.5		
3.4.4	38.5		
3.4.5	75		
3.4.6	50		

	Design Stress (ksi), Slenderness $<S_1$	S_1	Design Stress (ksi), $S_1 <$ Slenderness $< S_2$	S_2	Design Stress (ksi), Slenderness $> S_2$
3.4.8	24.5	5.5	$30.7 - 1.13\,(b/t)$	14	$210/(b/t)$
3.4.8.1	24.5	5.5	$30.7 - 1.13\,(b/t)$	18	$3,400/(b/t)^2$
3.4.9	24.5	18	$30.7 - 0.350\,(b/t)$	43	$660/(b/t)$
3.4.9.2	24.5	7.1	$25.7 - 0.169\,\lambda_s$	101	$87,000/\lambda_s^2$
3.4.10	24.5	17	$29.4 - 1.19\,\sqrt{R/t}$	(1)	$5,100 \Big/ \left[(R/t)\left(1 + \dfrac{\sqrt{R/t}}{35}\right)^2\right]$
3.4.11	24.5	8.5	$25.7 - 0.141\,(L_b/r_y)$	121	$126,000/(L_b/r_y)^2$
3.4.12	29.0	29	$44.1 - 2.79\,\sqrt{R/t}$	84	See 3.4.10
3.4.13	32.0	11	$40.9 - 0.780\left(\dfrac{d}{t}\sqrt{\dfrac{L_b}{d}}\right)$	35	$16,500/[(d/t)^2(L_b/d)]$
3.4.14	24.5	20	$25.7 - 0.270\,\sqrt{\dfrac{L_b S_c}{0.5\sqrt{I_y J}}}$	3,980	$34,000/(L_b S_c/0.5\sqrt{I_y J})$
3.4.15	24.5	5.5	$30.7 - 1.13\,(b/t)$	14	$210/(b/t)$
3.4.16	24.5	18	$30.7 - 0.350\,(b/t)$	43	$660/(b/t)$
3.4.16.1	29.0	0.1	$29.4 - 1.19\,\sqrt{R/t}$	(1)	$5,500 \Big/ \left[(R/t)\left(1 + \dfrac{\sqrt{R/t}}{35}\right)^2\right]$
3.4.16.3	24.5	7.1	$25.7 - 0.169\,\lambda_s$	101	$87,000/\lambda_s^2$
3.4.17	32.0	7.5	$40.9 - 1.19\,(b/t)$	23	$7,100/(b/t)^2$
3.4.18	32.0	39	$40.9 - 0.230\,(h/t)$	90	$1,830/(h/t)$
3.4.19	32.0	90	$40.9 - 0.099\,(h/t)$	207	$4,200/(h/t)$
3.4.20	17.0	34	$23 - 0.175\,(h/t)$	(1)	$53,000/(h/t)^2$
3.4.21	17.0	61	$31.6 - 0.240\,(a_e/t)$	(1)	$72,000/(a_e/t)^2$

Note

(1) Use the lesser of the design stresses for (a) slenderness between S_1 and S_2, and (b) slenderness greater than S_2.

ALLOY 5086-H34

F_{tu} =	44 ksi	F_{tuw} = 35 ksi	
F_{ty} =	34 ksi	F_{tyw} = 14 ksi	
F_{cy} =	32 ksi	F_{cyw} = 14 ksi	
F_{su} =	26 ksi	F_{suw} = 21 ksi	
F_{sy} =	20 ksi	k_t = 1.00	
E =	10,400 ksi		

Specification	Design Stress (ksi)		
3.4.1	32	Gross Area	
	37	Net Area	
3.4.2	32.5	Tubes	32.5
3.4.3	38.0		
3.4.4	42.0		
3.4.5	75		
3.4.6	50		

	Design Stress (ksi), Slenderness $<S_1$	S_1	Design Stress (ksi), $S_1 <$ Slenderness $< S_2$	S_2	Design Stress (ksi), Slenderness $> S_2$
3.4.8	30.5	5.1	$38.6 - 1.59\,(b/t)$	12	$230/(b/t)$
3.4.8.1	30.5	5.1	$38.6 - 1.59\,(b/t)$	16	$3,400/(b/t)^2$
3.4.9	30.5	16	$38.6 - 0.500\,(b/t)$	39	$740/(b/t)$
3.4.9.2	30.5	6.3	$32 - 0.240\,\lambda_s$	90	$87,000/\lambda_s^2$
3.4.10	30.5	15	$36.6 - 1.59\,\sqrt{R/t}$	(1)	$5,100 \left/ \left[(R/t)\left(1 + \dfrac{\sqrt{R/t}}{35}\right)^2 \right] \right.$
3.4.11	30.5	7.6	$32 - 0.197\,(L_b/r_y)$	108	$126,000/(L_b/r_y)^2$
3.4.12	35.5	27	$54.9 - 3.74\,\sqrt{R/t}$	73	See 3.4.10
3.4.13	39.5	11	$51.4 - 1.10\left(\dfrac{d}{t}\sqrt{\dfrac{L_b}{d}}\right)$	31	$16,500/[(d/t)^2(L_b/d)]$
3.4.14	30.5	16	$32 - 0.380\sqrt{\dfrac{L_b S_c}{0.5\sqrt{I_y J}}}$	3,160	$34,000/(L_b S_c/0.5\sqrt{I_y J})$
3.4.15	30.5	5.1	$38.6 - 1.59\,(b/t)$	12	$230/(b/t)$
3.4.16	30.5	16	$38.6 - 0.500\,(b/t)$	39	$740/(b/t)$
3.4.16.1	35.5	0.5	$36.6 - 1.59\,\sqrt{R/t}$	(1)	$5,500 \left/ \left[(R/t)\left(1 + \dfrac{\sqrt{R/t}}{35}\right)^2 \right] \right.$
3.4.16.3	30.5	6.3	$32 - 0.240\,\lambda_s$	90	$87,000/\lambda_s^2$
3.4.17	39.5	7.1	$51.4 - 1.68\,(b/t)$	20	$7,100/(b/t)^2$
3.4.18	39.5	37	$51.4 - 0.320\,(h/t)$	80	$2,100/(h/t)$
3.4.19	39.5	86	$51.4 - 0.139\,(h/t)$	185	$4,700/(h/t)$
3.4.20	19.0	33	$25.9 - 0.210\,(h/t)$	(1)	$53,000/(h/t)^2$
3.4.21	19.0	57	$35.6 - 0.290\,(a_e/t)$	(1)	$72,000/(a_e/t)^2$

Note

(1) Use the lesser of the design stresses for (a) slenderness between S_1 and S_2, and (b) slenderness greater than S_2.

ALLOY 5086-H111

				F_{tu} =	36 ksi		F_{tuw} = 35 ksi
				F_{ty} =	21 ksi		F_{tyw} = 14 ksi
				F_{cy} =	18 ksi		F_{cyw} = 13 ksi
				F_{su} =	21 ksi		F_{suw} = 21 ksi
	Design Stress (ksi)			F_{sy} =	12 ksi		k_t = 1.00
Specification				E =	10,400 ksi		
3.4.1	20	Gross Area					
	31	Net Area					
3.4.2	20.0	Tubes	20.0				
3.4.3	23.5						
3.4.4	26.0						
3.4.5	61						
3.4.6	41						

	Design Stress (ksi), Slenderness $<S_1$	S_1	Design Stress (ksi), $S_1 <$ Slenderness $< S_2$	S_2	Design Stress (ksi), Slenderness $> S_2$
3.4.8	17	5.8	$20.6 - 0.620\,(b/t)$	17	$171/(b/t)$
3.4.8.1	17	5.8	$20.6 - 0.620\,(b/t)$	22	$3,400/(b/t)^2$
3.4.9	17	19	$20.6 - 0.194\,(b/t)$	53	$540/(b/t)$
3.4.9.2	17	3.2	$17.3 - 0.094\,\lambda_s$	123	$87,000/\lambda_s^2$
3.4.10	17	18	$20 - 0.710\,\sqrt{R/t}$	(1)	$5,100 \Big/ \left[(R/t)\left(1 + \dfrac{\sqrt{R/t}}{35}\right)^2 \right]$
3.4.11	17.0	3.8	$17.3 - 0.078\,(L_b/r_y)$	148	$126,000/(L_b/r_y)^2$
3.4.12	20.0	36	$30 - 1.67\,\sqrt{R/t}$	109	See 3.4.10
3.4.13	22.0	13	$27.4 - 0.430\left(\dfrac{d}{t}\sqrt{\dfrac{L_b}{d}}\right)$	43	$16,500/[(d/t)^2(L_b/d)]$
3.4.14	17.0	4.0	$17.3 - 0.150\sqrt{\dfrac{L_b S_c}{0.5\sqrt{I_y J}}}$	5,910	$34,000/(L_b S_c/0.5\sqrt{I_y J})$
3.4.15	17.0	5.8	$20.6 - 0.620\,(b/t)$	17	$171/(b/t)$
3.4.16	17.0	19	$20.6 - 0.194\,(b/t)$	53	$540/(b/t)$
3.4.16.1	20.0	0	$20 - 0.710\,\sqrt{R/t}$	(1)	$5,500 \Big/ \left[(R/t)\left(1 + \dfrac{\sqrt{R/t}}{35}\right)^2 \right]$
3.4.16.3	17.0	3.2	$17.3 - 0.094\,\lambda_s$	123	$87,000/\lambda_s^2$
3.4.17	22.0	8.3	$27.4 - 0.650\,(b/t)$	28	$7,100/(b/t)^2$
3.4.18	22.0	43	$27.4 - 0.125\,(h/t)$	110	$1,500/(h/t)$
3.4.19	22.0	100	$27.4 - 0.054\,(h/t)$	250	$3,500/(h/t)$
3.4.20	11.5	37	$14.8 - 0.090\,(h/t)$	(1)	$53,000/(h/t)^2$
3.4.21	11.5	72	$20.3 - 0.123\,(a_e/t)$	(1)	$72,000/(a_e/t)^2$

Note

(1) Use the lesser of the design stresses for (a) slenderness between S_1 and S_2, and (b) slenderness greater than S_2.

ALLOY 5086-H116

F_{tu} = 40 ksi	F_{tuw} = 35 ksi
F_{ty} = 28 ksi	F_{tyw} = 14 ksi
F_{cy} = 26 ksi	F_{cyw} = 14 ksi
F_{su} = 24 ksi	F_{suw} = 21 ksi
F_{sy} = 16 ksi	k_t = 1.00
E = 10,400 ksi	

Specification	Design Stress (ksi)		
3.4.1	27	Gross Area	
	34	Net Area	
3.4.2	26.5	Tubes	26.5
3.4.3	31.0		
3.4.4	34.5		
3.4.5	68		
3.4.6	45		

	Design Stress (ksi), Slenderness $<S_1$	S_1	Design Stress (ksi), S_1 < Slenderness < S_2	S_2	Design Stress (ksi), Slenderness > S_2
3.4.8	24.5	5.5	$30.7 - 1.13\,(b/t)$	14	$210/(b/t)$
3.4.8.1	24.5	5.5	$30.7 - 1.13\,(b/t)$	18	$3{,}400/(b/t)^2$
3.4.9	24.5	18	$30.7 - 0.350\,(b/t)$	43	$660/(b/t)$
3.4.9.2	24.5	7.1	$25.7 - 0.169\,\lambda_s$	101	$87{,}000/\lambda_s^2$
3.4.10	24.5	17	$29.4 - 1.19\,\sqrt{R/t}$	(1)	$5{,}100 \Big/ \left[(R/t)\left(1 + \dfrac{\sqrt{R/t}}{35}\right)^2 \right]$
3.4.11	24.5	8.5	$25.7 - 0.141\,(L_b/r_y)$	121	$126{,}000/(L_b/r_y)^2$
3.4.12	29.0	29	$44.1 - 2.79\,\sqrt{R/t}$	84	See 3.4.10
3.4.13	32.0	11	$40.9 - 0.780\left(\dfrac{d}{t}\sqrt{\dfrac{L_b}{d}}\right)$	35	$16{,}500/[(d/t)^2(L_b/d)]$
3.4.14	24.5	20	$25.7 - 0.270\sqrt{\dfrac{L_b S_c}{0.5\sqrt{I_y J}}}$	3,980	$34{,}000/(L_b S_c/0.5\sqrt{I_y J})$
3.4.15	24.5	5.5	$30.7 - 1.13\,(b/t)$	14	$210/(b/t)$
3.4.16	24.5	18	$30.7 - 0.350\,(b/t)$	43	$660/(b/t)$
3.4.16.1	29.0	0.1	$29.4 - 1.19\,\sqrt{R/t}$	(1)	$5{,}500 \Big/ \left[(R/t)\left(1 + \dfrac{\sqrt{R/t}}{35}\right)^2 \right]$
3.4.16.3	24.5	7.1	$25.7 - 0.169\,\lambda_s$	101	$87{,}000/\lambda_s^2$
3.4.17	32.0	7.5	$40.9 - 1.19\,(b/t)$	23	$7{,}100/(b/t)^2$
3.4.18	32.0	39	$40.9 - 0.230\,(h/t)$	90	$1{,}830/(h/t)$
3.4.19	32.0	90	$40.9 - 0.099\,(h/t)$	207	$4{,}200/(h/t)$
3.4.20	15.0	37	$20.3 - 0.144\,(h/t)$	(1)	$53{,}000/(h/t)^2$
3.4.21	15.0	65	$27.9 - 0.198\,(a_e/t)$	(1)	$72{,}000/(a_e/t)^2$

Note

(1) Use the lesser of the design stresses for (a) slenderness between S_1 and S_2, and (b) slenderness greater than S_2.

ALLOY 5454-H111

			F_{tu} =	33 ksi	F_{tuw} = 31 ksi
			F_{ty} =	19 ksi	F_{tyw} = 12 ksi
			F_{cy} =	16 ksi	F_{cyw} = 12 ksi
			F_{su} =	19 ksi	F_{suw} = 19 ksi
Specification	Design Stress (ksi)		F_{sy} =	11 ksi	k_t = 1.00
			E =	10,400 ksi	

Specification	Design Stress (ksi)		
3.4.1	18	Gross Area	
	28	Net Area	
3.4.2	18.0	Tubes	18.0
3.4.3	21.0		
3.4.4	23.5		
3.4.5	56		
3.4.6	37		

	Design Stress (ksi), Slenderness $< S_1$	S_1	Design Stress (ksi), $S_1 <$ Slenderness $< S_2$	S_2	Design Stress (ksi), Slenderness $> S_2$
3.4.8	15	6.1	$18.1 - 0.510\,(b/t)$	18	$160/(b/t)$
3.4.8.1	15	6.1	$18.1 - 0.510\,(b/t)$	24	$3,400/(b/t)^2$
3.4.9	15	19	$18.1 - 0.161\,(b/t)$	56	$510/(b/t)$
3.4.9.2	15	3.8	$15.3 - 0.078\,\lambda_s$	131	$87,000/\lambda_s^2$
3.4.10	15	20	$17.7 - 0.600\,\sqrt{R/t}$	(1)	$5,100 \Big/ \left[(R/t)\left(1 + \dfrac{\sqrt{R/t}}{35}\right)^2\right]$
3.4.11	15.0	4.6	$15.3 - 0.065\,(L_b/r_y)$	157	$126,000/(L_b/r_y)^2$
3.4.12	18.0	36	$26.5 - 1.42\,\sqrt{R/t}$	118	See 3.4.10
3.4.13	20.0	12	$24.1 - 0.350\left(\dfrac{d}{t}\sqrt{\dfrac{L_b}{d}}\right)$	45	$16,500/[(d/t)^2(L_b/d)]$
3.4.14	15.0	5.8	$15.3 - 0.125\sqrt{\dfrac{L_bS_c}{0.5\sqrt{I_yJ}}}$	6,700	$34,000/(L_bS_c/0.5\sqrt{I_yJ})$
3.4.15	15.0	6.1	$18.1 - 0.510\,(b/t)$	18	$160/(b/t)$
3.4.16	15.0	19	$18.1 - 0.161\,(b/t)$	56	$510/(b/t)$
3.4.16.1	18.0	0	$17.7 - 0.600\,\sqrt{R/t}$	(1)	$5,500 \Big/ \left[(R/t)\left(1 + \dfrac{\sqrt{R/t}}{35}\right)^2\right]$
3.4.16.3	15.0	3.8	$15.3 - 0.078\,\lambda_s$	131	$87,000/\lambda_s^2$
3.4.17	20.0	7.6	$24.1 - 0.540\,(b/t)$	30	$7,100/(b/t)^2$
3.4.18	20.0	40	$24.1 - 0.103\,(h/t)$	117	$1,400/(h/t)$
3.4.19	20.0	91	$24.1 - 0.045\,(h/t)$	270	$3,200/(h/t)$
3.4.20	10.5	37	$13.4 - 0.078\,(h/t)$	(1)	$53,000/(h/t)^2$
3.4.21	10.5	74	$18.4 - 0.107\,(a_e/t)$	(1)	$72,000/(a_e/t)^2$

Note

(1) Use the lesser of the design stresses for (a) slenderness between S_1 and S_2, and (b) slenderness greater than S_2.

ALLOY 5454-H34

			F_{tu} =	39 ksi	F_{tuw} = 31 ksi
			F_{ty} =	29 ksi	F_{tyw} = 12 ksi
			F_{cy} =	27 ksi	F_{cyw} = 12 ksi
			F_{su} =	23 ksi	F_{suw} = 19 ksi
	Design Stress (ksi)		F_{sy} =	17 ksi	k_t = 1.00
Specification			E =	10,400 ksi	

Specification	Design Stress (ksi)		
3.4.1	28	Gross Area	
	33	Net Area	
3.4.2	27.5	Tubes	27.5
3.4.3	32.0		
3.4.4	36.0		
3.4.5	66		
3.4.6	44		

	Design Stress (ksi), Slenderness $<S_1$	S_1	Design Stress (ksi), $S_1 <$ Slenderness $< S_2$	S_2	Design Stress (ksi), Slenderness $> S_2$
3.4.8	25.5	5.4	$32 - 1.21\,(b/t)$	13	$210/(b/t)$
3.4.8.1	25.5	5.4	$32 - 1.21\,(b/t)$	18	$3,400/(b/t)^2$
3.4.9	25.5	17	$32 - 0.380\,(b/t)$	42	$680/(b/t)$
3.4.9.2	25.5	6.7	$26.7 - 0.180\,\lambda_s$	99	$87,000/\lambda_s^2$
3.4.10	25.5	17	$30.6 - 1.25\,\sqrt{R/t}$	(1)	$5,100 \left/ \left[(R/t)\left(1 + \dfrac{\sqrt{R/t}}{35} \right)^2 \right] \right.$
3.4.11	25.5	8.0	$26.7 - 0.150\,(L_b/r_y)$	119	$126,000/(L_b/r_y)^2$
3.4.12	30.0	29	$45.9 - 2.94\,\sqrt{R/t}$	82	See 3.4.10
3.4.13	33.5	11	$42.6 - 0.830 \left(\dfrac{d}{t}\sqrt{\dfrac{L_b}{d}} \right)$	34	$16,500/[(d/t)^2(L_b/d)]$
3.4.14	25.5	17	$26.7 - 0.290\sqrt{\dfrac{L_b S_c}{0.5\sqrt{I_y J}}}$	3,830	$34,000/(L_b S_c/0.5\sqrt{I_y J})$
3.4.15	25.5	5.4	$32 - 1.21\,(b/t)$	13	$210/(b/t)$
3.4.16	25.5	17	$32 - 0.380\,(b/t)$	42	$680/(b/t)$
3.4.16.1	30.0	0.2	$30.6 - 1.25\,\sqrt{R/t}$	(1)	$5,500 \left/ \left[(R/t)\left(1 + \dfrac{\sqrt{R/t}}{35} \right)^2 \right] \right.$
3.4.16.3	25.5	6.7	$26.7 - 0.180\,\lambda_s$	99	$87,000/\lambda_s^2$
3.4.17	33.5	7.2	$42.6 - 1.27\,(b/t)$	22	$7,100/(b/t)^2$
3.4.18	33.5	38	$42.6 - 0.240\,(h/t)$	88	$1,870/(h/t)$
3.4.19	33.5	87	$42.6 - 0.105\,(h/t)$	203	$4,300/(h/t)$
3.4.20	16.0	36	$21.7 - 0.160\,(h/t)$	(1)	$53,000/(h/t)^2$
3.4.21	16.0	63	$29.8 - 0.220\,(a_e/t)$	(1)	$72,000/(a_e/t)^2$

Note

(1) Use the lesser of the design stresses for (a) slenderness between S_1 and S_2, and (b) slenderness greater than S_2.

ALLOY 5456-H116

			F_{tu}	=	46 ksi	F_{tuw} = 42 ksi
			F_{ty}	=	33 ksi	F_{tyw} = 19 ksi
			F_{cy}	=	27 ksi	F_{cyw} = 19 ksi
			F_{su}	=	27 ksi	F_{suw} = 25 ksi
Specification	Design Stress (ksi)		F_{sy}	=	19 ksi	k_t = 1.00
			E	=	10,400 ksi	

Specification	Design Stress (ksi)		
3.4.1	31	Gross Area	
	39	Net Area	
3.4.2	31.5	Tubes	31.5
3.4.3	36.5		
3.4.4	41.0		
3.4.5	78		
3.4.6	52		

	Design Stress (ksi), Slenderness $<S_1$	S_1	Design Stress (ksi), $S_1 <$ Slenderness $< S_2$	S_2	Design Stress (ksi), Slenderness $> S_2$
3.4.8	25.5	5.4	$32 - 1.21\,(b/t)$	13	$210/(b/t)$
3.4.8.1	25.5	5.4	$32 - 1.21\,(b/t)$	18	$3,400/(b/t)^2$
3.4.9	25.5	17	$32 - 0.380\,(b/t)$	42	$680/(b/t)$
3.4.9.2	25.5	6.7	$26.7 - 0.180\,\lambda_s$	99	$87,000/\lambda_s^2$
3.4.10	25.5	17	$30.6 - 1.25\,\sqrt{R/t}$	(1)	$5,100 \left/ \left[(R/t)\left(1 + \dfrac{\sqrt{R/t}}{35}\right)^2 \right]\right.$
3.4.11	25.5	8.0	$26.7 - 0.150\,(L_b/r_y)$	119	$126,000/(L_b/r_y)^2$
3.4.12	30.0	29	$45.9 - 2.94\,\sqrt{R/t}$	82	See 3.4.10
3.4.13	33.5	11	$42.6 - 0.830\left(\dfrac{d}{t}\sqrt{\dfrac{L_b}{d}}\right)$	34	$16,500/[(d/t)^2(L_b/d)]$
3.4.14	25.5	17	$26.7 - 0.290\sqrt{\dfrac{L_b S_c}{0.5\sqrt{I_y J}}}$	3,830	$34,000/(L_b S_c/0.5\sqrt{I_y J})$
3.4.15	25.5	5.4	$32 - 1.21\,(b/t)$	13	$210/(b/t)$
3.4.16	25.5	17	$32 - 0.380\,(b/t)$	42	$680/(b/t)$
3.4.16.1	30.0	0.2	$30.6 - 1.25\,\sqrt{R/t}$	(1)	$5,500 \left/ \left[(R/t)\left(1 + \dfrac{\sqrt{R/t}}{35}\right)^2 \right]\right.$
3.4.16.3	25.5	6.7	$26.7 - 0.180\,\lambda_s$	99	$87,000/\lambda_s^2$
3.4.17	33.5	7.2	$42.6 - 1.27\,(b/t)$	22	$7,100/(b/t)^2$
3.4.18	33.5	38	$42.6 - 0.240\,(h/t)$	88	$1,870/(h/t)$
3.4.19	33.5	87	$42.6 - 0.105\,(h/t)$	203	$4,300/(h/t)$
3.4.20	18.0	34	$24.5 - 0.192\,(h/t)$	(1)	$53,000/(h/t)^2$
3.4.21	18.0	60	$33.7 - 0.260\,(a_e/t)$	(1)	$72,000/(a_e/t)^2$

Note
(1) Use the lesser of the design stresses for (a) slenderness between S_1 and S_2, and (b) slenderness greater than S_2.

ALLOY 6005-T5

$F_{tu} =$	38 ksi	$F_{tuw} =$ 24 ksi	
$F_{ty} =$	35 ksi	$F_{tyw} =$ 13 ksi	
$F_{cy} =$	35 ksi	$F_{cyw} =$ 13 ksi	
$F_{su} =$	24 ksi	$F_{suw} =$ 15 ksi	
$F_{xy} =$	20 ksi	$k_t =$ 1.00	
$E =$	10,100 ksi		

Specification	Design Stress (ksi)		
3.4.1	33	Gross Area	
	32	Net Area	
3.4.2	32.5	Tubes	33.0
3.4.3	39.0		
3.4.4	43.0		
3.4.5	65		
3.4.6	43		

	Design Stress (ksi), Slenderness $<S_1$	S_1	Design Stress (ksi), $S_1 <$ Slenderness $< S_2$	S_2	Design Stress (ksi), Slenderness $> S_2$
3.4.8	33	4.1	$38.3 - 1.30 \, (b/t)$	10	$260/(b/t)$
3.4.8.1	33	4.1	$38.3 - 1.30 \, (b/t)$	12	$3,300/(b/t)^2$
3.4.9	33	13	$38.3 - 0.410 \, (b/t)$	33	$810/(b/t)$
3.4.9.2	33	2.4	$33.5 - 0.210 \, \lambda_s$	66	$85,000/\lambda_s^2$
3.4.10	33	7.9	$36.7 - 1.32 \, \sqrt{R/t}$	132	$5,000 \Big/ \left[(R/t)\left(1 + \dfrac{\sqrt{R/t}}{35}\right)^2 \right]$
3.4.11	33.0	2.9	$33.5 - 0.174 \, (L_b/r_y)$	79	$122,000/(L_b/r_y)^2$
3.4.12	39.0	18	$55.1 - 3.79 \, \sqrt{R/t}$	55	See 3.4.10
3.4.13	43.0	11	$56.8 - 1.30 \left(\dfrac{d}{t}\sqrt{\dfrac{L_b}{d}}\right)$	29	$16,000/[(d/t)^2(L_b/d)]$
3.4.14	33.0	2.3	$33.5 - 0.330 \sqrt{\dfrac{L_b S_c}{0.5\sqrt{I_y J}}}$	1,700	$33,000/(L_b S_c/0.5\sqrt{I_y J})$
3.4.15	33.0	4.1	$38.3 - 1.30 \, (b/t)$	10	$260/(b/t)$
3.4.16	33.0	13	$38.3 - 0.410 \, (b/t)$	33	$810/(b/t)$
3.4.16.1	39.0	0	$36.7 - 1.32 \, \sqrt{R/t}$	140	$5,300 \Big/ \left[(R/t)\left(1 + \dfrac{\sqrt{R/t}}{35}\right)^2 \right]$
3.4.16.3	33.0	2.4	$33.5 - 0.210 \, \lambda_s$	66	$85,000/\lambda_s^2$
3.4.17	43.0	7.0	$56.8 - 1.98 \, (b/t)$	19	$6,900/(b/t)^2$
3.4.18	43.0	36	$56.8 - 0.380 \, (h/t)$	75	$2,100/(h/t)$
3.4.19	43.0	84	$56.8 - 0.164 \, (h/t)$	173	$4,900/(h/t)$
3.4.20	19.0	29	$23.2 - 0.147 \, (h/t)$	59	$51,000/(h/t)^2$
3.4.21	19.0	65	$31.9 - 0.200 \, (a_e/t)$	142	$70,000/(a_e/t)^2$

ALLOY 6061-T6

$F_{tu} = 38$ ksi	$F_{tuw} = 24$ ksi
$F_{ty} = 35$ ksi	$F_{tyw} = 15$ ksi
$F_{cy} = 35$ ksi	$F_{cyw} = 15$ ksi
$F_{su} = 24$ ksi	$F_{suw} = 15$ ksi
$F_{sy} = 20$ ksi	$k_t = 1.00$
$E = 10,100$ ksi	

Specification	Design Stress (ksi)		
3.4.1	33	Gross Area	
	32	Net Area	
3.4.2	32.5	Tubes	33.0
3.4.3	39.0		
3.4.4	43.0		
3.4.5	65		
3.4.6	43		

	Design Stress (ksi), Slenderness $<S_1$	S_1	Design Stress (ksi), $S_1 <$ Slenderness $< S_2$	S_2	Design Stress (ksi), Slenderness $> S_2$
3.4.8	33	4.1	$38.3 - 1.30\,(b/t)$	10	$260/(b/t)$
3.4.8.1	33	4.1	$38.3 - 1.30\,(b/t)$	12	$3,300/(b/t)^2$
3.4.9	33	13	$38.3 - 0.410\,(b/t)$	33	$810/(b/t)$
3.4.9.2	33	2.4	$33.5 - 0.210\,\lambda_s$	66	$85,000/\lambda_s^2$
3.4.10	33	7.9	$36.7 - 1.32\,\sqrt{R/t}$	132	$5,000 \Big/ \left[(R/t)\left(1 + \dfrac{\sqrt{R/t}}{35}\right)^2 \right]$
3.4.11	33.0	2.9	$33.5 - 0.174\,(L_b/r_y)$	79	$122,000/(L_b/r_y)^2$
3.4.12	39.0	18	$55.1 - 3.79\,\sqrt{R/t}$	55	See 3.4.10
3.4.13	43.0	11	$56.8 - 1.30\left(\dfrac{d}{t}\sqrt{\dfrac{L_b}{d}}\right)$	29	$16,000/[(d/t)^2(L_b/d)]$
3.4.14	33.0	2.3	$33.5 - 0.330\sqrt{\dfrac{L_b S_c}{0.5\sqrt{I_y J}}}$	1,700	$33,000/(L_b S_c/0.5\sqrt{I_y J})$
3.4.15	33.0	4.1	$38.3 - 1.30\,(b/t)$	10	$260/(b/t)$
3.4.16	33.0	13	$38.3 - 0.410\,(b/t)$	33	$810/(b/t)$
3.4.16.1	39.0	0	$36.7 - 1.32\,\sqrt{R/t}$	140	$5,300 \Big/ \left[(R/t)\left(1 + \dfrac{\sqrt{R/t}}{35}\right)^2 \right]$
3.4.16.3	33.0	2.4	$33.5 - 0.210\,\lambda_s$	66	$85,000/\lambda_s^2$
3.4.17	43.0	7.0	$56.8 - 1.98\,(b/t)$	19	$6,900/(b/t)^2$
3.4.18	43.0	36	$56.8 - 0.380\,(h/t)$	75	$2,100/(h/t)$
3.4.19	43.0	84	$56.8 - 0.164\,(h/t)$	173	$4,900/(h/t)$
3.4.20	19.0	29	$23.2 - 0.147\,(h/t)$	59	$51,000/(h/t)^2$
3.4.21	19.0	65	$31.9 - 0.200\,(a_e/t)$	142	$70,000/(a_e/t)^2$

ALLOY 6063-T5

F_{tu} =	22 ksi	F_{tuw} =	17 ksi
F_{ty} =	16 ksi	F_{tyw} =	8 ksi
F_{cy} =	16 ksi	F_{cyw} =	8 ksi
F_{su} =	13 ksi	F_{suw} =	11 ksi
F_{sy} =	9 ksi	k_t = 1.00	
E =	10,100 ksi		

Specification	Design Stress (ksi)		
3.4.1	15	Gross Area	
	19	Net Area	
3.4.2	15.0	Tubes	15.0
3.4.3	18.0		
3.4.4	20.0		
3.4.5	37		
3.4.6	25		

	Design Stress (ksi), Slenderness $<S_1$	S_1	Design Stress (ksi), $S_1 <$ Slenderness $< S_2$	S_2	Design Stress (ksi), Slenderness $> S_2$
3.4.8	15	4.3	$16.6 - 0.370\,(b/t)$	16	$168/(b/t)$
3.4.8.1	15	4.3	$16.6 - 0.370\,(b/t)$	18	$3,300/(b/t)^2$
3.4.9	15	14	$16.6 - 0.117\,(b/t)$	50	$540/(b/t)$
3.4.9.2	15	0	$14.7 - 0.061\,\lambda_s$	99	$85,000/\lambda_s^2$
3.4.10	15	8.3	$16.3 - 0.450\,\sqrt{R/t}$	(1)	$5,000 \left/ \left[(R/t)\left(1 + \dfrac{\sqrt{R/t}}{35} \right)^2 \right] \right.$
3.4.11	15.0	0	$14.7 - 0.051\,(L_b/r_y)$	119	$122,000/(L_b/r_y)^2$
3.4.12	18.0	25	$24.5 - 1.29\,\sqrt{R/t}$	95	See 3.4.10
3.4.13	20.0	11	$24.1 - 0.360\left(\dfrac{d}{t}\sqrt{\dfrac{L_b}{d}} \right)$	45	$16,000/[(d/t)^2(L_b/d)]$
3.4.14	15.0	0	$14.7 - 0.097\sqrt{\dfrac{L_b S_c}{0.5\sqrt{I_y J}}}$	3,830	$33,000/(L_b S_c/0.5\sqrt{I_y J})$
3.4.15	15.0	4.3	$16.6 - 0.370\,(b/t)$	16	$168/(b/t)$
3.4.16	15.0	14	$16.6 - 0.117\,(b/t)$	50	$540/(b/t)$
3.4.16.1	18.0	0	$16.3 - 0.450\,\sqrt{R/t}$	(1)	$5,300 \left/ \left[(R/t)\left(1 + \dfrac{\sqrt{R/t}}{35} \right)^2 \right] \right.$
3.4.16.3	15.0	0	$14.7 - 0.061\,\lambda_s$	99	$85,000/\lambda_s^2$
3.4.17	20.0	7.5	$24.1 - 0.550\,(b/t)$	29	$6,900/(b/t)^2$
3.4.18	20.0	39	$24.1 - 0.104\,(h/t)$	115	$1,380/(h/t)$
3.4.19	20.0	91	$24.1 - 0.045\,(h/t)$	270	$3,200/(h/t)$
3.4.20	8.5	34	$9.9 - 0.041\,(h/t)$	(1)	$51,000/(h/t)^2$
3.4.21	8.5	91	$13.6 - 0.056\,(a_c/t)$	(1)	$70,000/(a_c/t)^2$

Note
(1) Use the lesser of the design stresses for (a) slenderness between S_1 and S_2, and (b) slenderness greater than S_2.

ALLOY 6063-T6

F_{tu} = 30 ksi	F_{tuw} = 17 ksi
F_{ty} = 25 ksi	F_{tyw} = 8 ksi
F_{cy} = 25 ksi	F_{cyw} = 8 ksi
F_{su} = 19 ksi	F_{suw} = 11 ksi
F_{sy} = 14 ksi	k_t = 1.00
E = 10,100 ksi	

Specification	Design Stress (ksi)		
3.4.1	24	Gross Area	
	26	Net Area	
3.4.2	24.0	Tubes	24.0
3.4.3	28.0		
3.4.4	31.0		
3.4.5	51		
3.4.6	34		

	Design Stress (ksi), Slenderness $<S_1$	S_1	Design Stress (ksi), $S_1 <$ Slenderness $< S_2$	S_2	Design Stress (ksi), Slenderness $> S_2$
3.4.8	24	3.6	$26.7 - 0.760\,(b/t)$	12	$210/(b/t)$
3.4.8.1	24	3.6	$26.7 - 0.760\,(b/t)$	15	$3,300/(b/t)^2$
3.4.9	24	11	$26.7 - 0.240\,(b/t)$	39	$680/(b/t)$
3.4.9.2	24	0	$23.5 - 0.123\,\lambda_x$	78	$85,000/\lambda_x^2$
3.4.10	24	5.2	$25.9 - 0.830\,\sqrt{R/t}$	(1)	$5,000\left/\left[(R/t)\left(1 + \dfrac{\sqrt{R/t}}{35}\right)^2\right]\right.$
3.4.11	24.0	0	$23.5 - 0.102\,(L_b/r_y)$	94	$122,000/(L_b/r_y)^2$
3.4.12	28.0	21	$38.8 - 2.38\,\sqrt{R/t}$	70	See 3.4.10
3.4.13	31.0	11	$39.2 - 0.750\left(\dfrac{d}{t}\sqrt{\dfrac{L_b}{d}}\right)$	35	$16,000/[(d/t)^2(L_b/d)]$
3.4.14	24.0	0	$23.5 - 0.196\sqrt{\dfrac{L_b S_c}{0.5\sqrt{I_y J}}}$	2,380	$33,000/(L_b S_c/0.5\sqrt{I_y J})$
3.4.15	24.0	3.6	$26.7 - 0.760\,(b/t)$	12	$210/(b/t)$
3.4.16	24.0	11	$26.7 - 0.240\,(b/t)$	39	$680/(b/t)$
3.4.16.1	28.0	0	$25.9 - 0.830\,\sqrt{R/t}$	(1)	$5,300\left/\left[(R/t)\left(1 + \dfrac{\sqrt{R/t}}{35}\right)^2\right]\right.$
3.4.16.3	24.0	0	$23.5 - 0.123\,\lambda_x$	78	$85,000/\lambda_x^2$
3.4.17	31.0	7.3	$39.2 - 1.13\,(b/t)$	23	$6,900/(b/t)^2$
3.4.18	31.0	37	$39.2 - 0.220\,(h/t)$	90	$1,770/(h/t)$
3.4.19	31.0	87	$39.2 - 0.094\,(h/t)$	208	$4,100/(h/t)$
3.4.20	13.5	28	$15.8 - 0.083\,(h/t)$	(1)	$51,000/(h/t)^2$
3.4.21	13.5	72	$21.7 - 0.114\,(a_e/t)$	(1)	$70,000/(a_e/t)^2$

Note
   (1) Use the lesser of the design stresses for (a) slenderness between S_1 and S_2, and (b) slenderness greater than S_2.

APPENDIX J
Other Aluminum Structural Design Specifications

The Aluminum Association's *Specification for Aluminum Structures* isn't the only design standard for aluminum, although it does form the basis for many of the others. These other standards can be divided into two categories: general standards (like the *Specification for Aluminum Structures*, hereafter called the Aluminum *Specification*), and application-specific standards, which provide design rules for particular aluminum structural applications, such as pressure vessels. The important thing to remember is that all the U.S. specifications are based on the Aluminum *Specification*, which means they're the same as some part of some edition of the Aluminum *Specification*, or at least their authors thought so.

J.1 GENERAL DESIGN SPECIFICATIONS

J.1.1 International Standards

While the Aluminum *Specification* is the only general design specification used in the U.S., other nations have their own. The following list isn't all-inclusive, but it covers much of the English-speaking world:

Australia: The standards-writing body Australia Standards has produced AS 1664, *Aluminium Structures*, which is based on the Aluminum *Specification* with the exception of Section 8, Testing.

Canada: CSA CAN3-S157-M83, *Strength Design in Aluminum* (96), is an LRFD specification different from the LRFD Aluminum *Specification*. Published in 1983 by the Canadian Standards Association (CSA), an update has been drafted but not yet issued.

United Kingdom: British Standard BS 8118, *Structural Use of Aluminum*, Part I, Code of Practice for Design is also a limit state design specification.

Europe: Eurocode 9, *Design of Aluminium Structures* CEN/TC250/SC9 (102) is a limit state specification issued in 1997 as a prestandard by CEN, the European Committee for Standardization. (The CEN acronym is based on the French name for this committee, which is Comité Européen de Normalisation.) CEN members are the national standards bodies of Austria, Belgium, the Czech Republic, Denmark, Finland, France, Germany, Greece, Iceland, Ireland, Italy, Luxembourg, the Netherlands, Norway, Portugal, Spain, Sweden, Switzerland, and the United Kingdom. Eurocode 9 will be considered for adoption as a standard, but in the meantime, these nations may maintain their individual national standards.

Eurocode 9 includes:

Part 1.1 Members (including connections)
Part 1.2 Structural Fire Design
Part 2   Fatigue

Part 1.2 (Structural Fire Design) is the only code that addresses this issue.

International Organization for Standardization: ISO/TR 11069 *Aluminium Structures*—Material and Design—Ultimate Limit State under Static Loading (116) includes limit state design rules and a commentary. This document is a technical report rather than an international standard and was prepared by the ISO technical committee TC167, Steel and Aluminium Structures subcommittee SC 3, Aluminium Structures.

J.1.2   U.S. Building Codes

Three model building code organizations operate in the U.S. Of these, only the International Conference of Building Officials (ICBO) based in Whittier, CA, printed its own aluminum design specification as part of its model code, the *Uniform Building Code* (UBC) (115). (The "International" part of ICBO is something of an overstatement since the UBC hasn't even been adopted throughout the U.S.) The ICBO's aluminum specification was, however, only a regurgitation of parts of various previous editions of the Aluminum *Specification* and included no new or (intentionally) contradictory requirements.

Dealing with the U.S. model building codes appeared to be getting much simpler when all three codes of these were merged into the International Code Council's (ICC) *International Building Code* (IBC) in 2000, which adopted the Aluminum *Specification* by reference in Chapter 20. Then the National Fire Protection Association (NFPA) felt the need to promote a competing building code (NFPA 5000). As of this writing, NFPA 5000 is available in draft form only; Chapter 41 is intended for aluminum.

J.2   APPLICATION-SPECIFIC STANDARDS

When aluminum is used in structural applications, design rules unique to that application have often been codified by industry associations. Because steel is also used to make these products, aluminum and steel rules are often found side by side in the standards.

J.2.1 Piping

Aluminum's corrosion-resistance and good mechanical properties at low temperatures have led to its use in piping. The standard governing this application is the American Society of Mechanical Engineers (ASME) standard *B31.3 Process Piping* (84), formerly called *Chemical Plant and Petroleum Refinery Piping*. Its scope is piping in petroleum refineries, chemical, pharmaceutical, textile, paper, semiconductor, and cryogenic plants and terminals. Allowable stresses for aluminum piping alloys are given to 400°F.

A nonferrous flange standard (ASME B16.31), which addressed three aluminum alloys and other nonferrous materials, was published in 1971 but withdrawn in 1981 to avoid conflict with a nickel alloy flange standard (B16.5) published that year. ASME has drafted B16.46, *Aluminum Alloy Pipe Flanges*, but it has not been approved yet. Aluminum flanges are currently covered in Appendix L to B31.3.

J.2.2 Tanks and Pressure Vessels

For liquefied natural gas (LNG) storage tanks, a cryogenic application, aluminum is addressed in the American Petroleum Institute (API) Standard 620, *Design and Construction of Large, Welded, Low-Pressure Storage Tanks* (44), Appendix Q, Low-Pressure Storage Tanks for Liquefied Hydrocarbon Gases. Because this appendix includes steel alloys, its scope limits design metal temperatures to −270°F or above, but aluminum can also be used at lower temperatures.

For storage tanks with design temperatures up to 400°F, a more general design standard is ASME *B96.1 Welded Aluminum-Alloy Storage Tanks* (85). Its scope is vertical, cylindrical, flat-bottom tanks storing liquids at pressures at or below 1 psi. Design rules in B96.1 are consistent with the Aluminum *Specification* except that some safety factors are higher to account for the vagaries of field welding. B96.1 also accounts for creep effects at elevated temperatures. Used appropriately, this standard provides some allowable stresses for general structural applications of aluminum at elevated temperatures, which are otherwise hard to find.

Fixed aluminum dome roofs (as discussed in Section 10.2) and aluminum internal floating roofs are common on steel storage tanks. The design of these roofs is addressed in API Standard 650, *Welded Steel Tanks for Oil Storage* (45), Appendix G, Structurally Supported Aluminum Dome Roofs and Appendix H, Internal Floating Roofs, respectively. These applications are limited to design temperatures of 200°F or less.

Aluminum dome roofs are also used on steel water-storage tanks; structural design of these roofs is addressed in the American Water Works Association (AWWA) standard D100-96, *Welded Steel Tanks for Water Storage*, Section 15, Structurally Supported Aluminum Dome Roofs. A useful provision of this standard is a formula for determining an overall buckling external pressure for such domes. This formula gives the same results as we did in Section 10.2.3 and the example there. AWWA standard D103-87, *Factory-Coated Bolted Steel Tanks for Water Storage*, Appendix A, also addresses aluminum dome roofs.

J.2.3   Highway Bridges

The American Association of State Highway and Transportation Officials (AASHTO) maintains standard specifications for highway structures. States usually don't adopt these specifications word for word, but use them as models. For allowable stress design, AASHTO offers *Guide Specifications for Aluminum Highway Bridges* (1991) (36), based on the ASD Aluminum *Specification* provisions with the safety factors for bridges. Additional considerations unique to bridges, such as deflection and slenderness limits and minimum thicknesses, are provided. Fatigue provisions are in tabular form, rather than presented as equations. For redundant load path structures, allowable fatigue stresses are based on the 5th edition of the Aluminum *Specifications* (22); for nonredundant load path structures, allowable fatigue stresses are reduced by calculating them for a higher number of cycles than are actually anticipated.

For load and resistance factor design, AASHTO provides *LRFD Bridge Design Specifications* (1994) (35), Section 7, aluminum structures. This standard has lower resistance factors than those appearing in the LRFD version of the Aluminum *Specification* because bridges are designed to have greater reliability than building structures. The fatigue provisions are based on the 5th (1986) edition of the Aluminum *Specifications*, but they are in the form of an equation rather than a table and are slightly different from the current Aluminum *Specification*. The AASHTO LRFD *Specifications* also account for the variable amplitude loads experienced on bridges by providing a floor on allowable fatigue stresses for variable amplitude loads equal to one-half of the constant amplitude fatigue thresholds.

J.2.4   Supports for Highway Signs, Luminaires, and Traffic Signals

AASHTO also publishes *Standard Specifications for Structural Supports for Highway Signs, Luminaires, and Traffic Signals*. (Everyone else calls luminaires light poles.) Section 5, Aluminum Design, uses the Aluminum *Specification* allowable-stress design. A revised version of this standard has been drafted; it uses allowable stress design and covers static design of aluminum in Section 6 and fatigue design in Section 11. Fatigue details in Section 11 address situations beyond those shown in the Aluminum *Specification*, but they are rather conservative for aluminum.

TABLE J.1 ASTM Corrugated Pipe Specifications

Type of Pipe	Material Specification	Fabrication Specification	Installation Specification	Design Specification
Factory made	B744 (M197)	B745 (M196)	B788	B790
Structural plate	B746 (M219)	B746 (M219)	B789	B790

Note
  AASHTO specifications are given in parentheses.

J.2.5 Corrugated Pipe for Culverts

The American Society for Testing and Materials (ASTM) Subcommittee B07.08 writes a number of standards on corrugated aluminum pipe, pipe arches, and arches for culverts, storm sewers, and other buried conduits. Very similar versions of these standards are also available from AASHTO, but cleverly concealed with different designation numbers. A summary of the ASTM specifications is given in Table J.1. These ASTM specifications are included in the ASTM volume *Aluminum and Magnesium Alloys.*

J.2.6 Transmission Towers

Aluminum transmission towers, usually of 6061-T6 extrusions with 2024-T4 bolts, have been used at sites where access is difficult because they can be delivered by helicopter and are corrosion resistant. The "Guide for the Design of Aluminum Transmission Towers" appeared in the ASCE *Journal of the Structural Division*, December 1972, issue. Allowable stresses in this guide are higher than those in the Aluminum *Specification*. Fastener material specifications are covered by ASTM F901, *Aluminum Transmission Tower Bolts and Nuts.*

J.2.7 Railroad Cars

Aluminum welding used in rail cars is addressed in the American Welding Society's (AWS) D15.1 *Railroad Welding Specification—Cars and Locomotives.* Unfortunately, Table J2 of this document, which is titled "Allowable Weld Stresses (Aluminum)" lists shear *strengths*, not *allowable stresses*, for various aluminum filler alloys. So, don't forget to add a safety factor before using the table values for allowable stress.

APPENDIX K
Buckling Constants

Buckling Constants for Aluminum Alloys

Alloy Temper	Product*	Compression in Columns			Compression in Flat Plates			Compression in Round Tubes			Bending in Rectangular Bars			Bending in Round Tubes			Shear in Flat Plates		
		B_c (ksi)	D_c (ksi)	C_c	B_p (ksi)	D_p (ksi)	C_p	B_t (ksi)	D_t (ksi)	C_t	B_{br} (ksi)	D_{br} (ksi)	C_{br}	B_{tb} (ksi)	D_{tb} (ksi)	C_{tb}	B_s (ksi)	D_s (ksi)	C_s
1100-H12	Sheet and Plate	11.0	0.044	165	12.8	0.056	153	12.7	0.372	573	17.0	0.085	133	19.1	0.875	160	8.2	0.029	190
1100-H14	Drawn Tube	14.5	0.067	144	17.0	0.086	133	16.7	0.536	446	22.6	0.131	115	25.1	1.260	133	10.7	0.043	167
2014-T6	Sheet	68.6	0.544	52	79.1	0.674	48	74.3	3.132	94	119.4	1.530	52	109.5	8.754	39	45.1	0.290	64
2014-T651	Plate	67.3	0.529	52	77.7	0.656	49	73.0	3.059	95	117.1	1.486	53	109.5	8.754	41	45.9	0.298	63
2014-T6,T6510,T6511	Extrusions	59.9	0.444	55	69.0	0.549	52	65.2	2.629	105	103.6	1.238	56	97.8	7.523	44	40.9	0.250	67
2014-T6, T651	Cold-Finished Rod and Bar, Drawn Tube	61.1	0.458	55	70.5	0.567	51	66.5	2.699	103	105.9	1.278	55	99.7	7.724	44	42.6	0.266	66
Alclad 2014-T6	Sheet (0.039)	64.8	0.502	53	74.8	0.622	49	70.4	2.922	98	112.6	1.408	53	103.6	8.157	40	42.6	0.267	65
Alclad 2014-T6	Sheet (0.249)	67.3	0.531	52	77.7	0.659	48	73.0	3.068	94	117.1	1.493	52	107.6	8.571	39	44.2	0.283	64
Alclad 2014-T651	Plate	64.8	0.502	53	74.8	0.622	49	70.4	2.922	98	112.6	1.408	53	105.6	8.363	42	44.2	0.283	64
3003-H12	Sheet and Plate	11.0	0.044	165	12.8	0.056	153	12.7	0.372	573	17.0	0.085	133	19.1	0.875	160	9.1	0.033	182
3003-H14	Sheet and Plate	15.7	0.075	138	18.4	0.096	127	18.1	0.594	416	24.5	0.147	111	27.1	1.397	127	13.2	0.058	151
3003-H16	Sheet	20.4	0.112	121	24.2	0.145	111	23.5	0.843	327	32.2	0.222	96	35.3	1.984	106	16.6	0.083	134
3003-H18	Sheet	22.8	0.133	114	27.1	0.172	105	26.3	0.977	295	36.1	0.264	91	39.4	2.298	99	19.2	0.103	125
3003-H12	Drawn Tube	12.2	0.052	157	14.2	0.065	145	14.1	0.424	523	18.8	0.100	126	21.1	0.999	150	9.1	0.033	182
3003-H14	Drawn Tube	18.0	0.093	129	21.3	0.120	119	20.8	0.715	366	28.3	0.183	103	31.2	1.683	115	13.2	0.058	151
3003-H16	Drawn Tube	21.6	0.123	118	25.7	0.159	108	24.9	0.909	310	34.1	0.243	94	37.4	2.140	102	16.6	0.083	134
3003-H18	Drawn Tube	24.0	0.144	112	28.6	0.187	102	27.7	1.046	282	38.1	0.286	89	41.5	2.461	96	19.2	0.103	125
Alclad 3003-H12	Sheet and Plate	9.9	0.038	174	11.5	0.047	162	11.4	0.321	635	15.2	0.072	140	17.1	0.756	172	8.2	0.029	190
Alclad 3003-H14	Sheet and Plate	14.5	0.067	144	17.0	0.086	133	16.7	0.536	446	22.6	0.131	115	25.1	1.260	133	12.4	0.053	156
Alclad 3003-H16	Sheet	19.2	0.103	125	22.8	0.132	115	22.2	0.779	345	30.2	0.202	100	33.2	1.832	111	15.8	0.076	138
Alclad 3003-H18	Sheet	21.6	0.123	118	25.7	0.159	108	24.9	0.909	310	34.1	0.243	94	37.4	2.140	102	18.4	0.096	128
Alclad 3003-H14	Drawn Tube	16.8	0.084	133	19.9	0.108	123	19.4	0.654	389	26.4	0.165	107	29.2	1.538	121	12.4	0.053	156
Alclad 3003-H18	Drawn Tube	22.8	0.133	114	27.1	0.172	105	26.3	0.977	295	36.1	0.264	91	39.4	2.298	99	18.4	0.096	128
3004-H32	Sheet and Plate	20.4	0.112	121	24.2	0.145	111	23.5	0.843	327	32.2	0.222	96	35.3	1.984	106	16.6	0.083	134
3004-H34	Sheet and Plate	25.3	0.155	109	30.1	0.201	100	29.0	1.116	269	40.0	0.309	86	43.6	2.626	92	20.1	0.110	122
3004-H36	Sheet	29.0	0.190	102	34.6	0.248	93	33.2	1.334	238	46.1	0.381	81	49.8	3.140	85	22.8	0.132	115
3004-H38	Sheet	33.9	0.241	94	40.7	0.317	86	38.8	1.643	206	54.2	0.487	74	58.2	3.865	76	25.4	0.156	108
3004-H34	Drawn Tube	27.7	0.178	104	33.1	0.232	95	31.8	1.260	247	44.1	0.356	82	47.7	2.966	87	20.1	0.110	122
3004-H36	Drawn Tube	31.4	0.215	98	37.7	0.282	89	36.0	1.486	220	50.1	0.433	77	54.0	3.497	80	22.8	0.132	115
Alclad 3004-H32	Sheet	19.2	0.103	125	22.8	0.132	115	22.2	0.779	345	30.2	0.202	100	33.2	1.832	111	15.8	0.076	138

Alloy	Form																		
Alclad 3004-H34	Sheet	24.0	0.144	112	28.6	0.187	102	27.7	1.046	282	38.1	0.286	89	41.5	2.461	96	19.2	0.103	125
Alclad 3004-H36	Sheet	27.7	0.178	104	33.1	0.232	95	31.8	1.260	247	44.1	0.356	82	47.7	2.966	87	21.9	0.125	117
Alclad 3004-H38	Sheet	32.7	0.228	96	39.2	0.299	87	37.4	1.564	213	52.2	0.459	76	56.1	3.680	78	24.5	0.148	110
Alclad 3004-H131, H241, H341	Sheet	25.3	0.155	109	30.1	0.201	100	29.0	1.116	269	40.0	0.309	86	43.6	2.626	92	21.0	0.117	119
Alclad 3004-H151, H261, H361	Sheet	32.7	0.228	96	39.2	0.299	87	37.4	1.564	213	52.2	0.459	76	56.1	3.680	78	24.5	0.148	110
3005-H25	Sheet	22.8	0.133	114	27.1	0.172	105	26.3	0.977	295	36.1	0.264	91	39.4	2.298	99	17.5	0.089	131
3005-H28	Sheet	29.0	0.190	102	34.6	0.248	93	33.2	1.334	238	46.1	0.381	81	49.8	3.140	85	21.9	0.125	117
3105-H25	Sheet	19.2	0.103	125	22.8	0.132	115	22.2	0.779	345	30.2	0.202	100	33.2	1.832	111	14.9	0.070	142
5005-H12	Sheet and Plate	14.5	0.067	144	17.0	0.086	133	16.7	0.536	446	22.6	0.131	115	25.1	1.260	133	10.7	0.043	167
5005-H14	Sheet and Plate	16.8	0.084	133	19.9	0.108	123	19.4	0.654	389	26.4	0.165	107	29.2	1.538	121	13.2	0.058	151
5005-H16	Sheet	20.4	0.112	121	24.2	0.145	111	23.5	0.843	327	32.2	0.222	96	35.3	1.984	106	15.8	0.076	138
5005-H32	Sheet and Plate	12.2	0.052	157	14.2	0.065	145	14.1	0.424	523	18.8	0.100	126	21.1	0.999	150	9.1	0.033	182
5005-H34	Sheet and Plate	15.7	0.075	138	18.4	0.096	127	18.1	0.594	416	24.5	0.147	111	27.1	1.397	127	11.5	0.048	161
5005-H36	Sheet	18.0	0.093	129	21.3	0.120	119	20.8	0.715	366	28.3	0.183	103	31.2	1.683	115	14.1	0.064	146
5050-H32	Sheet	15.7	0.075	138	18.4	0.096	127	18.1	0.594	416	24.5	0.147	111	27.1	1.397	127	12.4	0.053	156
5050-H34	Sheet	20.4	0.112	121	24.2	0.145	111	23.5	0.843	327	32.2	0.222	96	35.3	1.984	106	15.8	0.076	138
5050-H32	Drawn Tube	16.8	0.084	133	19.9	0.108	123	19.4	0.654	389	26.4	0.165	107	29.2	1.538	121	12.4	0.053	156
5050-H34	Drawn Tube	21.6	0.123	118	25.7	0.159	108	24.9	0.909	310	34.1	0.243	94	37.4	2.140	102	15.8	0.076	138
5052-O	Sheet and Plate	10.4	0.041	170	12.1	0.051	158	12.1	0.345	608	16.1	0.078	137	18.1	0.812	167	7.0	0.023	207
5052-H32	Sheet and Plate	24.0	0.143	112	28.6	0.186	103	27.7	1.042	284	38.1	0.285	89	41.5	2.453	96	18.4	0.095	128
5052-H34	Drawn Tube	27.7	0.177	104	33.1	0.231	96	31.8	1.256	250	44.1	0.355	83	47.7	2.956	88	21.0	0.117	120
5052-H36	Sheet	30.2	0.201	100	36.1	0.263	91	34.6	1.405	231	48.1	0.405	79	51.9	3.306	83	23.7	0.139	113
5083-O	Extrusions	18.0	0.092	131	21.3	0.118	120	20.8	0.708	376	28.3	0.181	104	31.2	1.667	118	12.4	0.052	158
5083-H111	Extrusions	24.0	0.142	113	28.6	0.184	104	27.7	1.036	289	38.1	0.282	90	41.5	2.437	97	19.2	0.101	127
5083-O	Sheet and Plate	20.4	0.111	123	24.2	0.143	113	23.5	0.835	336	32.2	0.219	98	35.3	1.965	108	14.1	0.063	148
5083-H116, H321	Sheet and Plate (1.500)	30.2	0.199	101	36.1	0.261	92	34.6	1.396	235	48.1	0.401	80	51.9	3.285	84	25.4	0.154	110
5083-H116, H321	Plate (3.000)	27.7	0.175	105	33.1	0.229	96	31.8	1.248	254	44.1	0.351	84	47.7	2.937	89	23.7	0.138	114
5086-O	Extrusions, Sheet and Plate	15.7	0.074	140	18.4	0.095	129	18.1	0.588	427	24.5	0.145	112	27.1	1.384	129	10.7	0.042	170
5086-H111	Extrusions	20.4	0.111	123	24.2	0.143	113	23.5	0.835	336	32.2	0.219	98	35.3	1.965	108	16.6	0.081	136
5086-H112	Plate (0.500)	19.2	0.101	127	22.8	0.130	116	22.2	0.771	355	30.2	0.199	101	33.2	1.814	113	14.1	0.063	148
5086-H112	Plate (1.000)	18.0	0.092	131	21.3	0.118	120	20.8	0.708	376	28.3	0.181	104	31.2	1.667	118	12.4	0.052	158
5086-H112	Plate (3.000)	16.8	0.083	135	19.9	0.106	125	19.4	0.647	400	26.4	0.163	108	29.2	1.523	123	10.7	0.042	170
5086-H116	Sheet and Plate	30.2	0.199	101	36.1	0.261	92	34.6	1.396	235	48.1	0.401	80	51.9	3.285	84	22.8	0.130	116
5086-H32	Sheet, Plate, Drawn Tube	30.2	0.199	101	36.1	0.261	92	34.6	1.396	235	48.1	0.401	80	51.9	3.285	84	22.8	0.130	116
5086-H34	Sheet, Plate, Drawn Tube	37.7	0.278	90	45.4	0.367	82	43.0	1.867	192	60.5	0.565	71	64.6	4.394	73	28.2	0.180	105

Buckling Constants for Aluminum Alloys

Alloy Temper	Product*	Compression in Columns			Compression in Flat Plates			Compression in Round Tubes			Bending in Rectangular Bars			Bending in Round Tubes			Shear in Flat Plates		
		B_c (ksi)	D_c (ksi)	C_c	B_p (ksi)	D_p (ksi)	C_p	B_t (ksi)	D_t (ksi)	C_t	B_{br} (ksi)	D_{br} (ksi)	C_{br}	B_{tb} (ksi)	D_{tb} (ksi)	C_{tb}	B_s (ksi)	D_s (ksi)	C_s
5154-H38	Sheet	39.0	0.294	88	46.9	0.388	81	44.4	1.956	184	62.6	0.597	70	66.7	4.602	71	29.1	0.189	102
5454-O	Extrusions	13.3	0.058	152	15.6	0.074	140	15.4	0.474	495	20.7	0.113	122	23.1	1.116	144	9.1	0.033	184
5454-H111	Extrusions	18.0	0.092	131	21.3	0.118	120	20.8	0.708	376	28.3	0.181	104	31.2	1.667	118	14.9	0.069	144
5454-H112	Extrusions	14.5	0.066	146	17.0	0.084	135	16.7	0.530	458	22.6	0.129	117	25.1	1.248	136	9.1	0.033	184
5454-O	Sheet and Plate	13.3	0.058	152	15.6	0.074	140	15.4	0.474	495	20.7	0.113	122	23.1	1.116	144	9.1	0.033	184
5454-H32	Sheet and Plate	27.7	0.175	105	33.1	0.229	96	31.8	1.248	254	44.1	0.351	84	47.7	2.937	89	21.0	0.115	121
5454-H34	Sheet and Plate	31.4	0.212	99	37.7	0.278	90	36.0	1.472	227	50.1	0.426	78	54.0	3.463	82	23.7	0.138	114
5456-O	Sheet and Plate	21.6	0.121	119	25.7	0.156	110	24.9	0.900	499	34.1	0.239	95	37.4	2.119	104	14.9	0.069	144
5456-H116, H321	Sheet and Plate (1.250)	31.4	0.212	99	37.7	0.278	90	36.0	1.472	227	50.1	0.426	78	54.0	3.463	82	27.3	0.171	106
5456-H116, H321	Plate (1.500)	29.0	0.187	103	34.6	0.245	94	33.2	1.321	244	46.1	0.376	82	49.8	3.110	86	25.4	0.154	110
5456-H116, H321	Plate (3.000)	29.0	0.187	103	34.6	0.245	94	33.2	1.321	244	46.1	0.376	82	49.8	3.110	86	23.7	0.138	114
6005-T5	Extrusions	39.4	0.246	66	45.0	0.301	61	43.2	1.558	141	66.8	0.665	67	64.8	4.458	55	26.1	0.133	81
6105-T5	Extrusions	39.4	0.246	66	45.0	0.301	61	43.2	1.558	141	66.8	0.665	67	64.8	4.458	55	26.1	0.133	81
6061-T6, T651	Sheet and Plate	39.4	0.246	66	45.0	0.301	61	43.2	1.558	141	66.8	0.665	67	64.8	4.458	55	26.1	0.133	81
6061-T6, T6510, T6511	Extrusions	39.4	0.246	66	45.0	0.301	61	43.2	1.558	141	66.8	0.665	67	64.8	4.458	55	26.1	0.133	81
6061-T6, T651	Cold Fin. Rod and Bar	39.4	0.246	66	45.0	0.301	61	43.2	1.558	141	66.8	0.665	67	64.8	4.458	55	26.1	0.133	81
6061-T6	Drawn Tube	39.4	0.246	66	45.0	0.301	61	43.2	1.558	141	66.8	0.665	67	64.8	4.458	55	26.1	0.133	81
6061-T6	Pipe	39.4	0.246	66	45.0	0.301	61	43.2	1.558	141	66.8	0.665	67	64.8	4.458	55	26.1	0.133	81
6063-T5	Extrusions (0.500)	17.3	0.072	99	19.5	0.086	93	19.2	0.529	275	28.3	0.183	103	28.8	1.513	95	11.3	0.038	122
6063-T5	Extrusions (1.000)	16.2	0.065	102	18.2	0.078	96	18.0	0.484	290	26.4	0.165	107	26.9	1.384	99	10.6	0.034	127
6063-T52	Extrusions	17.3	0.072	99	19.5	0.086	93	19.2	0.529	275	28.3	0.183	103	28.8	1.513	95	11.3	0.038	122
6063-T6	Extrusions and Pipe	27.6	0.145	78	31.4	0.175	74	30.5	0.978	189	46.1	0.381	81	45.7	2.800	70	18.2	0.077	97
6066-T6, T6510, T6511	Extrusions	51.4	0.366	57	59.0	0.451	54	56.1	2.207	112	88.2	1.010	58	84.1	6.315	47	34.3	0.199	70
6070-T6, T62	Extrusions	51.4	0.366	57	59.0	0.451	54	56.1	2.207	112	88.2	1.010	58	84.1	6.315	47	34.3	0.199	70
6351-T5	Extrusions	39.4	0.246	66	45.0	0.301	61	43.2	1.558	141	66.8	0.665	67	64.8	4.458	55	26.1	0.133	81
6351-T6	Extrusions	41.7	0.268	64	47.8	0.329	60	45.8	1.682	134	71.0	0.729	65	68.6	4.815	53	27.7	0.145	78
6463-T6	Extrusions	27.6	0.145	78	31.4	0.175	74	30.5	0.978	189	46.1	0.381	81	45.7	2.800	70	18.2	0.077	97
7005-T53	Extrusions	48.9	0.334	60	56.2	0.411	56	53.5	2.045	121	83.9	0.918	61	80.2	5.853	49	33.4	0.189	73

*Maximum thickness indicated in parentheses.

Buckling Constants for Welded Aluminum Alloys

Alloy Temper	Product*	Compression in Columns B_c (ksi)	D_c (ksi)	C_c	Compression in Flat Plates B_p (ksi)	D_p (ksi)	C_p	Compression in Round Tubes B_t (ksi)	D_t (ksi)	C_t	Bending in Rectangular Bars B_{br} (ksi)	D_{br} (ksi)	C_{br}	Bending in Round Tubes B_{tb} (ksi)	D_{tb} (ksi)	C_{tb}	Shear in Flat Plates B_s (ksi)	D_s (ksi)	C_s
1100-H12, H14	All	3.7	0.0087	284	4.2	0.010	267	4.3	0.087	1375	5.5	0.016	232	6.4	0.204	332	2.4	0.0046	351
3003-H12, H14, H16, H18	All	5.4	0.015	236	6.1	0.018	221	6.2	0.142	1060	8.1	0.028	192	9.3	0.334	259	3.5	0.0082	290
Alclad 3003-H12, H14, H16, H18	All	4.8	0.013	250	5.5	0.016	234	5.5	0.123	1145	7.2	0.024	203	8.3	0.289	279	3.2	0.0069	307
3004-H32, H34, H36, H38	All	9.3	0.034	180	10.8	0.043	167	10.7	0.297	772	14.3	0.066	145	16.1	0.698	179	6.3	0.019	219
Alclad 3004-H32, H34, H36, H38	All	8.7	0.031	185	10.1	0.039	172	10.1	0.273	795	13.4	0.060	150	15.1	0.642	187	5.9	0.017	226
3005-H25	Sheet	7.0	0.023	206	8.1	0.028	192	8.1	0.204	875	10.7	0.043	167	12.2	0.481	216	4.7	0.012	253
5005-H12, H14, H32, H34	All	5.4	0.015	236	6.1	0.018	221	6.2	0.142	1065	8.1	0.028	192	9.3	0.334	259	3.5	0.0082	290
5050-H32, H34	All	6.5	0.020	215	7.4	0.025	201	7.5	0.183	930	9.8	0.038	175	11.2	0.430	228	4.3	0.011	264
5052-O, H32, H34	All	10.4	0.041	170	12.1	0.051	158	12.1	0.345	732	16.1	0.078	137	18.1	0.812	167	7.0	0.023	207
5083-O, H116, H321	Sheet and Plate (1.500)	20.4	0.111	123	24.2	0.143	113	23.5	0.835	515	32.2	0.219	98	35.3	1.965	118	14.1	0.063	148
5083-O	Extrusions	18.0	0.092	131	21.3	0.118	120	20.8	0.708	552	28.3	0.181	104	31.2	1.667	118	12.4	0.052	158
5083-H111	Extrusions	16.8	0.083	135	19.9	0.106	125	19.4	0.647	573	26.4	0.163	108	29.2	1.523	123	12.4	0.052	158
5086-O, H112, H116, H32, H34	Plate (3.000)	19.2	0.101	127	22.8	0.130	116	22.2	0.771	532	30.2	0.199	101	33.2	1.814	113	13.2	0.058	153
5086-O, H112, H116, H32, H34	All	15.7	0.074	140	18.4	0.095	129	18.1	0.588	596	24.5	0.145	112	27.1	1.384	129	10.7	0.042	170
5086-H111	Extrusions	14.5	0.066	146	17.0	0.084	135	16.7	0.530	622	22.6	0.129	117	25.1	1.248	136	10.7	0.042	170
5154-H38	Sheet	12.2	0.051	158	14.2	0.065	147	14.1	0.422	680	18.8	0.099	127	21.1	0.992	152	8.2	0.029	192
5454-O, H112, H32, H34	All	13.3	0.058	152	15.6	0.074	140	15.4	0.474	651	20.7	0.113	122	23.1	1.116	144	9.1	0.033	184
5454-H111	Extrusions	12.2	0.051	159	14.2	0.064	147	14.1	0.420	683	18.8	0.098	128	21.1	0.989	153	9.1	0.033	184
5456-O	Sheet and Plate	21.6	0.121	123	25.7	0.156	110	24.9	0.900	499	34.1	0.239	95	37.4	2.119	104	14.9	0.069	144
5456-H116, H321	Sheet and Plate (1.500)	20.4	0.111	123	24.2	0.143	113	23.5	0.835	515	32.2	0.219	98	35.3	1.965	108	14.9	0.069	144
5456-H116, H321	Plate (3.000)	19.2	0.101	127	22.8	0.130	116	22.2	0.771	532	30.2	0.199	101	33.2	1.814	113	14.1	0.063	148
6005-T5	Extrusions	14.5	0.067	144	17.0	0.086	133	16.7	0.536	446	22.6	0.131	115	25.1	1.260	133	9.9	0.038	174
6061-T6, T651, T6510, T6511†	All	16.8	0.084	133	19.9	0.108	123	19.4	0.654	389	26.4	0.165	107	29.2	1.538	121	11.5	0.048	161
6061-T6, T651, T6510, T6511‡	All	12.2	0.052	157	14.2	0.065	145	14.1	0.424	523	18.8	0.100	126	21.1	0.999	150	8.2	0.029	190
6063-T5, T52, T6	All	8.7	0.031	185	10.1	0.039	172	10.1	0.273	795	13.4	0.060	150	15.1	0.642	187	5.9	0.017	226
6351-T5, T6†	Extrusions	16.8	0.084	133	19.9	0.108	123	19.4	0.654	389	26.4	0.165	107	29.2	1.538	121	11.5	0.048	161
6351-T5, T6‡	Extrusions	12.2	0.052	157	14.2	0.065	145	14.1	0.424	523	18.8	0.100	126	21.1	0.999	150	8.2	0.029	190
6463-T6	Extrusions	8.7	0.031	185	10.1	0.039	172	10.1	0.273	795	13.4	0.060	150	15.1	0.642	187	5.9	0.017	226
7005-T53	Extrusions	27.7	0.174	106	33.1	0.228	97	31.8	1.244	427	44.1	0.350	84	47.7	2.928	89	19.2	0.101	127

*Maximum thickness indicated in parentheses.

†Values when welded with 5183, 5356, or 5556 alloy filler wire, regardless of thickness. Values also apply to thicknesses less than or equal to 0.375 in. when welded with 4043, 5554, or 5654 filler.

‡Values when welded with 4043, 5554, or 5654 alloy filler wire.

APPENDIX L
Metric Conversions

If you're like us, your problem with metric conversion charts is that they provide too much information. In structural design, you don't need to know how many becquerels are in a curie; you're just trying to remember if a newton is a measure of force or some kind of junk food. The conversions given here are what you need—but first a few comments.

No topic—other than LRFD, perhaps—elicits so many opinions from engineers besides metrication. That's probably because most people figure that you don't have to know much to comment on the topic. That might be worth rethinking. Before questioning conversions, let's review. There are two types of conversions from the foot-pound system (commonly referred to as the "U.S. customary units"; ironic considering that it's from England) to SI (the International System of Units, known by its French acronym and commonly called the metric system):

Soft conversion is conversion of foot-pound units to metric units without modifying the result to an even, convenient round number in metric units. For example, converting 1 in. to 25.4 mm is a soft conversion.

Hard conversion is conversion of foot-pound units to metric units and modifying the result to an even, convenient round number in metric units. Converting 1 in. to 25 mm is a hard conversion.

Most U.S. standards, such as ASTM material specifications, use hard conversions. For example, ASTM B221M lists the tensile ultimate strength of 6061 extrusion as 260 MPa, even though ASTM B221 shows it as 38 ksi, which is actually 262.0024 . . . MPa. Just as most aluminum strengths are rounded to the nearest 1 ksi in the foot-pound system, most strengths are rounded to the nearest 5 MPa in the SI system. In this book, where quantities are taken from standards, we've used the SI version of the standard for the SI quantity. The round-off usually is less than the accuracy in the foot-pound quantity; if you think it's worth worrying about, consider how accurate most structural design is. We have not given SI dimensions for fastener and pipe sizes because the schedules of available sizes for these are completely different from the U.S. schedules and, frankly, we don't often deal with them.

For simplicity's sake, you might want to try this in your calculations: convert in. to mm, lb to N, and ksi to MPa. That should take care of 80% of the conversions in common structural design. This approach works well because the resulting SI quantities are generally convenient numbers (not too large,

Table L.1   SI Unit Prefixes

Prefix	Name	Multiplier	Example
M	mega	10^6	MPa
k	kilo	10^3	kg
m	milli	10^{-3}	mm
μ	micro	10^{-6}	μm

not too small); an exception is moment of inertia in mm⁴, which becomes rather large for most shapes since millimeters are small in relation to inches and the effect is magnified by raising to the fourth power. Consequently, some prefer cm⁴ for properties like I, J, and C_w.

Length

1 in. = 25.4 mm
1 ft = 0.3048 m
1 mil = 0.001 in. = 25.4 μm

Volume

1 gal = 3.785 L

Mass

1 lb = 0.4536 kg

Force

1 lbf = 4.448 N

Moment or Torque

1 lbf-in. = 113.0 N-mm

Pressure

1 k/in.² = 6.895 MPa = 6.895 N/mm²
1 lbf/in.² = 6.895 kPa
1 lbf/ft.² = 0.04788 kPa
1 Pa = 1 N/m²

Speed

1 mi/h = 1.609 km/h

Temperature

(°F − 32)/1.8 = °C

Energy

1 ft-lbf = 1.356 J

APPENDIX M
Statistics

We increasingly find ourselves facing statistical expressions that we are expected to decipher. For many of us, however, our recall with respect to statistics is limited to the adage, "There are liars, damned liars, and statisticians." Another adage proclaims, "If you torture data long enough, it will confess to anything." While it is useful to invoke these sayings as reminders that we should regard statistical claims with a healthy dose of skepticism, they do not help us come to an understanding of the statistical issue at hand.

The statistical questions that we most frequently encounter as engineers deal in one way or another with the evaluation of test results. We love to collect data and make some assertions. In order for the validity of our assertions to be quantified in some way, however, we are obligated to assign some statistical measure to our results.

We may use statistics to describe a characteristic of a population by a single value when, in reality, values for this characteristic vary within the population. For example, the actual height of trees within a population varies from tree to tree, but we may wish to characterize the entire population in terms of the average height of the individual trees. Similarly, we may wish to characterize all production of a given aluminum alloy in terms of a specified minimum strength. Statisticians have developed their own jargon for discussing these issues, so let's begin by learning to talk the talk.

TALK THE TALK

Distribution refers to the relationship of individuals within a population to the average for that population. Some individuals will be above average, and others will be below average. The characteristic of the population that is measured or observed is the *variable*. We plot the distribution with the x-axis showing the measured values of our variable, and the y-axis showing the portion of the population that is observed to match a given value. The distribution may be described by means of two *parameters*. The plot of the distribution is located along the x-axis by its *mean*, and the spread of the distribution may be described by either its *variance* or its *standard deviation*.

Mean is the arithmetic average of the values determined for each individual within the population for the *variable* in question. When the distribution is plotted, the mean determines its location along the x-axis. It is considered a *parameter* of the distribution.

Normal distribution refers to a class of distribution that meets certain criteria, including symmetry of the distribution about the *mean.* A plot of the normal distribution resembles the familiar bell curve. It is routinely described in terms of two *parameters*, the *mean* and the *standard deviation.*

Parameter refers to some measure of the *distribution* of the *variable.* Paameters may be estimates of specified points, such as the *mean* or a minimum expected value. Parameters may also be used to describe the spread of the distribution, in terms of its *variance* or its *standard deviation.*

Population refers to the entire universe of individuals that may fit our category. For example, when we are evaluating the response of medical patients to a given drug, the population is every patient who may ever take that drug. Similarly, when we are determining the growth rate of a given species of tree, the population is every tree of that species that may ever exist. Correspondingly, when we are determining the strength of a given aluminum alloy, the population includes every ingot of that alloy that might ever be produced.

Standard deviation is the square root of the *variance* and is a preferred measure of the spread of the *distribution* in that it is expressed in the same units as the *variable.*

Variable refers to the characteristic of the *population* that we measure or observe, such as the height of the tree or the strength of the metal.

Variance is a measure of the spread of the distribution. This refers to the extent to which individual values deviate from the average, or *mean,* value. The spread is also referred to as the variability of the data.

WALK THE TALK

Let's use the measured speed of the vehicles passing a highway checkpoint to illustrate the concept of distribution. Our hypothetical checkpoint measured the speed of 100 vehicles, and recorded the following results:

Speed (miles per hour)	Number of Vehicles	Portion of the Population
62	4	0.04
64	11	0.11
66	22	0.22
68	26	0.26
70	22	0.22
72	11	0.11
74	4	0.04
	100	

These results are plotted in the scattergraph of Figure M-1. Our variable is the speed of the vehicles, and we assign it to the x-axis. The y-axis indicates the portion of the population that was measured at a given speed.

We may fit a curve to these points and approximate our data as a continuum, as shown in Figure M-2. This curve represents the distribution of values for the variable in question. In other words, it shows how individual values are distributed within the measured range of values.

The curve representing the distribution may be described in terms of its horizontal spread and its location along the x-axis. The parameters of *variance* and *mean* may be used to measure the spread and the location of the distribution, respectively. These concepts are illustrated in Figure M-3.

Now that we're up to speed on the terminology, we offer the following maxim: *When we know the value of a given variable for individuals selected at random from a population, we may use statistics to estimate the average value of the variable for the entire population and the distribution of individual values about the average.*

THE NORMAL DISTRIBUTION

The first thing we want to do with a real-life problem is to assume that our data is normally distributed. Statisticians will go on and on about other ways

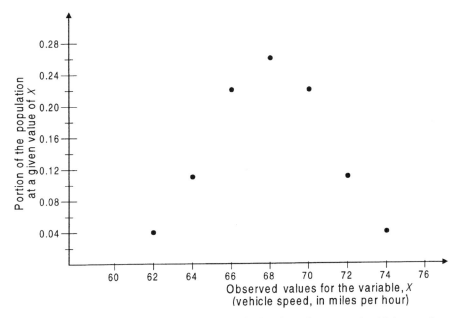

Figure M-1   Scattergraph illustrating the distribution of measured vehicle speeds.

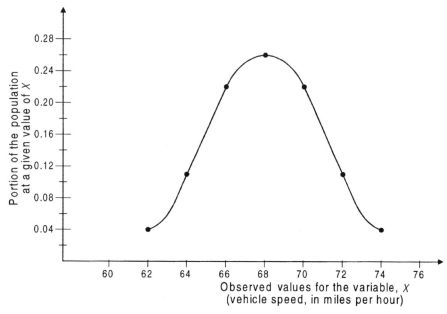

Figure M-2 Curve fit to the scattergraph of measured vehicle speeds.

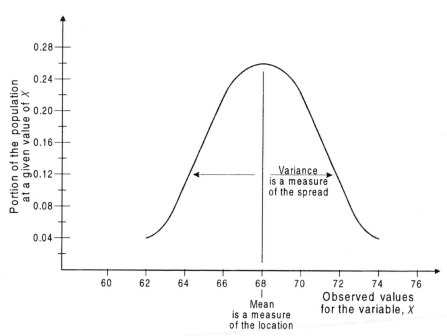

Figure M-3 Variance and mean as measures of the spread and location.

that the data might be distributed, and may question how we know that our data is normally distributed. But we don't want to engage in that discussion because the tools that we want to use depend upon the assumption of a normal distribution. So we mumble something about the "sample approaching a continuous Gaussian distribution in accordance with the central limit theorem." If pressed to explain ourselves, we will pretend to look down our noses with contempt at the apparent ignorance of the questioners, and proceed to go on about our business with our assumption of a normal distribution securely in hand.

A useful measure of the spread of the normal distribution is the standard deviation, which is the square root of the variance. We sometimes evaluate a normal distribution in terms of the proximity of a given percent of the population to the mean. We express this proximity in terms of multiples of the standard deviation. Figure M-4 shows percents of the population falling within selected intervals of $+/- z$ about the mean, where z has units of standard deviations. The symbol for the mean is μ, and the standard deviation is represented by σ. The intervals are, then, expressed as $[\mu +/- z\sigma]$.

A common rule of thumb is that 95% of the population falls within an interval of $+/- 2$ standard deviations. We can see from Figure M-4 that this is an approximation. It would be more accurate to state that 95.4% of the population falls within an interval of $+/- 2$ standard deviations, or to state that 95% of the population falls within an interval of $+/- 1.96$ standard deviations. We're not particularly concerned with this rule of thumb. We mention it only to illustrate how to read the figure.

An important qualifier shown in the title to Figure M-4 is the statement that the variance is known. When you collect data in the real world, however, it is almost a certainty that neither you nor anyone else knows what the variance is. Nevertheless, we will tarry awhile in the blissful domain of known variance long enough to illustrate some basic concepts.

The parameters typically used to describe a normal distribution are the mean and the standard deviation. The mean (μ) and standard deviation (σ) of the population are determined as follows:

$$\mu = \frac{\sum_{i=1}^{N} X_i}{N}$$

and

$$\sigma = \sqrt{\frac{\sum_{i=1}^{N} (X_i - \mu)^2}{N}}$$

where:

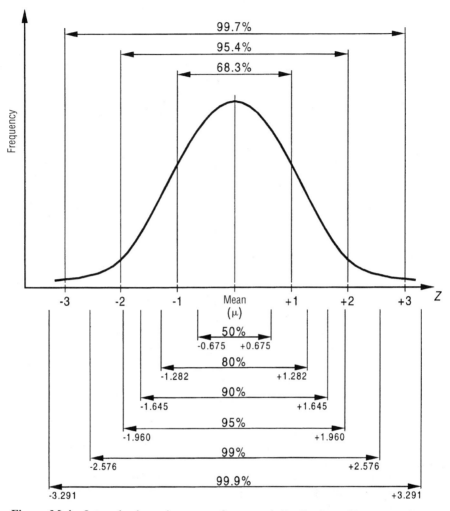

Figure M-4   Intervals about the mean of a normal distribution of known variance.

μ = mean for a population of N individuals
N = number of individuals in the population
X_i = value for individual i
 i = an individual in the population
σ = standard deviation of the population.

Again, the standard deviation is a measure of the variability, or spread, among individual values within the population. It is evident from the equation that the greater the deviation of individual values (X_i) from the average value (μ), the larger the standard deviation.

TWO-SIDED INTERVALS

An interval or range is often expressed in terms of percentiles. A *central percentage*, which is balanced about the mean of the population, is expressed as a $(1 - \alpha)$ interval, where α is the decimal fraction of the population that is not included in the interval. Because the central interval is bounded by limits on both sides, with the portion that is not included in the interval tailing off beyond the limits, it is commonly referred to as a "two-sided" or "two-tailed interval."

Because both the population distribution and the central interval are balanced about the mean, the portion of the population that is not covered by the interval is evenly distributed between the region above the upper limit and the region below the lower limit. Therefore, a $(\alpha/2)$ portion of the population resides beyond each limit. The lower limit of the $(1 - \alpha)$ interval is the $(\alpha/2)$ percentile, and the upper limit is the $(1 - \alpha/2)$ percentile.

As we saw in Figure M-4, the portion of the population that is included in the interval may be expressed in terms of the product of a coefficient (z) and the standard deviation. The use of $z_{(\alpha/2)}$ or $z_{(1-\alpha/2)}$ as the coefficient for the standard deviation is simply a matter of convention, in that:

$$z_{(\alpha/2)} = z_{(1-\alpha/2)}$$

Thus, the $(1 - \alpha)$ interval lies between the lower limit of $[\mu - (z_{(\alpha/2)}\,\sigma)]$ and the upper limit of $[\mu + z_{(1-\alpha/2)}\,\sigma)]$, and it may be expressed either as:

$$\mu +/- (z_{(\alpha/2)}\,\sigma),$$

or as:

$$\mu +/- (z_{(1-\alpha/2)}\,\sigma).$$

These relationships are illustrated in Figure M-5.

Selected values for z are shown in Figure M-4, and complete tables for values of z may be obtained from published text.[1] Use of these tables can be tricky. Recalling that the $(1 - \alpha)$ interval may be expressed as $[\mu +/- (z_{(1-\alpha/2)}\,\sigma)]$, we must take care to differentiate between the use of α and $\alpha/2$. While the interval in question covers $(1 - \alpha)$, we want the value of z at $z_{(1-\alpha/2)}$.

The nuance of α versus $\alpha/2$ may be illustrated with an example. Let's find the value of z for the interval that covers 95% of the population. We might restate the given information as the $(1 - \alpha)$ interval covering a 0.95 portion of the population. We then see that $(1 - \alpha)$ is equal to 0.95, α is equal to

[1]Remington, R. D., and M. A. Schork. *Statistics with Applications to the Biological and Health Sciences.* Englewood Cliffs: Prentice-Hall, 1969.

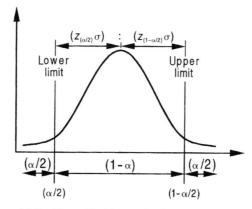

Figure M-5   Two-sided interval of a normal distribution.

0.05, $\alpha/2$ is equal to 0.025, and $(1 - \alpha/2)$ is equal to 0.975. This means that the limits of the 95% interval are at the 0.025 and 0.975 percentiles, corresponding to $z_{0.025}$ and $z_{0.975}$, respectively. We know that, for a normal distribution, $z_{0.025}$ is equal to $z_{0.975}$. We find from tables that the value of $z_{0.975}$ is 1.960, and, therefore, we express the 95% interval as

$$[\mu +/- (1.96\ \sigma)].$$

This example is shown in Figure M-6.

One-Sided Intervals

Percentiles may also be used to describe a single limit beyond which a $(1 - \alpha)$ portion of the population lies. If the limit is a lower limit, then the

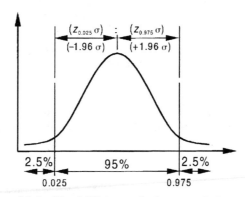

Figure M-6   The 95% interval of a normal distribution.

$(1 - \alpha)$ portion of the population resides above this limit, and the limit is the α percentile. We thereby mitigate the nomenclature difficulty that we just discussed for the two-sided interval. The lower limit for the one-sided interval covering $(1 - \alpha)$ is expressed as

$$[\mu - (z_{(\alpha)} \sigma)],$$

which is equal to

$$[\mu - (z_{(1-\alpha)} \sigma)].$$

The one-sided limit for a 95% interval, for example, is the 0.05 percentile and may be determined from

$$[\mu - (z_{0.95} \sigma)].$$

These relationships are shown in Figure M-7.

DISTRIBUTION OF THE MEAN

We have thus far been discussing the distribution of values for individuals within a population about the mean for the entire population. If we were to randomly select a group of individuals from the population—i.e., a subset of the population—the mean for that group is likely to deviate somewhat from the mean for the entire population. If we were to select many groups and determine the mean for each, we would generate a new data set. This data set is the mean for groups within the population. It would also be distributed symmetrically about the mean for the entire population. In fact, if the size of each group were one individual, then the distribution of the mean for the groups would be the same as the distribution of the individual values. As soon

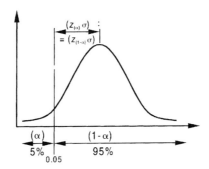

Figure M-7   One-sided interval of a normal distribution.

as we increase the group size beyond one, however, the value for the most deviant individual in the group is averaged with the values for less deviant individuals. We, therefore, expect the distribution of mean values to have less spread than the distribution of individual values. Both distributions, however, would be balanced about the population mean, μ.

The more compact distribution of the mean as compared to the distribution of individual values is illustrated in Figure M-8. The narrower curve of the mean is quantified by an adjustment to the standard deviation. Whereas the standard deviation for the individual values is σ, the standard deviation for the mean is $\sigma/\sqrt{N}$.

MEASURING (SAMPLING) STATISTICS

The preceding discussion on the distribution of the mean introduced the concept of selecting groups from the population. We call these groups *samples.*

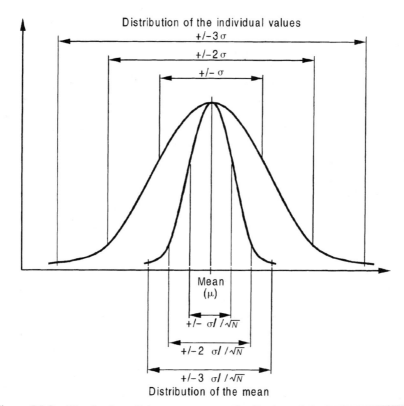

Figure M-8 Distribution of the mean versus distribution of the individual values.

In the real world, we never know the true characteristics of the entire theoretically possible population. We may generate a limited amount of data by performing tests on a sample of the population. Next, we apply statistics to estimate the characteristics of the entire population from our sample data. This process of attempting to infer the characteristics of a population from measurements of a portion (sample) of that population is referred to as *statistical inference.*

We may think of statistical inference as a systematic method for expressing not only what we know, but also how well we know it. We estimate values for the parameters in question, and we additionally calculate the level of *confidence* that we have in our estimates.

We will see that this concept of confidence is critical to our application of statistics to real-world problems. We felt it best to keep it out of sight, however, until after we had introduced you to the basic concepts of a normal distribution. Since we have neglected it for this long, let's ignore it a little bit longer while we wrap up our presentation of the mean and the standard deviation.

The best estimates of the true values of the parameters of the population are the values that we measure from our sample. Thus, we will estimate that the population mean (μ) is the same as the sample mean ($\bar{x}$), and that the population standard deviation (σ) is the same as the sample standard deviation (s). Values for $\bar{x}$ and s are obtained from the test data as follows:

$$\bar{x} = \frac{\sum\limits_{i=1}^{n} x_i}{n}$$

$$s = \sqrt{\frac{\sum\limits_{i=1}^{n} (x_i - \bar{x})^2}{n - 1}}$$

where:

$\bar{x}$ = mean for a sample of n individuals
n = number of individuals in the sample
x_i = value for individual i
i = an individual in the sample
s = standard deviation of the sample.

As with the population parameters, the standard deviation of the mean is related to the standard deviation of the individual values by a factor of $(1/\sqrt{n})$.

We wrap up our introduction to statistics with the following table, which summarizes the relationships that we have established.

Parameter	Population	Sample
Mean	$\mu = \dfrac{\sum\limits_{i=1}^{N} x_i}{N}$	$\bar{x} = \dfrac{\sum\limits_{i=1}^{n} x_i}{n}$
Standard deviation (of the individual values)	$\sigma = \sqrt{\dfrac{\sum\limits_{i=1}^{N} (X_i - \mu)^2}{N}}$	$s = \sqrt{\dfrac{\sum\limits_{i=1}^{n} (x_i - \bar{x})^2}{n - 1}}$
Standard deviation of the mean	$\dfrac{\sigma}{\sqrt{N}}$	$\dfrac{s}{\sqrt{n}}$

READY FOR THE REAL WORLD

Now that we've established the basic relationships that we will employ, we are ready to consider real-world problems. In our sheltered world of structural engineering, the parameters that we are most frequently interested in are the typical and minimum values. We will become better acquainted with the concept of *confidence* in the problem of determining typical values, and we will introduce the concept of *tolerance* in the problem of minimum values.

CONFIDENCE AT LAST

A quick explanation of confidence is that it is almost the same as probability. When we estimate that the value of a population parameter falls within a certain range, which we call the *confidence interval*, we are tempted to say that we have calculated the probability that the value of this parameter will fall within the specified range. For any given population, however, the parameter has some fixed value. We estimated the value of the parameter by testing a randomly selected portion of the population. Because variability among the individuals in the population exists, the value of our estimate will vary depending upon the group that we select for our sample. We see, then, that it is our estimate that is subject to chance, rather than the value of the population parameter itself. Therefore, we should speak of how confident we are in our estimate and not in terms of the probability of the value of the parameter. We feel obligated to maintain this distinction just in case a statistician looks over our shoulders, but you may feel free to read confidence as being synonymous with probability.

   There are two basic premises to the expression of a level of confidence. The first is that the more data we have, the more confident we are in our estimates. The second is that the more the results of our tests agree, the more confidence we have in them.

DETERMINATION OF TYPICAL VALUES

When we speak of the typical value of a variable, we are referring to the average. Our parameter, then, is the mean. The sample mean is our estimate of the population mean, and the confidence interval is a range around the sample mean that we predict will include the population mean.

We discussed earlier that we may describe an interval in terms of multiples of the standard deviation about the mean. In the hypothetical scenario of known variance, we have presented the following expression to describe an interval about the mean for the distribution of the individual values:

$$\mu +/- z_{(1-\alpha/2)}\sigma$$

For the discussion of the mean, we adjust the standard deviation by the $(1/\sqrt{N})$ factor, and an interval about the mean may be written as follows:

$$\mu +/- z_{(1-\alpha/2)} \frac{\sigma}{\sqrt{N}}$$

We replace the population parameters μ, σ, and N with the sample parameters $\bar{x}$, s, and n, respectively, to apply this expression to our sample data. We recall, however, that in our real-world problem the variance is not known, and the sample mean $(\bar{x})$ and sample standard deviation (s) are only estimates of the population mean (μ) and population standard deviation (σ). The coefficient z must, therefore, be replaced with a multiplier that incorporates the additional uncertainty accruing from the unknown variance.

A very clever person who worked for a beer company figured out what the distribution for this modified coefficient should be. He wrote his findings under the pseudonym "Student" and, although it is now known that his name was W. S. Gossett and he worked for the Guinness brewery, we still refer to this distribution by the quirky moniker "Student's t." We will, then, replace z with the Student's t-distribution (or just t-distribution, for brevity). The confidence interval for the sample mean may then be written as follows:

$$\bar{x} +/- t_{(1-\alpha/2)} \frac{s}{\sqrt{n}}$$

TROUBLES WITH THE t-DISTRIBUTION

We would be ready now to express the confidence interval for our estimate of the mean, or typical, value if only we could read the t-distribution chart. Both the rows and the columns can be tricky to navigate.

The difficulty with the columns is that they are often labeled in terms of the percentile. As we mentioned when we introduced the concept of intervals, the limit of the one-sided (1-α) interval is the (1-α) percentile, but the limit of the two-sided (1-α) interval is the (1-α/2) percentile. This means that the distribution for the one-sided 95% confidence interval is found in the column labeled $t_{0.95}$, but the limit of the two-sided 95% confidence interval is found in the column labeled $t_{0.975}$. The same table is used for both one-sided and two-sided confidence intervals, but users must be careful to correctly identify the appropriate column for a given case.

The difficulty with the rows is that they are labeled as degrees of freedom rather than as number of tests in the sample. For our case of the distribution of the sample mean, the degrees of freedom will always be (n-1), or one less than the number of tests in the sample. This means that, if we performed six tests, we use the row labeled 5 degrees of freedom.

The texts that publish these tables will indulge you with more detailed explanations[1], but we're ready to plow into an example to show how this works.

ILLUSTRATION OF TYPICAL VALUES, CONFIDENCE INTERVALS, AND THE *t*-DISTRIBUTION

Suppose that we wish to estimate, with a 90% level of confidence, the typical number of cycles that a certain connection detail will withstand before failure. We test six assemblies and find the following results.

Test Number	Number of Cycles to Failure
1	97
2	86
3	103
4	111
5	85
6	91

We understand "typical" to refer to the mean. The confidence interval for the mean has both a lower and an upper limit, so we will be computing a two-sided interval. We recall that the confidence interval for the sample mean is expressed as follows:

$$\bar{x} +/- t_{(1-\alpha/2)} \frac{s}{\sqrt{n}}$$

and, making the appropriate substitutions:

$$\bar{x} = \frac{\sum\limits_{i=1}^{n} x_i}{n} = \frac{97 + 86 + 103 + 111 + 85 + 91}{6} = 95.5 \text{ cycles}$$

$$s = \sqrt{\frac{\sum\limits_{i=1}^{n} (x_i - \bar{x})^2}{n - 1}}$$

$$= \sqrt{\frac{(97 - 95.5)^2 + (86 - 95.5)^2 + (103 - 95.5)^2 + (111 - 95.5)^2 + (85 - 95.5)^2 + (91 - 95.5)^2}{(6 - 1)}}$$

$$s = 10.2 \text{ cycles}$$

$$n = 6$$

We look up the value of the *t*-distribution for a two-sided confidence interval at the $1 - \alpha/2$) percentile, and we select the row corresponding to $(n - 1)$ degrees of freedom. Our confidence level is 90% and thus $(1 - \alpha)$ equals 0.90, α equals 0.10, $\alpha/2$ equals 0.05, and $(1 - \alpha/2)$ equals 0.95. The number of tests (n) is 6, so the degrees of freedom $(n - 1)$ is 5. The value we want, then, is $t_{0.95}$ at 5 degrees of freedom. We find from tables that it is 2.0150.

We can now solve for our confidence interval as follows:

$$\left[95.5 +/- 2.0150 \left(\frac{10.2}{\sqrt{6}} \right) \right] \text{ cycles}$$

or

$$95.5 +/- 8.39 \text{ cycles.}$$

That is, we have estimated with a 90% level of confidence that the typical number of cycles that our connection detail can withstand before failure is 95.5 cycles, plus or minus 8.39 cycles.

ABOVE ALL, TOLERANCE

A close relative of the confidence interval is the *tolerance interval.* The tolerance interval embraces the concept of a confidence level, with the additional concept of a portion of the population that may fall within the interval. Rather than saying that there is a $(1 - \alpha)$ level of confidence that a point (such as the mean) lies within a given interval, we might say that we have a (γ) level of confidence that a (P) portion of the population falls within that interval.

When we apply the notion of a tolerance interval to a one-sided problem, we are estimating either a minimum or a maximum tolerance limit. In the case of a minimum tolerance limit, we would say that we have a (γ) level of confidence that a (P) portion of the population falls above the minimum. This is the statistical evaluation that we apply when determining minimum expected strengths of materials.

DETERMINATION OF MINIMUM VALUES

When we estimate a minimum expected value for a population, we are dealing with the distribution of the individuals within the population, rather than with the mean. We have expressed the population distribution as follows, for the case of the variance being known:

$$\mu +/- z_{(1-\alpha/2)}\sigma$$

We need to adapt this expression for the population distribution to the distribution of a sample. As for the confidence interval, we replace the population parameters μ and σ with the sample parameters $\bar{x}$ and s, respectively. The replacement for the multiplier, z, is the tolerance limit factor, k. The tolerance interval for the distribution of a sample is, then, expressed as follows:

$$\bar{x} +/- ks$$

The one-sided tolerance limits take the form of:

$$\bar{x} + ks \text{ for a upper limit}$$
$$\bar{x} - ks \text{ for a lower limit.}$$

Whereas with the t-distribution, the same table could be used for either a one-sided or a two-sided interval, the tolerance limit factor is presented in separate tables for one-sided versus two-sided intervals. This makes each table somewhat easier to read, but many texts include tables of tolerance limit factors for only the two-sided case. You may have to visit a university library to find the one-sided case, but extensive tables are available.[2] A limited table is presented in Aluminum *Specification* Section 8.3.1, for the special case of complying with the testing provisions of that section.

The minimum mechanical properties published in the *Aluminum Design Manual* for the design of aluminum structure are defined as the strength which

---

[2]Odeh, R. E., and D. B. Owen. *Tables for Normal Tolerance Limits, Sampling Plans, and Screening.* New York and Basel: Marcel Dekker, 1980.

99% of the material is expected to exceed with a confidence of 95%. This is a straightforward application of the one-sided tolerance limit.

ILLUSTRATION OF MINIMUM EXPECTED VALUES AND THE ONE-SIDED TOLERANCE LIMIT

Suppose that we have developed a new aluminum alloy, and we wish to determine its minimum tensile ultimate strength. We would perform pull tests to generate sample data and then compute the mean ($\bar{x}$) and the standard deviation (s). These parameters would be estimated in the same manner as for the previous example of the confidence interval, so we won't repeat the mathematical steps. Let us assume that we found the sample mean to be 65.4 ksi and the sample standard deviation to be 3.13 ksi. The expression for the lower one-sided tolerance limit would, then, be:

$$(65.4 \text{ ksi}) - k(3.13 \text{ ksi})$$

We now need to look up the one-sided tolerance limit (k) that will be exceeded by 99% of the population with a 95% confidence level. We see from the table that the tolerance limit is additionally a function of the number of tests in the sample. The following table presents selected values of the $k_{(99,95)}$ one-sided tolerance limit factors, as well as the resulting minimum tensile ultimate strengths that we would compute for our example:

Number of Tests n	One-Sided Tolerance Limit k	Minimum Tensile Ultimate Strength (ksi)
3	10.55	32.4
8	4.353	51.8
15	3.520	54.4
30	3.064	55.8
100	2.684	57.0

In each case, the average strength of the samples was 65.4 ksi. It is evident that the deduction from the mean to the lower tolerance limit increases markedly when the sample size becomes small.

We observed earlier that the confidence level increases with additional data. We might observe as a corollary that additional data decreases the spread of the tolerance interval.

APPENDIX N
Technical Organizations

Aluminum Association (AA)
900 19th Street, N.W., Suite 300
Washington, DC 20006
202-862-5100, 301-645-0756 (publications)
202-862-5164 (fax)
www.aluminum.org
The Aluminum Association, founded in 1933, represents aluminum primary producers, recyclers, and producers of semi-fabricated products. The Aluminum Association is the main source of information, standards, and statistics concerning the U.S. aluminum industry.

Aluminum Association of Florida (AAF)
Association Headquarters
1650 S. Dixie Highway, Suite 500
Boca Raton, FL 33432
561-362-9019
561-395-8557 (fax)
www.aaof.org
The Aluminum Association of Florida is a trade association for aluminum construction contractors and fabricators and their suppliers.

Aluminum Extruders' Council (AEC)
1000 N. Rand Road, Suite 214
Wauconda, IL 60084
847-526-2010
847-526-3993 (fax)
www.aec.org
The Aluminum Extruders Council, founded in 1950, is an international association of aluminum extruders. Most North American extruders are members, representing nearly 200 plants and nearly 100 companies.

American Architectural Manufacturers Association (AAMA)
1827 Walden Office Square, Suite 104
Schaumberg, IL 60173-4268
847-303-5664

847-303-5774 (fax)
www.aamanet.org
The American Architectural Manufacturers Association is a trade association for the windows, doors, and skylights industry.

American Association of State Highway and Transportation Officials (AASHTO)

444 North Capital Street, N.W., Suite 249
Washington, DC 20001
202-624-5800, 800-231-3475 (publications)
202-624-5806 (fax)
www.transportation.org

American Concrete Institute (ACI)

P.O. Box 9094
Farmington Hills, MI 48333-9094
248-848-3800
248-848-3801 (fax)
www.aci-int.org
The American Concrete Institute is a technical and educational society dedicated to improving the design, construction, manufacture, and maintenance of concrete structures.

American Galvanizers Association (AGA)

6881 South Holly Circle, Suite 108
Englewood, CO 80112
800-468-7732
720-554-0909 (fax)
www.galvanizeit.org
The American Galvanizers Association is a not-for-profit trade association whose members represent the after-fabrication, hot-dip galvanizing industry throughout the U.S., Canada, and Mexico.

American Institute of Steel Construction (AISC)

One East Wacker Drive, Suite 3100
Chicago, IL 60601-2001
312-670-2400, 800-644-2400 (publications)
312-670-5403 (fax)
www.aisc.org
The American Institute of Steel Construction serves the structural steel industry in the U.S. Its purpose is to expand the use of fabricated structural steel through research, education, technical assistance, standardization, and quality control.

American Iron and Steel Institute (AISI)

1101 17th Street, N.W., Suite 1300

Washington, DC 20036
202-452-7100, 800-277-3850 (publications)
www.steel.org

American National Standards Institute (ANSI)
1819 L Street, NW, 6th Floor
Washington, DC, 20036
202-293-8020
202-293-9287 (fax)
www.ansi.org

American Petroleum Institute (API)
1220 L Street, N.W.
Washington, DC 20005
202-682-8000
www.api.org

American Society of Civil Engineers (ASCE)
1801 Alexander Bell Drive
Reston, VA 20191-4400
800-548-2723
www.asce.org

American Society of Mechanical Engineers (ASME)
Three Park Avenue
New York, NY 10016-5990
212-591-7722, 800-843-2763 (publications)
212-591-7674 (fax)
www.asme.org

American Society for Testing and Materials (ASTM)
100 Barr Harbor Drive
West Conshohocken, PA 19428-2959
610-832-9585
610-832-9555 (fax)
www.astm.org

American Welding Society (AWS)
550 N.W. LeJeune Road
Miami, FL 33126
800-443-9353 or 305-443-9353
305-443-7559 (fax)
www.amweld.org

ASM International
Materials Park, OH 44073-0002

800-336-5152
440-338-4634 (fax)
www.asminternational.org
ASM International serves the technical interests of metals and materials professionals.

Building Officials and Code Administrators (BOCA)
4051 West Flossmoor Road
Country Club Hills, IL 60478
708-799-2300
708-799-4981 (fax)
www.bocai.org
Building Officials and Code Administrators developed the model building codes used primarily in the northeastern part of the U.S. prior to the development of the International Building Code.

Canadian Standards Association (CSA)
178 Rexdale Boulevard.
Toronto, Ontario, Canada M9W 1R3
416-747-4000, 800-463-6727
416-747-4149 (fax)
www.csa.ca

Industrial Fastener Institute (IFI)
1717 East Ninth Street, Suite 1105
Cleveland, OH 44114-2879
216-241-1482
216-241-5901 (fax)
www.industrial-fasteners.org

International Code Council (ICC)
5203 Leesburg Pike, Suite 600
Falls Church, VA 22041
703-931-4533
703-379-1546 (fax)
www.intlcode.org
The International Code Council was established in 1994 as a nonprofit organization dedicated to developing a single set of comprehensive and coordinated national model construction codes. The founders of the ICC are Building Officials and Code Administrators International, Inc. (BOCA), International Conference of Building Officials (ICBO), and Southern Building Code Congress International, Inc. (SBCCI).

International Conference of Building Officials (ICBO)
5360 Workman Mill Road
Whittier, CA 90601

310-699-0541
310-692-3853 (fax)
www.icbo.org
The International Conference of Building Officials developed the model building codes used primarily in the western part of the U.S. prior to the development of the International Building Code.

Metal Building Manufacturers Association (MBMA)

1300 Sumner Avenue
Cleveland, OH 44115-2851
216-241-7333
216-241-0105 (fax)
www.mbma.com

Metal Construction Association (MCA)

104 South Michigan Avenue, Suite 1500
Chicago, IL 60603
312-201-0193
www.mca1.org
Founded in 1983, the MCA seeks to expand the use of metal in construction through marketing, technology, and education.

National Association of Aluminum Distributors (NAAD)

1900 Arch Street
Philadelphia, PA 19103-1498
215-564-3484
215-963-9784 (fax)
www.naad.org
The National Association of Aluminum Distributors is a trade association of the North American Service Centers and principal suppliers engaged in marketing aluminum products.

National Association of Architectural Metal Manufacturers (NAAMM)

8 South Michigan Avenue, Suite 1000
Chicago, IL 60603
312-332-0405
312-332-0706 (fax)
www.naamm.org
The National Association of Architectual Metal Manufacturers is a trade association representing manufacturers of metal products used chiefly in commercial and industrial building construction. These products include metal stairs and railings, flagpoles, expanded metal, hollow metal doors and frames, steel and aluminum bar grating, and metal lathing and furring.

National Association of Corrosion Engineers (NACE)

1440 S. Creek Drive

Houston, TX 77084-4906
281-228-6200
www.nace.org

National Fire Protection Association (NFPA)
1 Batterymarch Park
P.O. Box 9101
Quincy, MA 02269-9101
617-770-3000, 800-344-3555 (publications)
617-770-0700 (fax)
www.nfpacatalog.org

Non-Ferrous Founders' Society (NFFS)
1480 Renaissance Drive, Suite 310
Park Ridge, IL 60068
847-299-0950
www.nffs.org
The Non-Ferrous Founders' Society represents the non-ferrous metal casting industry.

Research Council on Structural Connections (RCSC)
www.boltcouncil.org
The Research Council on Structural Connections' purpose is to support investigations of the suitability and strength of structural connections, to promote the knowledge of economical and efficient practices relating to such structural connections, and to publish standards and other documents necessary to achieving its purpose. The primary document the Council publishes is the *Specification for Structural Joints Using ASTM A325 or A490 Bolts.*

Society of Automotive Engineers (SAE)
400 Commonwealth Drive
Warrendale, PA 15096-0001
724-776-4841, 877-606-7323
724-776-5760 (fax)
www.sae.org

The Society for Protective Coatings (SSPC)
40 24th Street, 6th Floor
Pittsburgh, PA 15222-4656
877-281-7772
412-281-9992 (fax)
www.sspc.org
The Society for Protective Coatings was founded in 1950 as the Steel Structures Painting Council, a professional society concerned with the use of coatings to protect industrial steel structures.

Southern Building Codes Congress International (SBCCI)
900 Montclair Road
Birmingham, AL 35213
205-591-1853
205-592-7001 (fax)
www.sbcci.org
The Southern Building Codes Congress International developed the model building codes used primarily in the southeastern part of the U.S. prior to the development of the International Building Code.

Specialty Steel Industry of North America (SSINA)
3050 K Street, N. W.
Washington, DC 20007
202-342-8630, 800-982-0355
202-342-8631 (fax)
www.ssina.com
The Specialty Steel Industry of North America is a voluntary trade association representing almost all the producers of specialty steel in North America. Members make a variety of products, including bar, rod, wire, angles, plate, sheet and strip, in stainless steel and other specialty steels.

Structural Stability Research Council (SSRC)
University of Florida
Department of Civil Engineering
P.O. Box 116580
Gainesville, FL 32611-6580
352-846-3874
352-846-3978 (fax)
www.ce.ufl.edu
The Structural Stability Research Council, formerly the Column Research Council, publishes the *Guide to Stability Design Criteria for Metal Structures*.

GLOSSARY

The definitions in this glossary are based on usage common in the aluminum industry. Terms italicized in the text and in these definitions are defined here.

alclad: an adjective applied to *aluminum alloys* given a thin aluminum or aluminum alloy coating metallurgically bonded to the surface. This coating electrolytically protects the core alloy against corrosion in a manner similar to the way zinc galvanizing protects steel.

allowable stress: a stress to which *stresses* produced by *nominal loads* are limited in *allowable stress design*.

allowable stress design (ASD): a method of proportioning structural components that limits *stresses* under *nominal loads* to the *yield* or *ultimate strength* divided by a *safety factor*.

alloy: a material with metallic properties and composed of two or more elements, at least one of which is a metal.

aluminum: a silvery, lightweight, easily worked metallic element resistant to corrosion and abundant in nature; its atomic number is 13. Commercially pure aluminum (99% or greater) is designated by an *alloy* number of 1xxx.

Aluminum *Specification*: the 7th edition of the *Specification for Aluminum Structures* published by the *Aluminum Association* (AA).

anchor bolt: a bolt used to anchor members to a foundation.

annealing: a thermal treatment that reduces the *yield strength* and softens a metal by relieving *stresses* induced by *cold work* or by coalescing precipitates from solid solution.

anodizing: forming an oxide coating on a metal by electrochemical treatment.

artificial aging: a rapid precipitation from solid solution at elevated temperatures to produce a change in *mechanical properties*. Tempers T5 through T10 are artificially aged.

ASD: see *allowable stress design*.

aspect ratio: the ratio of width to thickness for a rectangular plate *element*.

austenitic stainless steel: AISI 200 Series (chromium, nickel, and manganese alloys) and AISI 300 Series (chromium and nickel) *stainless steels*. These steels are nonmagnetic.

bar: a solid *wrought product* with a cross-section square or rectangular or a regular hexagon or octagon and which may have rounded corners, and with at least one perpendicular distance between parallel faces of 0.375 in. or more [more than 10 mm]. See Table G.1.

TABLE G.1 Wire, Rod, and Bar

Width or Diameter	Square □, Rectangular, Hexagon, or Octagon	Circular ○
< 0.375 in. ≤ 10 mm	wire	wire
≥ 0.375 in. > 10 mm	bar	rod

bar, cold-finished: *bar* made by *cold-working* to obtain better surface *finish* and dimensional *tolerances*.

beam: a member that supports loads that act perpendicular to its longitudinal axis. These loads exert bending and shear on the beam cross section.

beam-column: a member subjected to both bending and axial compression.

bearing connection: see *connection, bearing*.

billet: a *hot-worked* unfinished product that may be subsequently worked by *forging* or *extruding* or other methods.

blind rivet: see *rivet, blind*.

bow, lateral: a deviation from a straight line along a longitudinal edge.

bow, longitudinal: a deviation from flat of a *sheet* or *plate* with curvature in the direction of rolling.

bow, transverse: a deviation from flat of a *sheet* or *plate* with curvature perpendicular to the direction of rolling.

brace: for a *column*, a point of support against lateral translation. For a *beam*, a point of support against lateral translation or support preventing twisting about the beam's longitudinal axis.

brazing: the process of joining metals by fusion using filler metals with a melting point above 840°F [450°C], but lower than the melting point of the base metals being joined.

bridge-type structures: a class of structures that includes highway and railroad bridges. See also *building-type structures*.

buckling: the deflection and/or rotation of a structural component under a compressive load.

buckling, elastic: *buckling* that occurs at a *stress* less than the *yield strength* of a member in compression and predicted by the *Euler buckling* formula for overall member buckling. Upon removal of the load causing compression, the member returns to its initial shape.

buckling, flexural: *buckling* that is manifested as a lateral displacement similar to what a beam experiences.

buckling formula constants: the slope and y-intercept (stress) of a line predicting *inelastic buckling stress* as a function of *slenderness ratio*, and the slenderness at its intersection with the *elastic buckling* curve.

buckling, inelastic: *buckling* that occurs at a *stress* greater than the *yield strength* of a member in compression. The deformation that occurs is permanent and is not reversed by removing the load.

buckling, local: *buckling* of an *element* of a cross section (for example, a flange or a web) over a distance on the order of the width of the element. In its broadest sense, local buckling includes both *inelastic* and *elastic buckling*, but in the Aluminum *Specification* usually refers to elastic buckling.

buckling, overall: *buckling* of a *column* as a member, either as *flexural buckling*, *torsional buckling*, or *torsional-flexural buckling*.

buckling, torsional: *buckling* that is manifested as a twisting or corkscrewing of the member, without lateral translation occurring at any point along the member. An example of a member that tends to buckle in this manner is a cruciform section *column*.

buckling, torsional-flexural: a combination of *torsional* and *flexural buckling*. An example is buckling of an I-beam subjected to a major axis bending moment, which causes the beam to twist, as well as to displace laterally.

buffing: mechanical finishing done with rotating wheels with abrasives.

building-type structures: a class of structures that includes buildings, highway signs, and light poles, and other structures of the kind to which the AISC *Specification for Structural Steel Buildings* and AISI *Specification for the Design of Cold-Formed Steel Structural Members* would apply if these structures were made of steel. See also *bridge-type structures*.

built-up member (section): a member composed of metal plates or shapes joined together by mechanical fasteners or *welding*.

butt joint: a joint between two parts butted end to end. See also *groove weld*.

camber: curvature in the plane of the web of an unloaded beam.

casting: a metal object made by pouring molten metal into a mold and allowing it to solidify.

casting, permanent mold: a metal object made by introducing molten metal by gravity or low pressure into a mold made of durable material, usually iron or steel, and allowing it to solidify.

casting, sand: a metal object made by pouring molten metal into a sand mold and allowing it to solidify. The sand mold is typically discarded after one use.

checkered plate: see *tread plate*.

cladding: see *alclad*.

coefficient of thermal expansion: see *thermal expansion, coefficient of*.

coefficient of variation: see *variation, coefficient of*.

coil set: longitudinal bowing in unwound coil in the same direction as that of the *coiled sheet*, which is also called *coil curvature.*

coiled sheet: *sheet* that has been rolled into a coil.

cold-finished: cold-worked to improve finish and dimensional tolerances.

cold-formed members: members formed without the application of heat.

cold-working: permanent deformation of a metal that produces *strain hardening*, which results in an increase in strength and a loss in *ductility*. Working done at elevated temperatures does not result in strain hardening, hence the term *cold-working.*

column: a member subjected to axial compression.

compact section: a cross section composed of *elements* that have dimensions so that overall member *buckling* occurs before *local buckling* of any of the elements.

component: see *element.*

composite beam: a *beam* composed of two or more materials structurally connected so that the two act together under load.

connection, bearing: a connection designed to resist shear by bearing of a fastener on the sides of its hole.

connection, slip-critical: a connection designed to resist shear by friction between the *faying surfaces*, formerly referred to as a *friction connection.*

corrosion, exfoliation: a delamination parallel to the metal surface caused by the formation of corrosion product.

corrosion, galvanic: corrosion that occurs when two conductors with different electric potential are electrically connected by an electrolyte.

corrosion, pitting: corrosion resulting in small pits in a metal surface.

corrosion, stress: localized directional cracking caused by a combination of tensile *stress* and corrosive environment.

corrugating: forming *sheet* into a regular series of parallel longitudinal ridges and valleys.

coupon: a piece of metal from which a test specimen may be taken.

creep: an increase in deformation of a part under constant stress. Metals in tension may creep at elevated temperatures.

crippling stress: the average *stress* at maximum load in a member in compression. See *postbuckling.*

curtain wall: a perimeter wall of a building designed to carry wind loads to the building frame.

design strength: the product of the *resistance factor* and the *nominal strength*. This product must be less than the factored loads in *load and resistance factor design (LRFD).*

design stress: the product of the *resistance factor* and the *nominal strength* expressed as stress.

diaphragm: a flat element that resists shear loads acting in its own plane.

doubly symmetric section: a cross section that is symmetric about each of two and only two perpendicular axes, for example, an Aluminum Association standard I-beam.

draft: taper on the sides of a die or mold to allow removal of *forgings*, *castings*, or patterns from the die or mold.

drawing: pulling material through a die to reduce its size, change its cross section, or harden it.

ductility: the ability of a material to withstand *plastic strain* before rupture. A comparison of the *modulus of resilience* to the *modulus of toughness* is a measure of ductility.

duralumin: name used for the early aluminum-copper alloys that turned out to be not very durable.

dye penetrant testing: an inspection method that uses a penetrating dye or oil for the detection of surface cracks or defects.

edge stiffener: see *stiffener, edge*.

effective length: see *length, effective*.

effective net area: see *net area, effective*.

effective width: see *width, effective*.

elastic: pertaining to the behavior of material under load at *stresses* below the *proportional limit*, where deformations under load are not permanent.

elastic analysis: determination of the effects of loads on members based on the assumption that the member behavior is elastic.

elastic buckling: see *buckling, elastic*.

elastically supported flange: see *flange, elastically supported*.

element: a plate, either rectangular or curved in cross section and connected to other elements only along its longitudinal edges to other elements. See Figure 5.6 for examples. In the Aluminum *Specification*, elements of shapes are also called *components* of shapes.

elongation: the percentage increase in the distance between two gauge marks of a specimen tensile-tested to rupture. The elongation is dependent on the original gauge length (2 in. [50 mm] is often used), and original dimensions, such as thickness, of the specimen. Elongation is considered to give a general indication of *ductility*.

embossing: a raised relief pattern on a surface, typically used on aluminum *sheet* to enhance appearance, also called *patterned sheet*.

endurance limit: the fatigue strength of a plain specimen for a very large number of cycles (500,000,000) of stress reversals, measured by a standard test. For this constant amplitude loading, the fatigue strength will be no less than the endurance limit, even at a greater number of cycles. See also *fatigue limit*.

Euler buckling stress: the *elastic buckling stress* of a *column* predicted by Euler's formula $\pi^2 E/(L/r)^2$. See also *buckling, elastic*.

exfoliation: see *corrosion, exfoliation*.

extrusion: a product made by pushing material through an opening called a *die*.

extrusion billet: a solid or hollow stock, commonly cylindrical, charged into an *extrusion* press.

failure mode: see *limit state*.

fatigue: *fracture* caused by the repeated application of stresses.

fatigue limit: the allowable stress range for a structural component subjected to 5 million cycles of constant amplitude cyclic load, as calculated by Equation 4.8.1-2 in the Aluminum *Specification*. See also *endurance limit*.

faying surface: either of the adjacent surfaces held closely together in a joint.

filler wire: see *weld filler wire*.

fillet: a rounded corner.

finish: the characteristics of a surface.

first order analysis: analysis based on satisfying equilibrium conditions on the undeflected structure.

flange: (1) general use: a plate *element* of a *shape*, oriented parallel to the bending axis and near the extreme fiber; (2) specific use in some provisions of the Aluminum *Specification:* that portion of the cross section that lies more than two-thirds the distance from the neutral axis to the extreme fiber.

flange, elastically supported: the *flange* that lies on the compression side of the neutral axis of a *beam* bent about its minor axis. An example is shown in Figure 10.11. The flange is said to be elastically supported because its tendency to *buckle* can be measured by a spring constant. A method for calculating the strength of an elastically supported flange is given in the Aluminum *Specification* Section 4.10.

flexural buckling: see *buckling, flexural*.

flexure: bending.

forging: a product worked to a predetermined shape using one or more processes, such as hammering, upsetting, pressing, rolling, etc.

fracture: a rupture or break due to the application of tension.

fracture toughness: the ability to resist cracking at a notch or crack. See also *modulus of toughness* and *notch-sensitive*.

friction connection: see *connection, slip-critical*.

galvanic corrosion: see *corrosion, galvanic*.

gauge: a number designating the thickness of *sheet*, *plate*, or *wire*. Alternately, the spacing between fasteners in a direction perpendicular to the direction of force.

grip: the total thickness of parts joined by a fastener.

groove weld: a *weld* made at a *butt joint* and sometimes referred to as a butt weld. There are several kinds of groove welds, such as bevel, vee, J, and U welds; they may be the full depth or only a partial penetration of the thickness.

gross section or area: the full cross-sectional area less the area of holes not filled with fasteners.

hard metric conversion: see *metric conversion, hard.*

heat affected zone (HAZ): the region with reduced strength in the vicinity of a weld, taken as 1 in. [25 mm] from the centerline of a groove weld or the heel of a fillet weld according to the Aluminum *Specification.*

heat treating: obtaining desired properties by heating or cooling an *alloy* under controlled conditions. This does not include heating solely to allow *hot-working.*

heat-treatable alloy: an *alloy* that may be heat-treated to increase strength.

hot-rolled shapes: *shapes* formed by rolling when the metal is in a semi-molten state.

hot-working: *plastic* deformation at a temperature high enough that prevents *strain-hardening.*

inelastic: without return to original shape upon removal of load.

inelastic buckling: see *buckling, inelastic.*

intermediate stiffener: see *stiffener, intermediate.*

lateral bow: deviation from straight of a longitudinal edge.

length, effective: the length, equal to kL, over which a member would *buckle* if it were pin-ended about the axis of *buckling* rotation.

limit, proportional: see *proportional limit.*

limit state: a design condition at which a structure or component becomes unfit for service because it no longer performs its intended function (i.e., it has reached its serviceability limit state), or it is unable to support loads (i.e., it has reached its strength limit state). Also referred to as a failure mode.

load and resistance factor design (LRFD): a method of proportioning structural components using *load factors* and *resistance factors* so that no *limit state* is exceeded when the structure is subjected to all appropriate load combinations. LRFD is also called *strength design.*

load factor (γ): a factor applied to a *nominal load* to account for uncertainties in predicting it accurately.

local buckling: see *buckling, local.*

lockbolt: a fastener composed of a pin and a collar, which is swaged onto a series of concentric ridges on the pin during installation. (See Figure 8.5.)

lot, heat-treat: material of the same mill form, *alloy, temper,* section, and size traceable to one heat-treated furnace load or 8-hour period for continuous furnaces.

lot, inspection: an identifiable quantity of material submitted for inspection at one time.

LRFD: see *load and resistance factor design.*

mechanical properties: properties related to the behavior of material when subjected to force, including *modulus of elasticity, yield strength, ultimate strength,* and others.

mechanical properties, minimum: the *mechanical properties* which 99% of the material is expected to exceed at a confidence level of 0.95.

metric conversion, hard: conversion of foot-pound units to metric units and modifying the result to an even, convenient round number in metric units. See *metric conversion, soft.*

metric conversion, soft: conversion of foot-pound units to metric units without modifying the result to an even, convenient round number in metric units. See *metric conversion, hard.*

mill finish: the *finish* on material as produced by the mill without any additional treatment.

minimum mechanical properties: see *mechanical properties, minimum.*

modulus of elasticity (*E*): the slope of the stress-strain curve up to the *proportional limit*, which is the linear part of the curve. The modulus, also called *Young's Modulus*, may be measured for tension or compression. The compressive modulus for various aluminum alloys is given in Aluminum *Specification* Table 3.3-1 and averages about 10,000 ksi [70,000 MPa].

modulus of resilience: the area under the linear portion of the stress-strain curve, considered to be a measure of the *elastic* strain energy of the material. See *ductility* and *modulus of toughness.*

modulus of rigidity (*G*): the ratio of shear *stress* in a torsion test to shear *strain* in the *elastic* range, also called the *shear modulus.* An approximate average value for aluminum alloys is 3,800 ksi [26,000 MPa]. It can be calculated from the *modulus of elasticity (E)* and *Poisson's ratio* by the formula $G = E/(2[1 + v])$.

modulus of toughness: the entire area under the stress-strain curve up to rupture, considered to be a measure of the total strain energy the material can withstand before rupture. See *ductility* and *modulus of resilience.*

moment gradient: the variation of bending moment over the length of the *beam*, a graph of which is called the *moment diagram.*

natural aging: the change in mechanical properties that takes place in *heat-treatable* aluminum *alloys* at ambient temperature over time due to the precipitation of alloying elements from solid solution.

NDT: see nondestructive testing.

net area: the area of a cross section exclusive of holes.

net area, effective: the portion of the *net area* that may be used in calculating tensile *stress* for a member in tension. *Elements* of the section that are not connected at joints may not be fully effective in carrying tensile *stress*, so such *elements* may not be fully included in the net effective area.

neutral axis: the line of zero fiber stress in a section of a member in bending.

nominal loads: the magnitude of loads as given in applicable codes before any factors are applied.

nominal strength: the load-carrying capacity of a structure or component as determined by calculations using the *minimum mechanical properties* and *rational analysis* or by testing conducted in accordance with the Aluminum *Specification*.

non-compact section: a cross section with an *element* of dimensions so that the element's strength is limited by *inelastic buckling*. In aluminum, such an element has a *slenderness ratio* between *slenderness limits* S_1 and S_2. See also *compact* and *slender element section*.

nondestructive testing (NDT): tests that do not alter the mechanical properties of the material inspected, sometimes referred to as nondestructive examination (NDE).

non-heat-treatable alloy: an *alloy* that can be strengthened only by *cold-working*.

notch-sensitive: a condition characterized by the ultimate stress on the net section attained in tests of specimens with sufficiently sharp notches being less than the specified minimum *tensile yield strength*. See also *fracture toughness*.

open die: a die, usually for an *extrusion*, that is available for use by any customer without a die charge.

open section: a cross section that can be drawn by a single continuous line, such as a wide flange, channel, tee, zee, or angle.

overall buckling: see *buckling, overall*.

penetrant testing: see *dye penetrant testing*.

permanent mold casting: see *casting, permanent mold*.

physical properties: properties of a material other than *mechanical properties*, such as density or electrical conductivity.

pin: a fastener about whose axis connected parts are not substantially restrained from rotation.

pipe: *tube* in standardized diameters and wall thicknesses given in an ANSI standard by nominal diameter and schedule number.

pitting: see *corrosion, pitting*.

plastic: pertaining to behavior of aluminum under load at *stresses* above the *proportional limit*, where deformations under load are permanent. Also referred to as *inelastic*.

plastic design: a design method based on members having the ability to maintain a full plastic moment through large rotations so that a mechanism can develop, allowing moment redistribution.

plate: a rolled product with a rectangular cross section and thickness of at least 0.25 in. [more than 6.3 mm] with sheared or sawed edges. The requirement for sheared or sawed edges distinguishes the product from an *extrusion.* See also *sheet.*

point symmetric section: a cross section that is symmetric about any axis in the cross section drawn through a common point, for example, a cruciform or zee section with equal length legs.

Poisson's ratio (ν): the negative of the ratio of the transverse *strain* to the longitudinal *strain* of a member subjected to axial force within the proportional limit. Poisson's ratio for aluminum *alloys* is approximately 0.33.

ponding: the collecting of water in low areas on a roof.

postbuckling: pertaining to behavior of an element in compression after *elastic buckling* has occurred, referring to the ability of some structural components to sustain loads greater than the *elastic buckling load.* See *crippling stress.*

precipitation-heat treatment: *artificial aging.*

prismatic: having a constant cross section with respect to length.

profile: a *wrought product* much longer than its width other than *sheet, plate, rod, bar, tube,* or *wire.* The term profile is preferred over "*shape,*" but the terms are synonymous.

profile, structural: a *profile* in standard *alloys, tempers,* sections, and dimensions, such as angles, channels, I-beams, H-beams, tees, and zees, commonly used for structural purposes. These are shown in *Aluminum Design Manual,* Part VI, Section Properties.

proportional limit: the *stress* at which the stress-strain curve begins to deviate from a straight line.

pull-out force (P_{not}): the force in the direction of the axis of a screw required to pull a screw out of the material into which it is threaded. See Figure 8.18.

pull-over force (P_{nov}): the force in the direction of the axis of a screw required to pull the connected part under the head of the screw over the head. See Figure 8.18.

punching: forming a hole in metal by pushing a male die through the material, which is supported below by a female die of slightly larger dimensions.

purlin: a horizontal structural member that supports *roofing.*

radiography: X-ray pictures.

radius of gyration (r): the square root of the quantity obtained by dividing the moment of inertia by the area of a cross section. The radius of gyration is identified with respect to the axis about which the moment of inertia is taken.

radius of gyration, effective (r_{ye}): a distance that may be used instead of radius of gyration to calculate the moment carrying capacity of a *beam*, or the *buckling* load of a *column*.

rational analysis: prediction of load-carrying capacity using accepted principles of structural mechanics.

reaming: fabricating a hole to final size by enlarging a smaller hole.

re-entrant cut: a cut into material similar to that shown in Figure 3.14.

reliability index: a measure of the reliability of a structure equal to the number (β) of *standard deviations* of the natural log of the R/Q distribution between the mean and zero (where R is the resistance and Q is the load effect).

residual stress: *stress* in an unloaded member after it has been formed. Residual *stresses* may arise in aluminum members from *welding*, cold bending, differential cooling after heating, finishing, and other fabrication methods.

resilience, modulus of: see *modulus of resilience*.

resistance factor: a factor less than one accounting for variations in material properties, fabrication, and design, and the consequences of failure applied to the *nominal strength* for *load and resistance factor design*.

ribbed siding: *siding* with a particular trapezoidal cross section with dimensions shown in *Aluminum Design Manual*, Part VI, Table 31.

rigidity, modulus of: see *modulus of rigidity*.

rivet: a one-piece fastener that is deformed during installation to acquire its final configuration.

rivet, blind: a *rivet* that can be installed with access to only one side of the parts being joined.

rod: a solid *wrought product* circular in cross section with a diameter not less than 0.375 in. [more than 10 mm]. See Table G.1.

roll-forming: a fabrication method of forming parts with a constant cross section and longitudinal bends using a series of cylindrical dies in male-female pairs to progressively form a *sheet* or *plate* to a final shape in a continuous operation. For example, a *corrugated sheet*.

roofing: *corrugated* or folded *sheet* used to clad the roofs of buildings.

rust: a form of corrosion product common to iron bearing materials. Rust never sleeps.

safety factor: a factor greater than one by which *nominal strength* is divided to calculate *allowable stress*. Safety factors vary depending on the type of aluminum structure (*bridge* or *building*), as well as the type of *limit state* (*yielding* or *ultimate strength*).

sand casting: see *casting, sand*.

sandwich panel: a panel assembly consisting of an insulating core material with aluminum skins on both sides. Also called *aluminum composite material* (ACM). See Section 3.1.5.

second-order analysis: analysis based on satisfying equilibrium conditions on the deflected structure.

self-drilling screw: a screw that drills and taps its own hole as it is being driven.

shape: see *profile*.

shape factor: for bending, the ratio of the moment carried by a continuously braced beam with the full section *yielded* to the moment carried by the beam when *yielding* first occurs. The shape factor is a function of the shape of the cross section, being higher for cross sections with more material concentrated near the neutral axis than for those sections with material dispersed farther away from the neutral axis.

shear center: that point in the plane of a cross section through which transverse (shear) load must act to produce no twisting of the cross section. The shear center is also called the *flexural center.*

shear modulus: see *modulus of rigidity*.

shear ultimate strength (F_{su}): the maximum *stress* sustainable in shear.

sheet: a rolled product with a rectangular cross section and a thickness less than 0.25 in. and at least 0.006 in. [more than 0.15 mm through 6.3 mm] with slit, sheared, or sawed edges. The edge requirement is to distinguish this product from *extrusions*. See also *plate*.

siding: *corrugated* or folded *sheet* used to clad building walls.

singly symmetric section: a cross section that is symmetric about one and only one axis. For example, an Aluminum Association standard channel.

slender element section: a cross section with an *element* of dimensions so that the element will undergo *elastic buckling*. In aluminum, such an element has a *slenderness ratio* greater than the S_2 *slenderness limit*. See also *compact* and *non-compact section*.

slenderness limit (S_1, S_2): a *slenderness ratio* below which the compression-carrying capacity is predicted by one equation and above which the capacity is predicted by another equation.

slenderness ratio: a measure of the tendency of a compression member to *buckle*. For overall member *buckling*, this is the ratio of the *effective length* to *radius of gyration*, both with respect to the same axis. For *local buckling*, this is the ratio of width (or midthickness radius, for curved *elements*) of an *element* to its thickness. See also *aspect ratio*.

soft metric conversion: see *metric conversion, soft*.

space frame: a three-dimensional structural frame.

spinning: a fabrication method of shaping material into a piece with an axis of revolution in a spinning lathe with a mandrel.

SSRC: see Structural Stability Research Council.

stabilizing: a low-temperature heat treatment used to prevent age-softening of certain strain-hardened aluminum alloys containing magnesium.

stainless steel: alloys of iron containing 10.5% or more chromium. Standards for the design of stainless steel structural members are provided in ASCE Standard 8-90, *Specification for the Design of Cold-Formed Stainless Steel Structural Members* (82). See also *austenitic stainless steel*.

standard deviation: in statistics, a measure of variability equal to the square root of the arithmetic average of the squares of the deviations from the mean in a frequency distribution.

standing seam: a seam, usually between *roofing* sheets, made by fitting together the upturned legs of two adjacent pieces. See Figure 10.2.

steel: an *alloy* of iron with a small quantity of carbon and other elements. Synonym: an archaic metal that rusts.

Steel Manual: the 9th edition (1989) of the *Manual of Steel Construction* published by the AISC and which includes the *Specification for Structural Steel Buildings, Allowable Stress Design* (39).

stiffener: material added to a flat *element* of a cross section to increase its capacity to carry compression.

stiffener, edge: a *stiffener* along the free edge of an *element*.

stiffener, intermediate: a *stiffener* not located along the edge of an *element*.

strain (ϵ): the deformation of a member under load, referred to its original dimensions. For example, the elongation per unit length is a tensile strain.

strain hardening: *cold-working* that results in an increase in strength and loss of *ductility*.

strength design: see *load and resistance factor design*.

strength, nominal: see *nominal strength*.

stress (f): force divided by area.

Structural Stability Research Council (SSRC): Formerly the *Column Research Council*, this organization fosters research on the behavior of compressive components of metal structures and writes the *Guide to Stability Design Criteria for Metal Structures* (107). In the 4th edition of the *Guide*, Section 3.9 discusses aluminum columns.

stub column: a full cross-section compression test specimen short enough so that it will not *buckle elastically* or *inelastically* as a *column*.

supported: for plate *elements*, the condition of an edge that is continuously attached to another plate element. For members, the condition at the end of the member that prevents transverse displacement but permits rotation and longitudinal displacement.

tangent modulus of elasticity (E_t): the slope of the stress-strain curve beyond the *proportional limit*. This slope varies depending on the *strain*.

tapping screw: a screw that taps a predrilled hole as it is being driven.

temper: a condition produced by mechanical or thermal treatment.

tensile ultimate strength: see *ultimate strength*.

tensile yield strength: see *yield strength.*

tension field action: *postbuckling* behavior of an *element* in shear, such as the *web* of a *beam*, that demonstrates diagonal *buckling* waves, and by which tension is carried by diagonal bands and compression by vertical stiffeners, like a truss.

thermal expansion, coefficient of (α): the measure of the change in *strain* in a material caused by a change in temperature. This *physical property* is a function of *alloy* and temperature range, and is approximately 13×10^{-6} per °F [23×10^{-6} per °C] for aluminum *alloys.*

tolerance: an allowable deviation from a nominal or specified dimension or property.

torsion constant (J): a measure of the stiffness of a section in pure twisting, in units of length to the fourth power.

torsional buckling: see *buckling, torsional.*

torsional-flexural buckling: see *buckling, torsional-flexural.*

toughness, modulus of: see *modulus of toughness.*

trapezoidal sheet: *sheet* that has been formed by bending to have a cross section similar to that shown in Figure 10.1. *Aluminum Design Manual,* Part VI, Table 31 shows examples of trapezoidal sheet called ribbed *siding* and V-beam *roofing* and *siding.*

tread plate: *sheet* or *plate* with a raised diamond pattern on one side for slip-resistance. Also called *checkered plate* or *diamond plate.* See Section 3.1.5.

tube: a hollow *wrought product* that is longer than its width, is symmetrical and round, hexagonal, octagonal, elliptical, square, or rectangular, and has uniform wall thickness.

ultimate strength: the maximum strength a material can withstand before rupture. Two ultimate strengths exist for each *alloy* and *temper:* tension (F_{tu}) and shear (F_{su}).

V-beam: *roofing* or *siding* with a particular trapezoidal cross section with dimensions shown in *Aluminum Design Manual,* Part VI, Table 31.

variation, coefficient of: in statistics, the ratio of the *standard deviation* to the mean.

warping: distortion of a member's cross section from a flat plane.

warping constant (C_w): a measure of the resistance to rotation that arises because of restraint of *warping* of the cross section, in units of length to the sixth power. Formulas for warping constants for various cross sections can be found in Sharp (Table 5.6) (133) and the *Aluminum Design Manual,* Part VI, Table 29.

water staining: oxidation of an aluminum surface in the presence of water between two closely held metal surfaces.

web: a connecting *element* between flanges.

web crippling: *local buckling* of a *web* at a point of support or concentrated load.

weld filler wire: wire used as filler metal in *welds*, also called *weld wire*. Recommended filler alloys for various combinations of alloys to be *welded* together are given in Aluminum *Specification* Table 7.2-1.

welded tube: *tube* produced by longitudinally forming and seam-*welding sheet*.

welding: joining pieces by applying heat, with or without filler metal, to unite the pieces by fusion.

width, effective (b_e): that portion of the width of an *element* in compression over which an assumed uniform *stress* will predict the same behavior (such as deflection and strength) as produced by the actual nonuniform *stress* distribution. The portion of the *element* not included in the effective width may be considered to be *buckled* and supporting no load.

wire: a solid *wrought product* whose diameter or greatest perpendicular distance between parallel faces is less than 0.375 in. [less than or equal to 10 mm], with a round, square, rectangular, or regular octagon or hexagon cross section. See Table G.1.

working: deforming a metal by mechanical action, such as rolling, *extruding*, or *forging*.

working load: a load, such as snow load or wind load, as given by the appropriate code or specification, without any factors applied.

wrought products: products that have been mechanically *worked* by a process, such as rolling, *extruding*, *forging*, or drawing.

yield strength: the *stress* at and above which loading causes permanent deformation. In aluminum, this is measured by the 0.2% offset method: A line is drawn parallel to the linear portion of the stress-strain curve and through 0.2% *strain*, and the yield strength is taken as the *stress* at the intersection of this line and the stress-strain curve. Three yield strengths exist for each *alloy* and *temper*: one for each type of *stress* (compression (F_{cy}), tension (F_{ty}), and shear (F_{sy})).

yielding: deforming permanently.

Young's modulus: See *modulus of elasticity*.

REFERENCES

1. Alcoa, *Alcoa Structural Handbook*, Pittsburgh, PA, 1960.
2. Aluminum Association, *Aluminum Alloys for Cryogenic Applications*, Washington, DC, 1999.
3. Aluminum Association, *Aluminum Automotive Extrusion Manual AT6*, Washington, DC, 1998.
4. Aluminum Association, *Aluminum Design Manual*, Washington, DC, 2000.
5. Aluminum Association, *Aluminum Drainage Products Manual*, Washington, DC, 1983.
6. Aluminum Association, *Aluminum for Automotive Body Sheet Panels AT3*, Washington, DC, 1996.
7. Aluminum Association, *Aluminum Forging Design Manual*, 2nd edition, Washington, DC, October 1995.
8. Aluminum Association, *Aluminum Forgings Application Guide*, Washington, DC, 1st edition, November 1975.
9. Aluminum Association, *Aluminum in Building and Construction*, Washington, DC, undated.
10. Aluminum Association, *Aluminum Standards and Data 1998 Metric SI*, Washington, DC, April 1998.
11. Aluminum Association, *Aluminum Standards and Data 2000*, Washington, DC, June 2000.
12. Aluminum Association, *Automotive Aluminum Crash Energy Management Manual AT5*, Washington, DC, 1998.
13. Aluminum Association, *Commentary on Specifications for Aluminum Structures*, 2nd edition, Washington, DC, December 1982.
14. Aluminum Association, *Designation System for Aluminum Finishes*, 7th edition, Washington, DC, September 1980; reaffirmed 1993.
15. Aluminum Association, *Engineering Data for Aluminum Structures*, 5th edition, Washington, DC, December 1986.
16. Aluminum Association, *Fire Resistance and Flame Spread Performance of Aluminum and Aluminum Alloys*, Washington, DC, 1997.
17. Aluminum Association, *Guidelines for Minimizing Water Staining of Aluminum*, AA TR3, 3rd edition, Washington, DC, 1990.
18. Aluminum Association, *Guidelines for the Use of Aluminum with Food and Chemicals*, Washington, DC, 1994.
19. Aluminum Association, *International Alloy Designations and Chemical Composition Limits for Wrought Aluminum and Wrought Aluminum Alloys*, Washington, DC, 2001.

519

20. Aluminum Association, *Metalworking with Aluminum*, 2nd edition, Washington, DC, December 1975.

21. Aluminum Association, *Specifications for Aluminum Sheet Metal Work in Building Construction*, 4th edition, Washington, DC, October 2000.

22. Aluminum Association, *Specifications for Aluminum Structures*, 5th edition, Washington, DC, December 1986.

23. Aluminum Association, *Standards for Aluminum Sand and Permanent Mold Castings*, 14th edition, Washington, DC, March 2000.

24. Aluminum Association, *Structural Design with Aluminum*, 1st edition, Washington, DC, May 1987.

25. Aluminum Association, *Tempers for Aluminum and Aluminum Alloy Products*, Washington, DC, 2001.

26. Aluminum Association, *The Aluminum Association Position on Fracture Toughness Requirements and Quality Control Testing, AA T-5*, Washington, DC, 1987.

27. Aluminum Association, *Welding Aluminum: Theory and Practice*, 3rd edition, Washington, DC, November 1997.

28. American Architectural Manufacturers Association, *AAMA 609-93 Voluntary Guide Specification for Cleaning and Maintenance of Architectural Anodized Aluminum*, Schaumberg, IL, 1993.

29. American Architectural Manufacturers Association, *AAMA 611-98 Voluntary Specification for Anodized Architectural Aluminum*, Schaumberg, IL, 1998.

30. American Architectural Manufacturers Association, *AAMA 2603-98 Voluntary Specification, Performance Requirements, and Test Procedures for Pigmented Organic Coatings on Aluminum Extrusions and Panels*, Schaumberg, IL, 1998.

31. American Architectural Manufacturers Association, *AAMA 2604-98 Voluntary Specification, Performance Requirements, and Test Procedures for High Performance Organic Coatings on Aluminum Extrusions and Panels*, Schaumberg, IL, 1998.

32. American Architectural Manufacturers Association, *AAMA 2605-98 Voluntary Specification, Performance Requirements, and Test Procedures for Superior Performing Organic Coatings on Aluminum Extrusions and Panels*, Schaumberg, IL, 1998.

33. American Architectural Manufacturers Association, *Aluminum Curtain Wall Design Guide Manual*, Palatine, IL, 1979.

34. American Architectural Manufacturers Association, *Metal Curtain Wall Fasteners*, AAMA TIR-A9-1991, Palatine, IL, 1991.

35. American Association of State Highway and Transportation Officials, *LRFD Bridge Design Specifications*, Washington, DC, 1994.

36. American Association of State Highway and Transportation Officials, *Guide Specifications for Aluminum Highway Bridges*, Washington, DC, 1991.

37. American Institute of Steel Construction, *Code of Standard Practice for Steel Buildings and Bridges*, Chicago, IL, 2000.

38. American Institute of Steel Construction, *Load and Resistance Factor Design Specification for Structural Steel Buildings*, 2nd edition, Chicago, IL, December 1993.

39. American Institute of Steel Construction, *Manual of Steel Construction, Allowable Stress Design*, 9th edition, Chicago, IL, 1989.

40. American Iron and Steel Institute, *Cold-Formed Steel Design Manual, Specification for the Design of Cold-Formed Steel Structural Members*, August 1986 edition with December 1989 addendum, Washington, DC, 1991.

41. American Iron and Steel Institute, *Load and Resistance Factor Design Specification for Cold-Formed Steel Structural Members*, Washington, DC, August 1991.

42. American National Standards Institute, *ANSI H35.1-1997, Alloy and Temper Designation Systems for Aluminum*, New York, 1997.

43. American National Standards Institute, *ANSI H35.2-1997, Dimensional Tolerances for Aluminum Mill Products*, New York, 1997.

44. American Petroleum Institute, *API Standard 620, Design and Construction of Large, Welded, Low-Pressure Storage Tanks,* 9th edition, Washington, DC, February 1996.

45. American Petroleum Institute, *API Standard 650, Welded Steel Tanks for Oil Storage,* 10th edition, Washington, DC, November 1998.

46. American Society for Testing and Materials, *A36 Standard Specification for Structural Steel*, Philadelphia, 1991.

47. American Society for Testing and Materials, *B26 Specification for Aluminum-Alloy Sand Castings*, West Conshohocken, PA, 1998.

48. American Society for Testing and Materials, *B108 Specification for Aluminum-Alloy Permanent Mold Castings*, West Conshohocken, PA, 1998.

49. American Society for Testing and Materials, *B209 Specification for Aluminum and Aluminum-Alloy Sheet and Plate*, West Conshohocken, PA, 1996.

50. American Society for Testing and Materials, *B210 Specification for Aluminum and Aluminum-Alloy Drawn Seamless Tubes*, West Conshohocken, PA, 1995.

51. American Society for Testing and Materials, *B211 Specification for Aluminum and Aluminum-Alloy Bar, Rod, and Wire*, West Conshohocken, PA, 1995.

52. American Society for Testing and Materials, *B221 Specification for Aluminum and Aluminum-Alloy Extruded Bars, Rods, Wire, Profiles, and Tubes*, West Conshohocken, PA, 1996.

53. American Society for Testing and Materials, *B241 Specification for Aluminum and Aluminum-Alloy Seamless Pipe and Seamless Extruded Tube*, West Conshohocken, PA, 1996.

54. American Society for Testing and Materials, *B247 Specification for Aluminum and Aluminum-Alloy Die Forgings, Hand Forgings, and Rolled Ring Forgings*, West Conshohocken, PA, 1995.

55. American Society for Testing and Materials, *B275 Standard Practice for Codification of Certain Nonferrous Metals and Alloys, Cast and Wrought,* West Conshohocken, PA, 1996.

56. American Society for Testing and Materials, *B308 Specification for Aluminum-Alloy 6061-T6 Standard Structural Profiles*, West Conshohocken, PA, 1996.

57. American Society for Testing and Materials, *B313 Specification for Aluminum and Aluminum-Alloy Round Welded Tubes*, West Conshohocken, PA, 1995.

58. American Society for Testing and Materials, *B316 Specification for Aluminum and Aluminum-Alloy Rivet and Cold-Heading Wire and Rods*, West Conshohocken, PA, 1996.

59. American Society for Testing and Materials, *B345 Specification for Aluminum and Aluminum-Alloy Seamless Pipe and Seamless Extruded Tube for Gas and Oil Transmission and Distribution Piping Systems*, West Conshohocken, PA, 1996.

60. American Society for Testing and Materials, *B361 Specification for Factory-Made Aluminum and Aluminum-Alloy Welding Fittings*, West Conshohocken, PA, 1995.

61. American Society for Testing and Materials, *B429 Specification for Aluminum-Alloy Extruded Structural Pipe and Tube*, West Conshohocken, PA, 1995.

62. American Society for Testing and Materials, *B483 Specification for Aluminum and Aluminum-Alloy Drawn Tubes for General Purpose Applications*, West Conshohocken, PA, 1995.

63. American Society for Testing and Materials, *B491 Specification for Aluminum and Aluminum-Alloy Extruded Round Tubes for General Purpose Applications*, West Conshohocken, PA, 1995.

64. American Society for Testing and Materials, *B547 Specification for Aluminum and Aluminum-Alloy Formed and Arc-Welded Round Tube*, West Conshohocken, PA, 1995.

65. American Society for Testing and Materials, *B548 Method for Ultrasonic Inspection of Aluminum-Alloy Plate for Pressure Vessels*, West Conshohocken, PA, 1997.

66. American Society for Testing and Materials, *B557 Test Methods of Tension Testing Wrought and Cast Aluminum- and Magnesium-Alloy Products*, West Conshohocken, PA, 1994.

67. American Society for Testing and Materials, *B565 Test Method for Shear Testing of Aluminum and Aluminum-Alloy Rivets and Cold-Heading Wire and Rods*, West Conshohocken, PA, 1994.

68. American Society for Testing and Materials, *B594 Practice for Ultrasonic Inspection of Aluminum-Alloy Wrought Products for Aerospace Applications*, West Conshohocken, PA, 1997.

69. American Society for Testing and Materials, *B632 Specification for Aluminum and Aluminum-Alloy Rolled Tread Plate*, West Conshohocken, PA, 1995.

70. American Society for Testing and Materials, *B645 Practice for Plane Strain Fracture Toughness Testing of Aluminum Alloys*, West Conshohocken, PA, 1998.

71. American Society for Testing and Materials, *B744 Specification for Aluminum Alloy Sheet for Corrugated Aluminum Pipe*, West Conshohocken, PA, 1995.

72. American Society for Testing and Materials, *B745 Specification for Corrugated Aluminum Pipe for Sewers and Drains*, West Conshohocken, PA, 1997.

73. American Society for Testing and Materials, *B746 Specification for Corrugated Aluminum Alloy Structural Plate for Field-Bolted Pipe, Pipe-Arches, and Arches*, West Conshohocken, PA, 1995.

74. American Society for Testing and Materials, *B788 Practice for Installing Factory-Made Corrugated Aluminum Culverts and Storm Sewers Pipe*, West Conshohocken, PA, 1997.

75. American Society for Testing and Materials, *B789 Practice for Installing Corrugated Aluminum Structural Plate Pipe for Culverts and Sewers*, West Conshohocken, PA, 1997.

76. American Society for Testing and Materials, *B790 Practice for Structural Design of Corrugated Aluminum Pipe, Pipe Arches and Arches for Culverts, Storm Sewers and Other Buried Conduits*, West Conshohocken, PA, 1997.

77. American Society for Testing and Materials, *B864 Specification for Corrugated Aluminum Box Culverts*, West Conshohocken, PA, 1995.

78. American Society for Testing and Materials, *E1592 Standard Test Method for Structural Performance of Sheet Metal Roof and Siding Systems by Uniform Static Air Pressure Difference*, Philadelphia, 1994.

79. American Society for Testing and Materials, *F467 Nonferrous Nuts General Use*, West Conshohocken, PA, 2001.

80. American Society for Testing and Materials, *F468 Nonferrous Bolts, Hex Cap Screws, and Studs for General Use*, West Conshohocken, PA, 2001.

81. American Society for Testing and Materials, *F593 Stainless Steel Bolts, Hex Cap Screws, and Studs*, West Conshohocken, PA, 1998.

82. American Society of Civil Engineers, ANSI/ASCE-8-90, *Specification for the Design of Cold-Formed Stainless Steel Structural Members*, New York, 1991.

83. American Society of Civil Engineers, ASCE 7-98, *Minimum Design Loads for Buildings and Other Structures*, New York, 2000.

84. American Society of Mechanical Engineers, ASME B31.3-1999 edition with 2000 addenda, *Process Piping*, New York, 1999.

85. American Society of Mechanical Engineers, ASME/ANSI B96.1-1999, *Welded Aluminum-Alloy Storage Tanks*, New York, 2000.

86. American Welding Society, AWS A2.4-98, *Standard Symbols for Welding, Brazing, and Nondestructive Examination*, Miami, FL, 1998.

87. American Welding Society, AWS A5.3/A5.3M: 1999, *Specification for Aluminum and Aluminum-Alloy Electrodes for Shielded Metal Arc Welding*, Miami, FL, 1999.

88. American Welding Society, AWS A5.10/A5.10M: 1999, *Specification for Bare Aluminum and Aluminum-Alloy Welding Electrodes and Rods*, Miami, FL, 2000.

89. American Welding Society, AWS B2.1.015-91, *Standard Welding Procedure Specification for Gas Tungsten Arc Welding of Aluminum, (M-22 or P-22), 10 through 18 Gauge, in the As-Welded Condition, With or Without Backing*, Miami, FL, 1991.

90. American Welding Society, AWS D1.1:2000, *Structural Welding Code—Steel*, Miami, FL, 2000.

91. American Welding Society, AWS D1.2-97, *Structural Welding Code—Aluminum*, Miami, FL, 1997.

92. American Welding Society, AWS D3.7-90, *Guide for Aluminum Hull Welding*, Miami, FL, 1990.

93. American Welding Society, AWS D8.14M/D8.14:2000, *Specification for Automotive and Light Truck Components Weld Quality—Aluminum Arc Welding*, Miami, FL, 2000.

94. American Welding Society, AWS D10.7-86R, *Recommended Practices for Gas Shielded Arc Welding of Aluminum and Aluminum Alloy Pipe*, Miami, FL, 1986.

95. Brockenbrough, R. L. and Johnston, B. G., *Steel Design Manual*, United States Steel Corp., Pittsburgh, PA, 1974.

96. Canadian Standards Association, *Strength Design in Aluminum*, CAN3-S157-M83, Rexdale, Ontario, December 1983.

97. Clark, John W., and Rolf, Richard L., "Design of Aluminum Tubular Members," ASCE *Journal of the Structural Division,* New York, December 1964.

98. Crawley, S. W., and Dillon, R. M., *Steel Buildings Analysis and Design*, 4th edition, John Wiley & Sons, New York, 1993.

99. Davis, J. R., and Associates, ed., *Aluminum and Aluminum Alloys*, ASM International, Materials Park, OH, 1993.

100. Ducotey, Steve, and Larsen, Curtis, "Comparison of ASD vs. LRFD for an Aluminum Clear Span Roof Structure," *ASCE Structures Congress XV*, Portland, OR, 1997.

101. Ellifritt, D. S., "The Mysterious 1/3 Stress Increase," *Engineering Journal*, 4th Quarter, 1977, American Institute of Steel Construction.

102. European Committee for Standardization (CEN), *ENV 1999-1-1 Eurocode 9: Design of aluminium structures*, Brussels, 1998.

103. Ferry, Robert L., and Kissell, J. Randolph, "The Aesthetics of Triangulated Latticed Structures," *Structures Congress '92 Compact Papers*, April 1992.

104. Fortin, Beaulieu, and Bastien, "Experimental Investigation of Aluminum Friction-Type Connections," *INALCO 9 Proceedings*, Munich, 2001.

105. Galambos, T. V., "History of Steel Beam Design," *Engineering Journal*, 4th Quarter, 1977, American Institute of Steel Construction.

106. Galambos, T. V., and Ellingwood, B., "Serviceability Limit States: Deflections," *Journal of the Structural Division, ASCE,* Vol. 112, January 1986.

107. Galambos, T. V., ed., *Guide to Stability Design Criteria for Metal Structures*, 4th edition, John Wiley & Sons, New York, 1988.

108. Galambos, T. V., *Load and Resistance Factor Design For Aluminum Structures*, Washington University Report No. 54, St. Louis, MO, May 1979.

109. Gaylord, E. H., and C. N. Gaylord, *Structural Engineering Handbook*, 2nd edition, McGraw-Hill, New York, 1979.

110. Hartmann, E. C., G. O. Hoglund, and H. A. Miller, "Joining Aluminum Alloys," *Steel*, August 7–September 11, 1944.

111. Hinkle, A. J., and M. L. Sharp, *Load and Resistance Factor Design of Aluminum Structures*, Alcoa Report No. 89-57-03, Alcoa Center, PA, February 1989.

112. Industrial Fasteners Institute, *Fastener Standards*, 6th edition, Cleveland, OH, 1988.

113. Industrial Fasteners Institute, *Metric Fastener Standards*, 2nd edition, Cleveland, OH, 1983.

114. International Code Council, *International Building Code*, Falls Church, VA, 2000.

115. International Conference of Building Officials, *Uniform Building Code*, 1991 edition, Whittier, CA, 1991.

116. International Organization for Standardization (ISO), *ISO TR 11069 Aluminium Structures—Material and design—Ultimate limit state under static loading*, Geneva, Switzerland, 1995.

117. Kaiser Aluminum, *Welding Kaiser Aluminum*, 2nd edition, Oakland, CA, 1978.

118. Kaufman, J. Gilbert, ed., *Properties of Aluminum Alloys: Tensile, Creep, and Fatigue Data at High and Low Temperatures*, ASM International, Materials Park, OH, 1999.

119. Kaufman, J. Gilbert, *Fracture Resistance of Aluminum Alloys*, ASM International, Materials Park, OH, 2001.

120. Kaufman, J. G., and Marshall Holt, *Fracture Characteristics of Aluminum Alloys*, Alcoa Research Laboratories Technical Paper No. 18, Alcoa, Pittsburgh, PA, 1965.

121. Kissell, J. R., and R. L. Ferry, "Aluminum Friction Connections," *ASCE Structures Congress XV*, Portland, OR, 1997.

122. Kulak, G. L., J. W. Fisher, and J. H. A. Struik, *Guide to Design Criteria for Bolted and Riveted Joints*, 2nd edition, John Wiley & Sons, New York, 1987.

123. Luttrell, C. R., "Thermal Cycling of Slip-Critical Aluminum Joints," *ASCE Structures Congress XVII*, New Orleans, 1999.

124. Luttrell, C. R., "Turn-of-Nut Method for Aluminum Joints," *ASCE Structures Congress XVII*, New Orleans, 1999.

125. Menzemer, Craig, "Failure of Bolted Connections in an Aluminum Alloy," *Journal of Materials Engineering and Performance*, ASM, Vol. 8, No. 2, April 1999.

126. Metal Building Manufacturers Association, *Low Rise Building Systems Manual*, Cleveland, OH, 1986.

127. Nelson, F. G., and R. L. Rolf, "Shear Strengths of Aluminum Alloy Fillet Welds," *Welding Research Supplement*, February 1966.

128. Nuernberger, H. H., *Alcoa Aluminum Alloys 6070 (Extrusions)*, Pittsburgh, PA, July 1982.

129. Pezze, F., J. Tang, and G. Fu, *Aluminum Bridges Versus Steel Bridges, Report 64*, Engineering Research and Development Bureau, New York State Department of Transportation, Albany, NY, 1992.

130. Report of the Task Committee on Lightweight Alloys, "Suggested Specifications for Structures of Aluminum Alloys 6061-T6 and 6062-T6," *Journal of the Structural Division, ASCE*, December 1962.

131. Research Council on Structural Connections, *Specification for Structural Joints Using ASTM A325 or A490 Bolts*, Chicago, IL, 2000.

132. Richter, D. L., "Temcor Space Structure Development," *Proceedings of the Third International Conference on Space Structures*, p. 1019, New York, 1984.

133. Sharp, M. L., *Behavior and Design of Aluminum Structures*, McGraw-Hill, New York, 1993.

134. Sharp, Maurice L., Glenn E. Nordmark, and Craig C. Menzemer, *Fatigue Design of Aluminum Components and Structures*, McGraw-Hill, Inc., New York, 1996.

135. Societé de Sauvegarde de la Citadelle des Baux, *Visite of the Baux de Provence Citadel*, Les Baux de Provence, France.

136. Specialty Steel Industry of the United States, *Design Guidelines for the Selection and Use of Stainless Steel*, Washington, DC, undated.

137. U.S. Department of Defense, MIL HDBK-5H, *Metallic Materials and Elements for Aerospace Vehicle Structures*, 2001.

138. White, R. N., P. Gergely, and R. G. Sexsmith, *Structural Engineering, Vol. 3, Behavior of Members and Systems*, John Wiley & Sons, New York, 1974.

139. Wright, D. T., "Membrane Forces and Buckling in Reticulated Shells," *Journal of the Structural Division, ASCE,* February 1965.

140. Young, W. C., *Roark's Formulas for Stress and Strain*, 6th edition, McGraw-Hill, New York, 1989.

141. Yu, W. W. *Cold-Formed Steel Design*, 2nd edition, John Wiley & Sons, New York, 1991.

INDEX

527